Advances in Intelligent Systems and Computing

Volume 775

The series "Advances in Intelligent Systems and Computing" contains publications on theory, applications, and design methods of Intelligent Systems and Intelligent Computing. Virtually all disciplines such as engineering, natural sciences, computer and information science, ICT, economics, business, e-commerce, environment, healthcare, life science are covered. The list of topics spans all the areas of modern intelligent systems and computing such as: computational intelligence, soft computing including neural networks, fuzzy systems, evolutionary computing and the fusion of these paradigms, social intelligence, ambient intelligence, computational neuroscience, artificial life, virtual worlds and society, cognitive science and systems, Perception and Vision, DNA and immune based systems, self-organizing and adaptive systems, e-Learning and teaching, human-centered and human-centric computing, recommender systems, intelligent control, robotics and mechatronics including human-machine teaming, knowledge-based paradigms, learning paradigms, machine ethics, intelligent data analysis, knowledge management, intelligent agents, intelligent decision making and support, intelligent network security, trust management, interactive entertainment, Web intelligence and multimedia.

The publications within "Advances in Intelligent Systems and Computing" are primarily proceedings of important conferences, symposia and congresses. They cover significant recent developments in the field, both of a foundational and applicable character. An important characteristic feature of the series is the short publication time and world-wide distribution. This permits a rapid and broad dissemination of research results.

More information about this series at http://www.springer.com/series/11156

Hasan Ayaz · Lukasz Mazur
Editors

Advances in Neuroergonomics and Cognitive Engineering

Proceedings of the AHFE 2018 International Conference on Neuroergonomics and Cognitive Engineering, July 21–25, 2018, Loews Sapphire Falls Resort at Universal Studios, Orlando, Florida USA

Editors
Hasan Ayaz
Drexel University
Philadelphia, PA, USA

Lukasz Mazur
University of North Carolina-Chapel Hill
Chapel Hill, NC, USA

ISSN 2194-5357 ISSN 2194-5365 (electronic)
Advances in Intelligent Systems and Computing
ISBN 978-3-319-94865-2 ISBN 978-3-319-94866-9 (eBook)
https://doi.org/10.1007/978-3-319-94866-9

Library of Congress Control Number: 2018947370

Printed on acid-free paper

This Springer imprint is published by the registered company Springer International Publishing AG part of Springer Nature
The registered company address is: Gewerbestrasse 11, 6330 Cham, Switzerland

Advances in Human Factors and Ergonomics 2018

AHFE 2018 Series Editors

Tareq Z. Ahram, Florida, USA
Waldemar Karwowski, Florida, USA

9th International Conference on Applied Human Factors and Ergonomics and the Affiliated Conferences

Proceedings of the AHFE 2018 International Conferences on Neuroergonomics, Cognitive Engineering, Cognitive Computing and Internet of Things, held on July 21–25, 2018, in Loews Sapphire Falls Resort at Universal Studios, Orlando, Florida, USA

Advances in Affective and Pleasurable Design	*Shuichi Fukuda*
Advances in Neuroergonomics and Cognitive Engineering	*Hasan Ayaz and Lukasz Mazur*
Advances in Design for Inclusion	*Giuseppe Di Bucchianico*
Advances in Ergonomics in Design	*Francisco Rebelo and Marcelo M. Soares*
Advances in Human Error, Reliability, Resilience, and Performance	*Ronald L. Boring*
Advances in Human Factors and Ergonomics in Healthcare and Medical Devices	*Nancy J. Lightner*
Advances in Human Factors in Simulation and Modeling	*Daniel N. Cassenti*
Advances in Human Factors and Systems Interaction	*Isabel L. Nunes*
Advances in Human Factors in Cybersecurity	*Tareq Z. Ahram and Denise Nicholson*
Advances in Human Factors, Business Management and Society	*Jussi Ilari Kantola, Salman Nazir and Tibor Barath*
Advances in Human Factors in Robots and Unmanned Systems	*Jessie Chen*
Advances in Human Factors in Training, Education, and Learning Sciences	*Salman Nazir, Anna-Maria Teperi and Aleksandra Polak-Sopińska*
Advances in Human Aspects of Transportation	*Neville Stanton*

(continued)

(continued)

Advances in Artificial Intelligence, Software and Systems Engineering	*Tareq Z. Ahram*
Advances in Human Factors, Sustainable Urban Planning and Infrastructure	*Jerzy Charytonowicz and Christianne Falcão*
Advances in Physical Ergonomics & Human Factors	*Ravindra S. Goonetilleke and Waldemar Karwowski*
Advances in Interdisciplinary Practice in Industrial Design	*WonJoon Chung and Cliff Sungsoo Shin*
Advances in Safety Management and Human Factors	*Pedro Miguel Ferreira Martins Arezes*
Advances in Social and Occupational Ergonomics	*Richard H. M. Goossens*
Advances in Manufacturing, Production Management and Process Control	*Waldemar Karwowski, Stefan Trzcielinski, Beata Mrugalska, Massimo Di Nicolantonio and Emilio Rossi*
Advances in Usability, User Experience and Assistive Technology	*Tareq Z. Ahram and Christianne Falcão*
Advances in Human Factors in Wearable Technologies and Game Design	*Tareq Z. Ahram*
Advances in Human Factors in Communication of Design	*Amic G. Ho*

Preface

This book brings together a wide-ranging set of contributed articles that address emerging practices and future trends in cognitive engineering and neuroergonomics —both aim to harmoniously integrate human operator and computational system, the former through a tighter cognitive fit and the latter a more effective neural fit with the system. The chapters in this book uncover novel discoveries and communicate new understanding and the most recent advances in the areas of workload and stress, activity theory, human error and risk, and neuroergonomic measures, cognitive computing as well as associated applications.

The book is organized into six main sections:

Section 1: Cognition and Performance
Section 2: Neurophysiological Sensing
Section 3: Brain—Computer Interfaces
Section 4: Systemic-Structural Activity Theory
Section 5: Cognitive Computing and Internet of Things
Section 6: Cognitive Design

Collectively, the chapters in this book have an overall goal of developing a deeper understanding of the couplings between external behavioral and internal mental actions, which can be used to design harmonious work and play environments that seamlessly integrate human, technical, and social systems.

Each chapter of this book was either reviewed or contributed by members of the Cognitive and Neuroergonomics Board. For this, our sincere thanks and appreciation to the board members listed below:

Neuroergonomics, Cognitive Engineering

Winston “Wink” Bennett, USA
Alexander Burov, Ukraine
P. Choe, Qatar
M. Cummings, USA
Frederic Dehais, France
X. Fang, USA
Chris Forsythe, USA
Qin Gao, China
Klaus Gramann, Germany
Y. Guo, USA
Peter Hancock, USA
David Kaber, USA
Kentaro Kotani, Japan
Ben Lawson, USA
S.-Y. Lee, Korea
Harry Liao, USA
Y. Liu, USA
Ryan McKendrick, USA
John Murray, USA
A. Ozok, USA
O. Parlangeli, Italy
Stephane Perrey, France
Robert Proctor, USA
A. Savoy, USA
K. Vu, USA
Thomas Waldmann, Ireland
Tomas Ward, Ireland
Brent Winslow, USA
G. Zacharias, USA
L. Zeng, USA
Matthias Ziegler, USA

Cognitive Computing and Internet of Things

Hanan A. Alnizami, USA
Thomas Alexander, Germany
Carryl Baldwin, USA
O. Bouhali, Qatar
Henry Broodney, Israel
Frederic Dehais, France
Klaus Gramann, Germany
Ryan McKendrick, USA
Stephane Perrey, France

Stefan Pickl, Germany
S. Ramakrishnan, USA
Duncan Speight, UK
Martin Stenkilde, Sweden
Ari Visa, Finland
Tomas Ward, Ireland
Matthias Ziegler, USA

It is our hope that professionals, researchers, and students alike find the book to be an informative and valuable resource—one that helps them to better understand important concepts, theories, and applications in the areas of cognitive engineering and neuroergonomics. Beyond basic understanding, the contributions are meant to inspire critical insights and thought-provoking lines of follow on research that further establish the fledgling field of neuroergonomics and sharpen the more seasoned practice of cognitive engineering. While we don't know where the confluence of these two fields will lead, they are certain to transform the very nature of human-systems interaction, resulting in yet to be envisioned designs that improve form, function, efficiency, and the overall user experience for all.

July 2018

Hasan Ayaz
Lukasz Mazur

Contents

Brain-Computer Interfaces

Systemic-Structural Activity Theory

Cognition and Performance

Beyond 2020 NextGen Compliance: Human Factors and Cognitive Loading Issues for Commercial and General Aviation Pilots

Mark Miller and Sam Holley(✉)

Embry-Riddle Aeronautical University Worldwide College of Aeronautics, Daytona Beach, FL, USA
{millmark, Sam.Holley}@erau.edu

Abstract. As previously identified by the authors, digitized flight decks have realigned SHELL model components and introduced cognitive overload concerns. Considering changes from implementing Next Generation air traffic management requirements in 2020, the authors assess digitized interfaces associated with cockpit displays of information integral to performance based navigation and similar operations. Focus is placed on Automatic Dependent Surveillance Broadcast, digitized communications, and expanded electronic flight bags. The ADSB (In) cockpit display will enable pilots to have flight visual awareness on aircraft, terrain, weather and hazards to flight through live satellite updates every second. Increased optical demands and cognitive loading are anticipated for general aviation and commercial pilots, beyond operational levels for those currently using advanced technologies. With nearly continuous cognitive processing and embedded information in the enhanced SHELL model by the authors, potential overload and concerns of situational awareness become likely candidates for human factors problems. Addressing these concerns, areas of emphasis for transition to NextGen 2020 operations are delineated, potential risks among increased cognitive disparities identified, and suggested foci recommended.

Keywords: SHELL · Digitized flight deck · Human factors · Working memory Crew resource management

1 Introduction

The next generation of Air Traffic Control (ATC) technological is slowly becoming a reality for controlling the airspace over the United States (U.S.). NextGen as it is called will modernize the old ATC system in the U.S. by switching it from land-based technologies to satellite based technologies. The increased gains in efficiency and safety have tremendous potential. NextGen aircraft technologies will not be interrupted by signal intervals like previous equipment, but instead will be constant. Pilots will now receive continuous output on other aircraft, terrain and weather from the more accurate satellite fed devices. They will be able to fly more direct and efficient routes by using Global Positioning Satellite (GPS) data from the satellites and will no longer have to rely on a ground system of antiquated Navigation Aides to keep them on

H. Ayaz and L. Mazur (Eds.): AHFE 2018, AISC 775, pp. 3–13, 2019.
https://doi.org/10.1007/978-3-319-94866-9_1

highways in the sky. NextGen will also boast of a better way of keeping track of all aircraft in airspace through ADSB (Out) technology that continuously puts out the aircrafts position every second through the use of satellites. This will act as a new form of transponder to let air traffic controllers and pilots know exactly where other aircraft are. The Federal Aviation Administration (FAA) has now mandated that all aircraft flying in the U.S. be equipped with ADSB (Out) equipment by 2020. The goal of this mandate is make the NextGen system fully functional. In the process of NextGen becoming fully functional, pilots will have options to use ADSB (In) and Datalink technologies to enhance their information. ADSB (In) will allow pilots to actually know where the other participating aircraft are along with being made aware of where the closest terrain is and how the weather will affect the flight. In addition, Datalink will allow digitalized text communication in flight. With NextGen imminent starting in 2020, there is currently only an outlook of relief as the skies over the United States are forecasted to get more crowded over the next 20 years and the current system cannot handle that forecasted growth. While NextGen is the long awaited ATC infrastructure for US airspace moving into the future, it is not without some serious questions to be answered in the area of computer information and automation concerning Nextgen cockpit technologies. How will these technologies affect both General Aviation and Commercial pilots flying in the future US airspace?

2 Analyzing the Increase of Computer Usage on Human Factors and Potential Human Error

The increase of computer information and automation in U.S. Commercial Cockpits has been well documented over the last 30 years. What started with automated flight controls, auto throttle and flight management systems designed to gain maximum fuel efficiency in flight soon shifted to improved flight navigation flight navigation systems and safety devices like Ground Proximity Warning Systems (GPWS), Traffic Collision Avoidance Systems (TCAS), and Onboard Weather Radar. All these technologies related to computer information and automation have greatly increased fuel efficiency, flight navigation, while at the same time greatly enhancing aviation safety. However, with the introduction of Electronic Flight Bags along with NextGen specific ADSB (In) and Datalink communications devices working their way into the cockpit, there has been a serious shift in how much computer information and automation is now available to pilots. All this addition of technology sounds good until it can have potential negative effects on the human performance in flight. Analysis of aviation accident reports that involved some kind of computer information and automation in the past 20 years indicates that the effects the computer information and automation on pilots show potential deficiencies in the four following areas: (1) Complacency in relying on computers, (2) Confusion and not understanding the computers, (3) Becoming overly focused on a computer and distracted from flying, and (4) Using the computer as optical inside only with little outward scanning. In Fig. 1, the SHELL diagram for human factors analysis was updated in 2017 [1] to take into account the increase of computer information and automation in the cockpit while including NextGen technologies. In this SHELL 2017 model, the potential for possible human factors issues that could lead to human error can be seen more clearly. The Hardware (H)-Liveware (L) and the Environment (E)-Liveware (L) linkages show the original areas of the computer being introduced in the cockpit. The Software (S)-Liveware (L), the Liveware (L)-Liveware (L) along with the Environment (E)-Liveware (L) show the new areas that computers have been introduced into the cockpit in the form of EFB, Datalink digitalized texting and ADSB (In) communications. The first important observation made clearly visible in this model is that the computer information and automation have become interfaces between what used to be direct linkages. In the evolution of flight, the SHELL interfaces were originally direct linkages to the human (Liveware (L)) at the center of the SHELL. However, in 2017 version of the SHELL, the evolution of infused computer technologies in aircraft cockpits has created clear computer interfaces in each linkage. Another important issue that is seen in the SHELL Model 2017 is that the new computers (EFB and Datalink) added in the Software (S)-Liveware (L) and Liveware (L)-Liveware (L) interfaces introduced in the last 15 years have now made every computer interface concatenated so that they can potentially overlap with one another. The most important observation in the SHELL Model 2017 is that those interface areas that have been newly created computer interfaces will grow in use in the Software (S)-Liveware (L)-Liveware (L)-Liveware (L) and the Environment (E)-Liveware (L) as NextGen becomes more functional in 2020 and beyond.

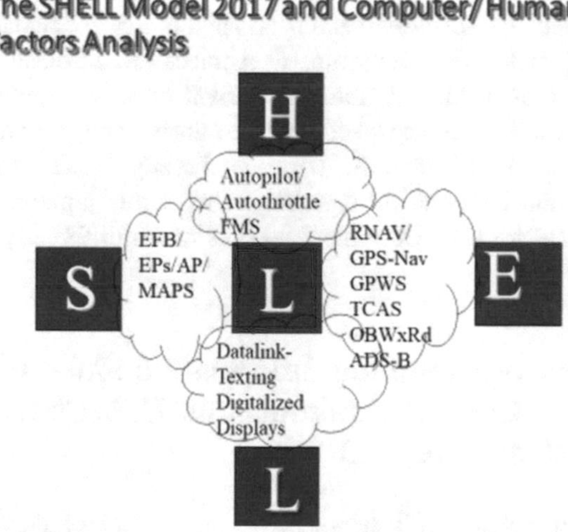

Fig. 1. The SHELL model 2017 and computer/human factors analysis [1]

3 NextGen Cockpit Equipment and the Effect of the Computer in the Cockpit in 2020 and Beyond

As shown in red in Fig. 1, within the Environment (E)-Liveware (L) linkage, the ADSB (In) will become a much greater tool for pilots in the cockpit as it will give constant visual updates to the pilots of the whereabouts of other terrain, other aircraft, and weather, all on a small screen built into the ADSB- (In) equipment. With that, navigation equipment will be linked to continuously updated satellite data to fly precise inputted automated routes that can also fly around updated weather. To support this the Software (S)-Liveware (L) linkage will continue to grow in use of the EFB. It is assumed at this juncture of the FAA's ADSB (Out) mandate by 2020 that the vast majority of U.S. pilots both Commercial and in General Aviation will switch to the efficiency gained from carrying an aviation oriented computerized tablet loaded with pubs, maps and procedures that will update electronically. As this trend continues to grow, there will also be another trend of EFBs and similar cockpit computer devices directly hooking up to cockpit displays in 2020 and beyond. Pilots of NextGen will use ADSB (In) data along with precision satellite navigation automation to fly with maps and instrument approach plates that are computer generated from their EFB. Where the NextGen equipment radically departs from normal aviation operations is in the area of communications. Where once pilots observed gauges together and discussed the readings with each other and the crew over cockpit radio, they have slowly changed to observing digital readouts of similar information over the computer screen. Pilots who once spoke to maintenance personnel or dispatchers via radio can now communicate with them via a digitized electronic text format. NextGen in 2020 and beyond will also favor more communications with ATC through a digitized text format called Datalink. Although there are many pros and cons as to how much and when Datalink will be ultimately used in the new ATC system, it has certainly already shown great promise by being used at selected over-crowded airports in the U.S. for Standardized Departures and Routing along with IFR Departure Clearances. Regardless of how digitalized communication evolves in flight within the cockpit or with communications from the cockpit to the ground, it will surely continue to grow. The growing use of computer information and automation in U.S. cockpits directly related to NextGen will be enormous in 2020 and beyond, but it is important at this juncture to note that it will affect the majority of the U.S. Commercial pilots much differently from their General Aviation counterparts.

4 The Influence of Computer Information and Automation from Nextgen Cockpit Equipment on U.S. Commercial Pilots in 2020 and Beyond

From a U.S. Commercial Pilot's perspective, the advent of NextGen equipment or related equipment like the EFB is inevitable and in many cases already happening. In the case of the ADSB (In) technology, it is presumed that all the Major Commercial Operators will be eventually equipped and gain the extra benefits for pilots that the ADSB (In) gives in the form of a visual display for the aircraft's relation to terrain, other aircraft and updated weather. The human factors ergonomic issue related to this is that U.S. CFR Part 121 carriers already have these three benefits in their cockpit layouts in the form of GPWS, TCAS and Onboard Weather Radar. In fact, the integration and improvement of these technologies over the last 30 years has brought the U.S. Commercial Industry to its highest levels of safe operations in the last decade. The problem with the addition of ADSB (In) for U.S. Commercial pilots is how to integrate the ADSB (In) visual display of terrain to work with the GPWS. Similarly, how to integrate the ADSB visual display of other aircraft into using the current TCAS system. Also important is how to use the ADSB (In) live weather display with the Onboard Weather Radar system. While the ADSB (In) seems like an immediate great addition of redundant systems in a visual form to boost safety margins in aviation, at the same time the three major technologies that ADSB (In) will support take a great deal of training and crew coordination to use properly. With ADSB (In), new human factors guidelines will need to be determined for the appropriate use of the ADSB (In) in U.S. commercial cockpits. In particular how to use ADSB (In) with each system optimally will also need to be determined. Will a priority still be given to respond to a TCAS alert if there is no threat observed on the ADSB-in or vice versa? Along with new human factors guidelines, training will also be imperative to integrate the use of ADSB (In) in simulators and Crew Resource Management. In the cases of the Software (S)-Liveware (L) linkage of the SHELL 2017, many companies have already been standardizing and upgrading their EFB devices for years. Though EFB and its informational software like maps and approach plates are not FAA mandated NextGen cockpit devices, they are certainly technologies that have been developing as strong supportive devices for NextGen flight. As these EFB devices become more powerful and integrate more into the cockpit displays, they will also call for more standardization and more training from each U.S. Commercial carrier. The last perspective from Liveware (L)-Liveware (L) linkage of the SHELL 2017 Model is related to communications and in the increase use of digitized texting in the cockpit. Many U.S. carriers have already installed Datalink and have capabilities of digitized texting in the cockpit. For the U.S. Commercial industry this will mean finding a consensus on when and where to use such digitized texting to communicate safely while at the same time finding where efficiencies and safer operations can be gained without jeopardizing efficient radio communications that already exist. The U.S. Commercial industry and their pilots should be the benefactor of the NextGen related computer technologies in the cockpit as long as the appropriate human factors guidelines are set along with the appropriate training for their integration to avoid human error.

5 The Influence of Computer Information and Automation from NextGen Cockpit Equipment on U.S. General Aviation Pilots in 2020 and Beyond

The General Aviator in the U.S. will be affected much differently through the implementation of NextGen cockpit technologies. Assuming that the ADSB (In) technology will someday be reasonably affordable, General Aviation enthusiasts will welcome the safety gains immediately attained by installing the ADSB (In) component in their cockpit to go with the mandatory ADSB-out component. Unlike their Commercial Airline counterparts, the vast majority of General Aviation enthusiasts do not have extra safety equipment in the cockpit to help them with deal with the Environment (E)-Liveware (L) linkage in the SHELL 2017 Model. In fact, very few General Aviation aircraft in the United States have GPWS, TCAS or Onboard Weather Radar in their cockpits. Most General Aviators simply are made aware of the terrain by looking at it or using maps. They keep separation from other aircraft by scanning more outward, while sometimes working with ATC. For in flight weather, General Aviators use the forecast and then are expected to use good judgement should the weather deteriorate. Coming from a standpoint of having nothing to enhance safety, to now having something that covers all three of the most dangerous parts of the aviation environment is certainly a great boost for the General Aviator, but this upwelling of new tools for the General Aviator could come with a human factors penalty. The penalty stems from the fact that most General Aviators fly single piloted and are not in team trained crews like their Commercial counterparts. Suddenly installing a magic video box in the form of ADSB (in) within a General Aviation cockpit will give pilots the visual tools immediately to help them avoid terrain, see other aircraft and work better with the weather, but this will come at a cost of looking visually more inside instead of having a primary scan outside. Experienced pilots with terrific scans could become overly focused while looking at the smaller ADSB (In) screen and less outside the cockpit where their eyes belong. Training General Aviators on how to integrate the ADSB (In) information into their flying properly will be ongoing in 2020 and beyond. In the Software (S)-Hardware (H) linkage in the SHELL 2017, General Aviators are not far behind their commercial counterparts when it comes to EFBs. In fact, companies like Jeppesen have come up with excellent EFB equipment for General Aviators that is far superior to the former method of carrying maps and approach plates. This has been a major enhancement for General Aviators over the past decade. This technology of cockpit computer information is a tremendous help as long as the General Aviator is able to operate the EFB device efficiently and not becoming overly focused on it while flying. Efficient ways of using such EFBs for General Aviators will be the key if they are also operating ADSB (In) equipment simultaneously. In the Liveware-Liveware (L) linkage of the SHELL 2017, the General Aviator will be at a disadvantage of being single piloted and trying to communicate through digitized text messaging while flying at the same time. Although some advantages could be gained in the form of using digitalized texting communications for copying taxi instructions, Standard Departures or copying IFR clearances, the General Aviator will have to exercise extreme caution while attempting to communicate digitally while taxing or in flight as the same human factors that have deemed texting dangerous while driving a car could also be at work in a single piloted aircraft as well. Although the General Aviation pilot will realize gains in safety and efficiency through NextGen cockpit equipment, without proper human factors standards and training, the General Aviator being often single piloted could fall prey to human error caused by NextGen cockpit equipment.

6 NextGen Computer Technology in the Cockpit Could Lead to Declines in Situational Awareness

The most important thing any pilot will learn related to situational awareness is to prioritize to fly the aircraft in safe parameters first, navigate the aircraft second and then communicate last. Aviate, navigate, communicate is an age old aviation adage that keeps pilots safe and alive by prioritizing situational awareness while flying. The first computers in aircraft aimed to increase situational awareness by helping keep that 'aviate' a priority by being directly integrated with the flight controls as depicted in the SHELL Model 2017 under the Liveware (L)-Liveware (L) linkage. Computers were added to the Liveware (L)-Environment (E) to help pilots improve their situational awareness to navigate better, avoid terrain, other aircraft and bad weather in the process. This was the older paradigm of using computers inflight to enhance efficiency in the cockpit and increase situational awareness around the "aviate and navigate' priority. The new paradigm introduces more efficient ways to communicate through computer technologies. In the new millennium cockpit, computer technologies have been introduced: in the Environment (E)-Liveware (L) linkage though ADSB (In) to communicate visually to pilots about terrain, other aircraft and weather, in the Software (S)-Liveware (L) linkage using EFB to communicate information visually to fly with and in Liveware (L)-Liveware (L) using Datalink and texting to communicate visually with digitized written language to others. This new paradigm of computers in the cockpit is about communicating visually with pilots. Referring back to the old adage of 'Aviate, Navigate and Communicate', the prioritizing situational awareness word of 'Communicate' was deemed the last priority in keeping overall situational awareness, but now could suddenly become a higher priority with these new NextGen cockpit computer technologies. Is it possible that 'Aviate and Navigate" could be affected by these new paradigm visual communication devices? Could these devices cause visual communications to sometimes interfere and overwhelm the 'Aviate and Navigate' situational awareness priorities?

Figure 2 is a simple Risk Assessment Matrix [2] that exposes the potential for problems with NextGen computer technologies in U.S. cockpits in 2020 and beyond. Across the top of the Matrix from right to left shows the slow increase of usage of computer information and automation in cockpits from 1980 to the 2020 FAA NextGen mandate and beyond. Once 2020 occurs, so begins the common use of all the new computer communications devices (ADSB (In), EFB and Datalink) in the cockpit. Due to cost affordability, the fast growth in these NextGen cockpit technologies will not happen immediately, but these communications computer tools for the cockpit will increase in usage beyond 2020 and eventually the sheer numbers of this growth will increase the probability of the occurrence of a loss of situational awareness related to Aviate, Navigate and Communicate; especially if human factors standards and training are not addressed. However, even with a herculean effort of human factors training and safety campaigning by the FAA, the most critical area of 'Aviate' (flying the aircraft safely) related to situational awareness could be left vulnerable on the left 'Severity' side of the Risk Assessment Matrix. This is because the lower priority of 'Communicate' in terms of the NextGen equipment could become visually overwhelming. The main reason why this should be concerning is because we are at a juncture in using all these new computerized cockpit tools together while having very little understanding of how they work with the human mind in flight cognitively.

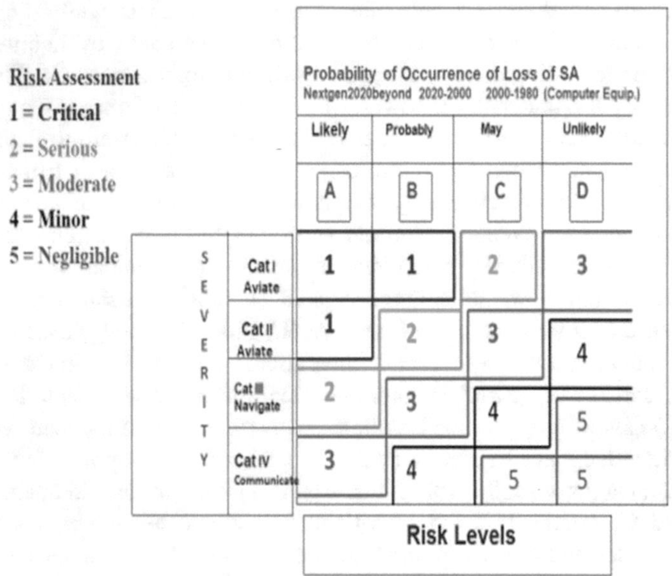

Risk Assessment
1 = Critical
2 = Serious
3 = Moderate
4 = Minor
5 = Negligible
Probability of Occurrence of Loss of SA
Nextgen2020beyond 2020-2000 2000-1980 (Computer Equip.)
Likely
Probably
May
Unlikely
A
B
C
D
SEVERITY
Cat I Aviate
Cat II Aviate
Cat III Navigate
Cat IV Communicate
Risk Levels

Fig. 2. Risk Assessment Matrix of the loss of situational awareness from the increased use of computer information and automation in modern cockpits versus (Aviate, Navigate and Communicate) [2]

7 NextGen 2020 Cognitive Processing Challenges

Previously discussed was the aspect that text information will replace audio, and that digitized information will require looking down (inviting other issues), including the tendency to turn off information feeds to reduce confusion and overload. Some potential deficiencies in cognitive processing that pilots are likely to encounter when adopting the new technology and procedures include confusion when interpreting the digital output, distraction and excessive loading in working memory, and reduced outward scanning for situational awareness. Since ADSB (Out) will be required in Airspace Classes A, B, C, and E (above 10,000 feet), services like TIS-B (traffic information) and FIS-B (flight information) will add to the cognitive processing requirements for pilots, some of whom may not be familiar with these flight demands. With regard to ADSB requirements now in effect in Europe, and required in 2020 within the U.S., potential cognitive differences may include latency in communications, alerts, symbology, colors, selection of traffic by crew, and integrating TCAS alert symbology. Other concerns [3] illuminate variations in electronic charting, e.g., terrain, airspace, approach paths, and landing systems. With requirements for digitized information and display increasingly mandatory in aircraft, the ability to discern similarities and differences accurately could present a challenge for some general aviation pilots. With standardization of display information elusive among manufacturers, this could well present a problem into the foreseeable future.

An earlier convention was that humans process information at a set rate, although later evidence showed the rate varies based on individual skills and type of information involved. Limited capacity theory suggests a limit to how much information can be allocated to performance, influenced by task complexity, and the allocation for primary and secondary tasks. The serial process is sequential, the parallel process provides for two or more channels operating simultaneously (although independently), and a hybrid variant that may process serially and in parallel with convergence, but can produce bottlenecks. In naturally occurring channels for vision or symbols these flow smoothly in parallel channels. However, where multiple visual signals are moving in the same channel, capacity is reached more readily and cognitive slowing may result [4]. This suggests that working memory might take a parallel processing track as opposed to sequential, which doubles the neural resources required and accelerates onset of compromised cognitive processing. As the growth in visualized digital data increases on the flight deck and in cockpits, the susceptibility for such delay in cognitive processing increases.

The term multitasking describes performing multiple tasks at once, although the evidence does not address adequately the issue of how people designate primary and secondary tasks. This has prompted a concept of task-switching to explain how multitasking is effective. Wickens [5] has determined that performance decrement rests on whether more than one task is performed simultaneously calling upon the same perceptual, cognitive, and psychomotor resources. Since most tasks are performed in stages, resources are adequate for demands made, however, an individual's load capacity may be reached where a single, large task becomes paramount. This might be the case where deconfliction decision making takes precedence with a less experienced operator in the NextGen 2020 environment.

8 Cognitive Loading and NextGen 2020

Cain [4] defined mental workload as measures that characterize task performance relative to operator capability. Earlier views that workload was principally additive, with demands on undifferentiated resources, has been replaced by the perspective that information processing comprises multiple resources operating differentially according to task complexities. The inference of cognitive loading initially was measured by direct observation of performance and use of rating scales and similar instruments to gauge decrements in task execution. As psychophysiological measures have entered the literature in greater emphasis, the point has been made that physiological methods do not measure imposed load, and instead provide information of individual responses to load. With less experienced operators entering the ADS-B environment, unaccustomed cognitive loading may tax some pilots and crewmembers.

Variable capacity theory [6] provides for operator intentions in setting task priorities and expanded channel capacity as workload increases, although fixed limits do not appear to be reliably predictable. Coping and resilience have been suggested as explanations for variable capacity and, along with several other proposed explanations for adaptive responses, have opened the investigation into variable capacities subject to situational relevance and individual states [7]. Fatigue, for instance, has been shown to reduce speed of output [8]. In high task situations this will be further exacerbated.

A comparison to studies of text messaging brings to bear a directly relevant issue with digitized cockpit communications. A comparison of heavy truck operation [9] and aircraft suggests that elevated risk for crash or near-crash increases 2300% over non-texting operations. Where ADS-B is initiated in single pilot operations, with no second crew member available, it is prudent to assess the potential for a similar elevated risk potential. Accompanying elevated risk invites increased anxiety, which increases attention to threat-related stimuli that can arise from confusion while attempting to comprehend the ADS-B information for less experienced pilots [10]. Recognition primed decision making applies where a learned optimal response has been successfully employed (often in emergencies) and can quickly be evaluated to meet a situation [11]. In such events, very experienced operators who regularly rehearse emergency events can evaluate a situation more rapidly with coherence and have the benefit of RPD, where general aviation pilots might be less able to develop a comparable level of skill.

Essentially, the adoption of ADS-B introduces several added degrees of freedom in cognitive processes as a result of NextGen procedures [12]. When considering Wickens' [5] concept for rearrangement of cognitive channels, where variable upper limits exist, the potential for elasticity is enhanced. In the case of general aviation pilots, however, not acquiring the associated task selection and sequencing skills actually reduces the degrees of freedom and invites potentially catastrophic outcomes. With added ATM and ADS-B requirements, general aviation pilots are more likely to reach a fixed upper limit of channel capacity and resulting notable decline in performance. A further consideration is the difficulty when bifurcation becomes imminent, as with a critical decision point. The influence of ensuing instability in cognitive processes suggests that pilots with high or excessive cognitive loading may alter a behavior pattern that could set into action an undesired sequence of events.

Further study of potential areas for concern in meeting the NextGen 2020 environment is needed. Task performance is situationally dependent, however, changes and response time are normally generalizable. A minus is that analysis may not indicate the unobserved part of the process. Still unresolved is the issue of whether mental workload is a scalar or vector quantity, particularly in regard to predictive modeling. As a scalar measure, cognitive loading is approached as a one-dimensional measurement (magnitude) of a single quantity. A vector approach, on the other hand, can have two measures (e.g., magnitude and direction) associated with a quantity. The relevance to cognitive workload is in determining relationships among vector measures and subsequent reliability and validity for prediction of cognitive processing and behaviors [4]. The effect is somewhat obvious when considering the added cognitive loading for NextGen 2020 ADS-B and allied demands for new or less experienced pilots.

Operators with poor understanding of a situation are prone to errors. Where this is attributable to lack of awareness rather than proficiency failure, a question of cognitive ability has been investigated. Working memory and spatial memory have been areas of special interest, along with cognitive style characteristics [13]. With added cognitive load, this could precipitate earlier onset of fatigue, which is reflected in the CUSP model, described operationally as a decrement in work capacity over time. With added cognitive workload associated with NextGen 2020, fatigue occurs earlier and memory and perception are the first to degrade [14]. This can be illustrated with the complexities the general aviation pilot might encounter when ADSB (In) becomes a requirement. The U.S. FAA has mapped this system and complexities in Advisory Circular 172-B.

In summary, the growing optical and cognitive workloads for pilots in a digitized environment, represented in SHELL Model 2017, will likely be challenged further with implementation of NextGen 2020 and ADS-B. That, and increasing use of electronic flight bags, accelerates risk of cognitive overload, confusion, fatigue, and loss of situational awareness. Regulators, manufacturers, operators, and pilots might take notice of these impending threats and address them in upgraded training and procedures.

References

1. Miller, M.D.: Aviation human factors: The SHELL model 2017 and computer/human factors analysis. In: Presentation to FAA Aviation Safety Conference, Honolulu (2016)
2. Miller, M.D.: Human Factors Computer Information/Automation Beyond 2020 NextGen Compliance: Risk Assessment Matrix of Situational Awareness (Cockpit Computer Use versus Aviate, Navigate, Communicate). Presentation to FAA Aviation Safety Conference, Honolulu (2017)
3. Chandra, D.C., Kendra, A.: Review of safety reports involving electronic flight bags. U.S. Department of Transportation, Report Number DOT-VNTSC-FAA-10-08, Washington DC (2010)
4. Cain, B: A Review of the Mental Workload Literature. Defense Research and Development Toronto (Canada) Conference Paper Number RTO-TR-HFM-121-Part II (2007)
5. Wickens, C.D.: Multiple resources and mental workload. Hum. Factors **50**, 449–455 (2008)
6. Ralph, J., Gray, W.D., Schoelles, M.J.: Squeezing the balloon: analyzing the unpredictable effects of cognitive overload. In: Proceedings of the Human Factors and Ergonomics Society, vol. 54, pp. 299–303. Sage, New York (2010)
7. Matthews, G., Campbell, S.E.: Sustained performance under overload: personality and individual differences in stress and coping. Theor. Issues Ergon. Sci. **10**, 417–443 (2009)
8. Lorist, M.M., Faber, L.G.: Consideration of the influence of mental fatigue on controlled and automatic cognitive processes. In: Ackerman, P. (ed.) Cognitive Fatigue, pp. 105–126. American Psychological Association, Washington, DC (2011)
9. Dingus, T.A., Hanowski, R.J., Klauer, S.G.: Estimating crash risk. Ergon. Des. **19**(4), 8–12 (2011)
10. Eysenck, M., Derakshan, N., Santos, R., Calvo, M.: Anxiety and cognitive performance: attentional control theory. Emotion **7**(2), 336–353 (2007)
11. Bond, S., Cooper, S.: Modeling emergency decisions: recognition-primed decision making. J. Clinical Nursing **15**, 1023–1032 (2006)
12. Hollis, J., Kloos, H., Van Orden, G.C.: Origins of order in cognitive activity. In: Guastello, S.J., Koopmans, M., Pincus, D. (eds.) Chaos and Complexity in Psychology: The Theory of Nonlinear Dynamical Systems, pp. 206–241. Cambridge University Press, Cambridge (2009)
13. Durso, F.T., Sethumadhavan, A.: Situation awareness: understanding dynamic environments. Hum. Factors **18**, 15–26 (2008)
14. Logie, R.H.: The functional organization and capacity limits of working memory. Curr. Dir. Psychol. Sci. **20**, 240–245 (2011)

Use of Dry Electrode Electroencephalography (EEG) to Monitor Pilot Workload and Distraction Based on P300 Responses to an Auditory Oddball Task

Zara Gibson[1(✉)], Joseph Butterfield[1], Matthew Rodger[1], Brian Murphy[1], and Adelaide Marzano[2]

[1] Queen's University Belfast, University Road, Belfast BT7 1NN, Northern Ireland
{zgibson01, j.butterfield, m.rodger, brian.murphy}@qub.ac.uk
[2] University of the West of Scotland, 1 County Place, Paisley PA1 1BN, Scotland
Adelaide.Marzano@uws.ac.uk

Abstract. This study aims to examine whether dry electrode EEG can detect and show changes in the P300, in a movement and noise polluted flight simulator environment with a view to using it for workload and distraction monitoring. Twenty participants completed take-off, cruise and landing flight phases in a flight simulator alongside an auditory oddball task. Dry EEG sensors monitored the participants' brain activity throughout the task and P300 responses were extracted from the resulting data. Results show that dry EEG can extract P300 responses as participants register oddball tone stimuli. The method can indicate workload for each condition based on the outputs from the EEG electrodes; landing (M = 287.5) and take-off (M = 484.6) procedures were more difficult than cruising (M = 636.6). With the differences between cruising and landing being statistically significant (p = .001). Outcomes correlate with participant NASA-TLX scores of workload that report landing to be the most difficult.

Keywords: P300 · Flight simulation · Workload · Dry EEG · Human factors

1 Introduction

1.1 The P300

Electroencephalography (EEG) is an electrophysiological monitoring method which is used to record electrical activity of the human brain from multiple electrodes/sensors placed on the scalp [1]. Event-related potentials (ERPs) are often extracted from the EEG signal. An ERP is a measured brain response that is generated by the brain as a result of and related to, a specific sensory, cognitive or motor event occurring internally or externally [2]. ERPs are particularly useful as they can be recorded noninvasively whilst also providing a range of useful information about cognitive processes such as

H. Ayaz and L. Mazur (Eds.): AHFE 2018, AISC 775, pp. 14–26, 2019.
https://doi.org/10.1007/978-3-319-94866-9_2

stimulus detection and attention. When recorded by EEG, the P300 (sometimes referred to as P3 or P3b in literature) surfaces as a positive deflection in voltage with a latency of roughly 250 to 500 ms (Fig. 1) [3].

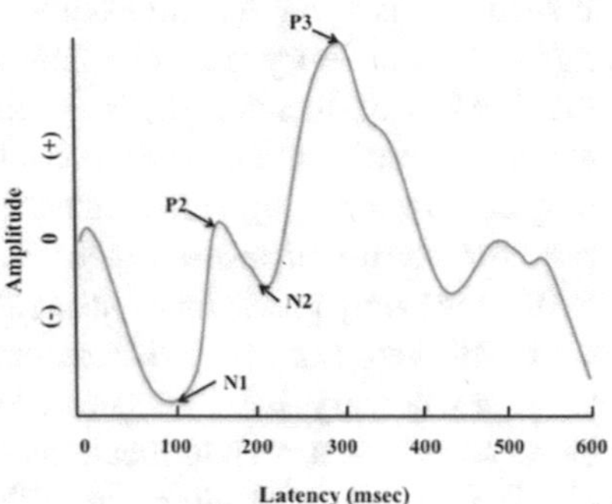

Fig. 1. An image of a P300 response to an unexpected stimuli (image from Waryasz 2017 [4])

The P300 amplitude is sensitive to the amount of attentional resources engaged during dual-task performance. Normally a primary task varying in cognitive demand is completed whilst the participant is engaged in a secondary task of mentally counting visual or auditory oddball stimuli. As the primary task increases in difficulty the P300 amplitude decreases. This is due to the primary task taking up more mental resources, thereby less are available to devote to the secondary oddball task. Hence, the amplitude of the P300 decreases to reflect the decrease in attentional resources devoted to the task. This occurs regardless of the modality of the primary task [5].

1.2 EEG and Flight Simulation

Since EEG emerged in the 20th century, there has been little variation in how EEG is measured. Currently manufacturers and researchers are moving towards developing wireless, mobile-based EEG and this has driven the development of alternative electrodes for physiological monitoring. The conventional wet adhesive Ag/AgCl electrodes used almost universally in clinical applications today provide an excellent signal but are not compatible for mobile use [6]. Technological advances in the area, such as a new generation of dry electrodes and wireless EEG caps, have opened the way for EEG to now be used in a much wider range of instances and places even more of a priority on developing aspects such as usability and signal quality [7]. In their review of the dry sensor EEG development, Lopez-Gordo et al. [7] suggest that although a broad diversity of approaches have been evaluated without a consensus in procedures and methodology, performance is not far from that obtained with wet electrodes. Hence, dry electrodes can be considered a useful tool in a variety of novel applications. One area of interest for dry electrode EEG is flight simulation. A wireless and portable EEG headset allows for more accurate flight-testing as pilots can interact normally without concern for wiring. A concern of utilising dry EEG however is that the movement of these dry electrodes against the scalp can severely interfere with the EEG signal, hence, research has been trying to resolve the issue [6].

Callan, Durantin, and Terzibas [8] utilised dry EEG in a motion platform-based flight simulator and compared this to an open cockpit biplane to determine if the technology can be used in such a noise-polluted environment. Their participant completed a passive task using random auditory presentations of a chirp sound (as this would not interfere with the flight task), to illicit ERPs. Their investigation suggested that dry EEG can be used in an environment with considerable vibrations, wind, acoustic noise and physiological artefacts and still achieve good single trial classification performance necessary for successful measurement.

However, Callan et al's [8] study only used one subject and used a 72-electrode cap (of which only a maximum of 46 were not rejected from each condition). The study proposed here aims to use a smaller dry EEG headset (with a smaller number of electrodes: 8) positioned based on prior research to locations showing the effectiveness of the parietal (P3, Pz and P4) sensors in recording the P300 and their neighbouring locations (C3, Cz and C4). Parietal locations have been established as areas more sensitive to the presentation of the P300 ERP and have been used in many P300 studies e.g. [9]. Additional measurement was taken at Oz and Fz as the P300 can be detected along the midline. This experiment will use more participants to gauge the effectiveness of the dry EEG across multiple individuals and will also utilise the oddball task (a task known for eliciting the P300 in neuroscientific research [10] and varying levels of flight complexity to manipulate workload). In other words this experiment is a development of Callan et al. [8] to evaluate the usefulness of portable dry EEG with multiple subjects in monitoring workload changes and attention allocation in a flight simulator setting.

2 Method

2.1 Equipment

Flight Simulator. The MP520 Engineering Flight Simulator configured to the default Cessna C172 was used in this experiment (Fig. 2(i)). The C172 is a single-engine, fixed wing aircraft. The model encapsulates a number of hardware systems within the main components, which include the cockpit housing, and the Instructor Operator Station (IOS). The simulator is equipped with flight and engine controls including; the control side-stick for pitch and roll control, rudder pedals for yaw and two throttle levers to control thrust. Switches operate flaps, spoilers, landing gear and brakes. As inexperienced participants were used, all flight performance data was displayed on a head-up display (HUD) overlaid on the visual scene. The communications system included a headset, which enables cockpit communication to the IOS intercom. Flight data, such as altitude, pitch and speed, were automatically recorded. The data logging rate was set to 1 Hz. The data was recorded as an Action Script Communication file (.ASC) which was exported to Microsoft Excel for analysis.

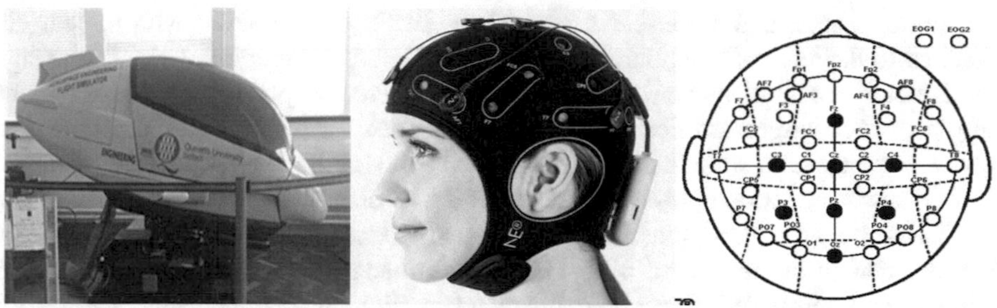

Fig. 2. Experimental design; (i) Merlin Flight Sim, (ii) ENOBIO set-up, (iii) Sensor locations

EEG Enobio 8. Enobio (Neuroelectrics, Barcelona, Spain; Fig. 2(ii)) is a wearable, wireless EEG sensor with 8 EEG channels and a triaxial accelerometer, for the recording and visualisation of 24-bit EEG data at 500 Hz. The Necbox is the core and the control unit of Enobio. The Necbox is a battery-operated device that connects through Bluetooth to the Neuroelectrics Instrument Controller (NIC) software running on a computer. The 10-electrode cable contains 8 channels, numbered from 1 to 8, for EEG monitoring, and two reference channels labelled with CMS and DRL. The Ear clip is an easy-to-use alternative to the sticktrode. It is a dual reference electrode because it is used to connect the two reference channels, CMS and DRL, to the same earlobe. A Neoprene Cap holds the sensors in place. In this study the locations Fz, Cz, C3, C4, Pz, P3, P4 and Oz were used (Fig. 2(iii)).

Paradigm Set-Up. Audacity software was used to cut a 1000 Hz soundtrack into a sound file 0.06 s long. A 0.01 s period of silence was played before 0.01 s of a fade in, 0.03 s of sound and 0.01 s of fade out (0.05 s sounds have been used in literature e.g. [11]. This fade-in and fade-out effect smoothed the tones. The same was done for a 2000 Hz soundtrack. PsychoPy software was used to create a two-tone oddball paradigm. A Psychopy database was set up to schedule the presentation sequencing and onsets of the normal and oddball stimuli. This way all participants will receive the same number of normal and oddball tones and at the same points of the trial. Three different sequences were created- one for each of the three trial types. In Psychopy a basic task was created with in introductory screen, an inter-stimulus interval (ISI), tone and end aspects. A loop was created around the ISI and tone to run for the length of the Excel file. Lastly, the Psychopy code was edited to incorporate a Lab-Streaming Layer which was linked to the Enobio recording software, NIC. This sent markers to the EEG recording to mark when a sound occurred and what type of sound occurred in the recording to isolate ERPs.

2.2 Procedure

Participants were pre-screened for the experiment by a self-exclusionary medical requirements sheet. Accepted participants were assigned a two-hour slot in which to complete the experiment. On the day, participants were briefed on the experiment, the EEG cap function and reminded to move their heads as little as possible throughout.

The participant signed a consent form and the EEG cap was fitted with the ear clip attached to the left ear as a reference. Lastly, an impedance reading was taken using a multi-meter (readings above 0 and below 20 kΩ were acceptable and all participants fulfilled this requirement prior to testing). Any issues, for instance signal problems or sensor-scalp contact, were identified and remedied at the beginning of the experiment.

Participants then completed a 3-trial learning session of their first procedure using the flight simulator. One trial with a verbal talk through of the procedure and one without, and an optional third trial if their performance could be improved. The participant then completed a recorded independent flight procedure. Next, participants were introduced to the oddball task and were given a 30-s practice session to aid familiarisation with the normal and oddball tones. The communications system operated via two headsets, one for the pilot and one for the instructor. When the oddball task was occurring, the instructor placed the headset at the speaker of the laptop to transfer the sound to the pilot's headset. There was no echoing of the sound due to the enclosure of the simulator. Participants then repeated the flight procedure with the listening task. This flight learning sequence was repeated for each of the three trials.

Trials were randomised between participants to help remove learning effects e.g. some started with take-off and others with cruising or landing. Furthermore, the first half of participants completed the control flight condition first and the second half completed the oddball flight condition first.

At the end of each trial a NASA-TLX questionnaire was given to participants i.e. after take-off, cruising and landing. The NASA-TLX is a useful self-report measure to gauge participants' own opinion of how difficult the task is [12]. Participants were instructed to rate each of the six dimensions a score from 0–20 and the raw scores were used for analysis purposes. Participants were also given additional questions asking how well they followed the instructions, how well they feel they performed and which of the three flight procedures was the most difficult in their opinion.

2.3 Data Analysis

EEG recordings were imported into the EEGLAB [13] add-on for Matlab (a software programme that enables analysis and design processes using Python coding) where they were pre-processed. Channel locations were then added and data was filtered using a 0.5 Hz high pass filter and a 40 Hz low pass filter. Using ERPLAB (a plug-in for the EEGLAB software [14]) the EEG file was epoched (the extraction of time windows) from 200 ms pre-stimulus (before the marker) to 800 ms post-stimulus (after the marker). The data was also baseline corrected to the period pre-stimulus.

Once data was epoched, artefact rejection techniques available in ERPLAB were utilised to identify 'bad' epochs. The techniques used were moving window peak-to-peak threshold, simple voltage threshold, blocking and flat lining, Sample-to-sample threshold and rate of change. If a bad channel was identified the channel would be removed from the artefact rejection phase as it would result in all epochs being flagged for rejection. Once 'bad' epochs were flagged, the data was viewed to manually identify any visual trials not picked up by the artefact rejections techniques. These epochs were then removed from the dataset. A median ERP was calculated for each sensor in each condition for each participant ($8 \times 3 \times 20$). Once this was done the mean across the sensors and conditions was found (8 × 3). From these mean waveforms the area under the curve was calculated using the Trapezium rule in Matlab to quantify the difference in P300 responses.

Flight performance was measured on the basis of how accurately participants followed the instructions e.g. percentage accuracy in the timing of meeting task requirements. For example, raising the landing gear at 250 ft following take-off. A further measure was the monitoring of participants pitch and altitude over the course of the trial and comparing to the average to gauge any deviations in flight path. The final measure participants self-reported workload and task difficulty in the NASA-TLX.

2.4 Hypotheses

Flight performance - performance will be consistent between the oddball and non-oddball flight conditions due to participants prioritising the manual flight task.

Oddball performance - accuracy of the oddball task will be affected by the type and complexity of flight task; the cruising flight task will show the greatest accuracy. A greater margin of error is more likely for landing and take-off as these are more difficult and place a greater load on working memory, landing will show the biggest error as this is the most difficult and time pressured task.

ERP analysis - outputs will show that with increased workload the P300 amplitude and area will decrease. Tones may be missed due to prioritising the flight task over the oddball when in conditions of higher workload. In the cruising task, the area will be the greatest as this is the least demanding task, followed by take-off and landing.

NASA-TLX - subjective workload measures will be higher for the landing and take-off tasks than for cruising. Landing will be perceived as the most workload-inducing task, followed by take-off and cruising the lowest workload task.

3 Results

3.1 Physiological Measures - Oddball Performance and ERP Outputs

Results for the oddball task showed that participants were most accurate in the cruising condition (Mean estimate error = −0.96%) and least accurate in the landing condition (Mean estimate error = +4.48%). Performance in the take-off condition fell in between (Mean estimate error = +1.40%). This could imply participants were not attending the oddball task as much in the landing scenario and were attending the task more in the cruising condition. Figure 3 shows participant's percentage error in the task.

Some participants also appeared to struggle with the take-off task. Three participants performed perfectly in the oddball task across all three conditions. As stated previously, gaining a score of less than 75% in the oddball task would mean the participant has not fulfilled the task adequately; from the information above, two participants (P14 and P18) failed the task during take-off, two participants (P17 and P18) failed the cruising oddball and three participants, (P17, 18 and 19) failed the landing task. P18 failed all three listening tasks by over- or under-estimating the number of oddball tones in the task.

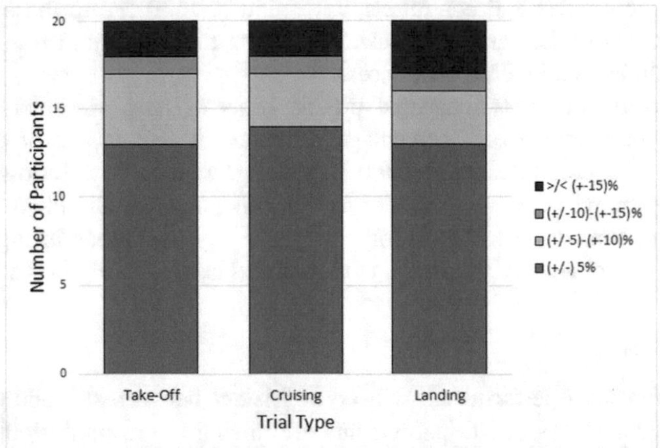

Fig. 3. Participant percentage error in the oddball counting task

From individual ERP outputs, a grand average was calculated for each procedure and the graphical outputs below show the wave pattern for each sensor. Fz, Cz and Pz are the core areas for measurement of the P300 and these are the sensors displayed below. The C4 and P3 sensors were more difficult to get reliable readings from due to variability in participant head shape and suffered from offset interference in cases. As a result, they are excluded from the results; below in Fig. 4 the average P300 measurements are displayed for each flight task.

From Fig. 4 it appears that the cruising and take-off conditions show more activation than the landing condition, indicating that more attention is devoted to the oddball task in these conditions. In order to analyse the relative activation differences between flight procedures a positive waveform analysis was carried out using the trapezoidal method on all positive inflections to quantify the area under the curve between 450 ms and 798 ms. The comparison between areas is shown in Fig. 5.

From Fig. 5 is it apparent that cruising shows the most activation at for all channels but shows similar activation to take-off at Pz. From the proposed theory of activation representing attentional resource allocation. This shows that participants devoted more attention to the oddball task in the cruising condition. This is also reflected in their oddball performance, as the performance in the task was more accurate. The reverse pattern is found in landing in which less attentional resources are allocation to the oddball task according to the area under the curve for the ERPs and this coincides with this being the task in which oddball performance was the poorest.

T-tests were used to measure any significant differences in area between conditions (Bonferroni correction, p = .0167) only the difference between cruising (M = 636.6, SD = 46489.5) and landing (M = 287.5, SD = 24997.7) was statistically significant (t (6) = 6.62, p = .001). There was no significant difference between take-off (M = 484.6, SD = 52212.5) and cruising (p = .160) or take-off and landing (p = .017).

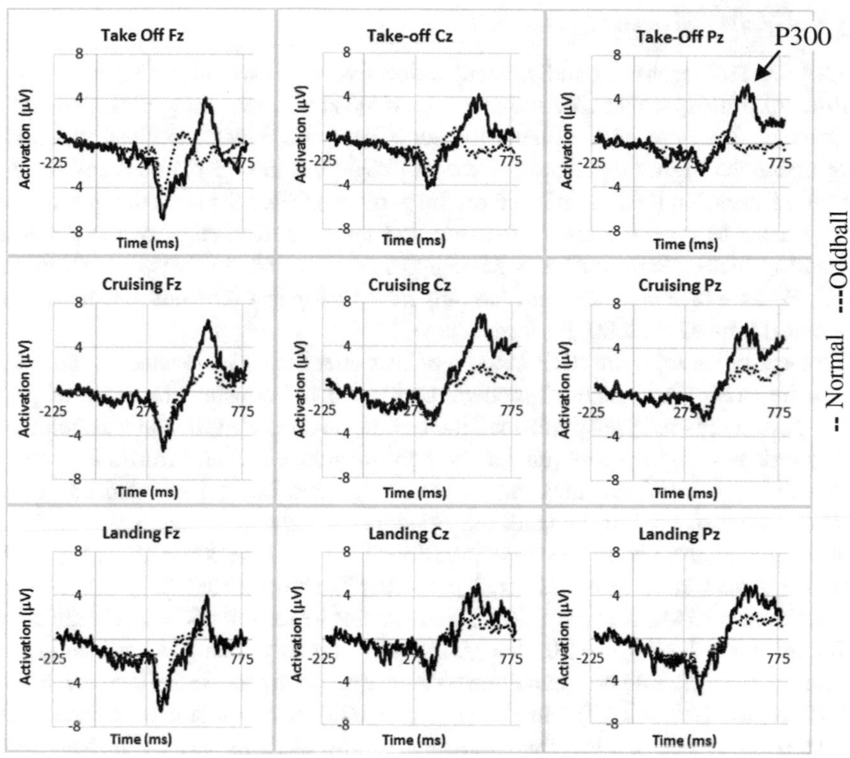

Fig. 4. ERP outputs for Fz, Cz and Pz sensor locations across the flight tasks

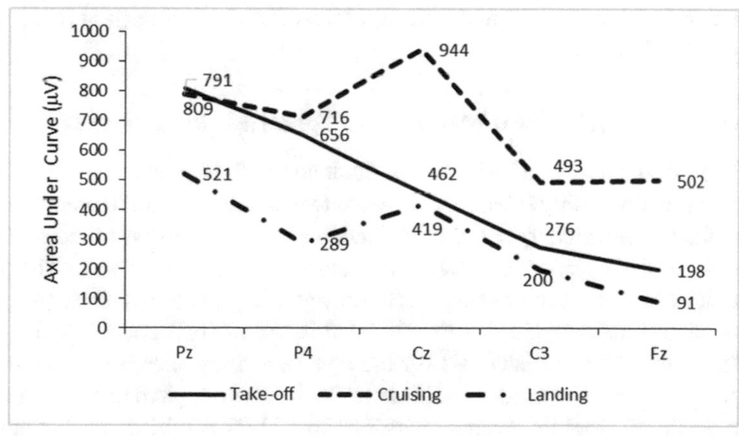

Fig. 5. Positive waveform analysis for each flight procedure with respect to the sensor location (parietal to frontal locations)

3.2 Subjective Measures - NASA-TLX

The NASA-TLX measures individuals' subjective opinions on six factors. For the purposes of this experiment, these have been divided into two categories, demand and performance. For demand the higher the score the more difficult the task was and for performance the higher the score the worse people felt they performed and the more effort and frustration the task inferred on the participant. Results from the questionnaire show cruising to have the lowest demand and lowest performance of the three tasks (Demand = 16.30, Performance = 21.45). Take-off came in between (Demand = 23.10, Performance = 23.25) and landing had the highest demand and highest performance (Demand = 26.80, Performance = 27.50).

This coincides with the ERP outputs as less attention was devoted to the oddball task in the landing and take-off procedures than to the cruising. In terms of performance, participants felt they performed the best in the cruising task and least well in the landing task as a higher score reflects the performance, effort and frustration involved in the task. Two Freidman rank tests were carried out, one for demand and one for performance rankings. For demand, the Friedman test showed there was a significant variation in demand across the three conditions (χ^2 (2, N = 20) = 12.342, p =.002). Wilcoxon signed rank tests (adjusted p = 0.0167) showed that only the difference between cruising (Mean Rank = 2.58) and landing (Mean Rank = 1.50) ranks were significant (W = 8.67, N = 20, p = .005) with landing being significantly more demanding than the cruising condition. There was no significant difference between take-off (Mean Rank = 1.93) and cruising (p = .025) or between take-off and landing (p = .163). The Friedman test for performance showed there was a significant difference in performance between the three conditions (χ^2 (2, N = 20) = 6.911, p = .032). Here the Wilcoxon signed rank test showed no significant interactions between take-off (Mean Rank = 1.98) and cruising (Mean Rank = 2.43, p = .205), cruising and landing (Mean Rank = 1.60, p = .024) and between take-off and landing (p = .115) when the adjusted p-value is used.

3.3 Performance/Behavioral Measures - Flight Task Performance

Take-Off. Participants' were instructed to pitch up to 15 degrees at a speed of 65 knots and maintain this until 1000 ft where they were to use pitch control to attempt to remain at 1000 ft for the remainder of the trial. Landing gear was to be retracted at 250 ft. Performance in the ascent was relatively consistent for participants. The range of maximum deviation from the average altitude was 76.23 ft to −64.85 ft from average for control conditions and 83.66 ft to −77.97 ft in the oddball condition (P9 data was excluded from the deviation analysis for the control as they veered off course halfway through the ascent under control conditions). The Regarding pitch control, the range of maximum deviation from the average was 7.55 to −11.07 for the control condition and 13.56 and −9.26 for the oddball condition. Lateral control of the plane can be an issue for some participants; one participant veered as far as 454 m left of the straight course in the control condition and one participant veered over 1000 m to the right in the oddball condition. For take-off, participants had to pitch upward at an indicated speed of 65 knots. Generally, participants followed this instruction very well. All participants took off very close to 65 knots, 75% were within 5% of the target speed for the control flight and 70% for the flight with the listening task, the remaining 15 and 20% respectively, fell within 10% of the target speed. Another instruction was to retract the landing gear of the aircraft at an altitude of 250 ft. Participants were much less accurate at raising the landing gear at 250 ft; two of the twenty participants completely forgot to perform the task in the dual-task condition. Two people fell outside 15% of the target altitude in the normal condition and one in the dual-task. One participant in each condition was between 10 and 15% of the target.

Cruising. For this procedure participants had simply to maintain their altitude at 1000 ft by controlling the pitch of the aircraft. Regarding staying at 1000 ft participants performed reasonably well, for the control condition participants flew with a deviation range of +9.61% and −11.2% of the target 1000 ft. For the oddball condition performance was more or less the same with a deviation range of +5.06% and −17.18% of the target 1000 ft range. Similar to take-off, lateral control was the most inaccurate part of the procedure. In the control condition deviation ranged from 410.53 m to the left rom start point to 241.00 m to the right, interestingly the oddball condition range was larger with one participant veering as far as 912.18 m off course yet another participant held the straight trajectory perfectly for the entirety of the trial.

Landing. In this procedure participants had to descend from an altitude of 2000 ft at a pitch of 3° to land on the runway and deploy the landing gear and flaps at 500 ft of altitude. Generally, participants performed consistently on their descent (135.13 ft above and 182.29 ft below the average in the control flight and 239.22 ft above and 239.47 ft below in the oddball condition). Once participants reached the point of switching on the flaps and landing gear, however, the flight path deviation was the greatest (455.03 ft above and 259.81 ft below the average path in the control and 312.16 ft above and 313.89 ft below in the oddball condition). The average pitch for the trial was −3.58 in the control with a range of +26.73 to −11.91 and a −3.58 average in the oddball with a range of +27.73 to −11.03. This shows participants tried to maintain the −3° pitch throughout, the deviations in range are mainly when the flaps are deployed as this causes lift in the aircraft, which is more difficult to control. All except one participant successfully landed on the runway in both conditions, one participant over shot due to not deploying the flaps completely and losing control of the descent. Only one participant deployed the landing flaps outside of a 15% error rate, all the others were ±15% in accuracy.

4 Discussion

As predicted in our hypotheses, the dry electrode EEG was sensitive enough to detect changes in the P300 in a movement and noise polluted environment. This agrees with the research carried out by Callan and colleagues in their 2015 study [8]. Further to this, the method used above allowed for further ERP analysis to determine positive waveform area. The results support the hypothesis in that largely cruising area was higher than take-off and landing had the smallest area overall.

The t-tests carried out on the positive waveform analysis showed that the difference between cruising and landing conditions was significant but not between take-off and cruising. and take-off and landing. This makes sense when the task requirements are considered. Take-off was considered an intermediate workload as it has some of the pressure and workload involved in landing and then becomes more similar to cruising as the trial continues. It follows that take-off is not significantly different from either of those tasks. Whereas, cruising is a consistent monitoring and marginal adjustment task, landing is a focused procedure that requires manual interaction with the simulator, thresholds to meet and a target. Hence, the tasks differed from each other on level of difficulty with cruising being easiest, landing being the most difficult and take-off falling somewhere in-between. The results showed that landing is significantly more difficult than cruising but neither are significantly different from take-off.

The results of the subjective NASA-TLX measure of this experiment showed Cruising to be the easiest task according to the demand questions and participants felt marginally more content with their performance in cruising than the other tasks. When t-tests were performed, the only significant difference was between the demand of cruising and of landing. This corresponds to the effect we see in the ERP analysis. Interestingly there were no significant differences in performance scores and this also reflects the effects we see in the flight data. This could reflect a task-prioritisation coping mechanism that has been seen throughout research and specific instructions can lead to prioritisation [15]. Yogev-Seligmann, et al. [15] found that adults significantly increased gait speed compared to the control condition when told to prioritise gait. Gait speed was reduced when priority was given to the cognitive task. In this study, participants were given no particular instructions on which of the two tasks to prioritise (flight/listening task), it could be that the lack of prioritisation meant participants prioritised the ecologically valid flight task; the task that in reality could have serious consequences if not performed adequately. In future work it may be interesting to see how participants cope when told to prioritise one task over another.

As predicted, flight performance did not vary significantly between normal and oddball conditions. This could support the theory that individuals when faced with dual task conditions prioritise the primary task [15]. In terms of oddball performance, the average percentage error did change with task in that landing had a higher average percentage error but this was still within 5% of the actual answer. Take-off average was within 2% accuracy and cruising average was within 1% average accuracy. What is interesting in the study is the diversity with which participants adapted to aircraft control. P1 struggled in lateral control of the simulation in the normal condition but not as much in the oddball condition for take-off. A possible explanation could be practise effects (as they completed the normal condition first and completed the take-off procedure first). A recent study [16] showed that both implicit and explicit knowledge help in dual task performance i.e. knowledge gained from single task conditions can benefit performance in dual task conditions. Alternatively, it could be reasoned that divided attention made performance better. Interestingly dual task training can improve the automatization of the primary task [17].

An interesting avenue for future research could be the effect of personality on novel task performance. The anxiety-performance relationship [18] relates to sporting achievement and the phenomenon of 'choking under pressure'. Anxiety can cause an impact on motor performance, mediated by the individual's confidence in the automaticity of performance under stress; this is termed skill establishment [18]. The more established and automatic the skill becomes, the less affected it will be by participant anxiety/ nerves. This could be a potential avenue for future investigation as well as the role of practice and how this interacts with personality.

Overall, cruising and landing show significant differences in workload with take-off workload somewhere between. This three-measurement approach to participant's workload (physiology, self-report and performance) has shown effectiveness in determining the level of workload flight conditions and procedures induce. Future work could apply the methodology to different scenarios and conditions using qualified pilots to gauge workload and the differences in how pilots cope with workload compared to inexperienced user used in this study and the methodological changes such as using varied cap sizes to help with better electrode placement and participant comfort.

In conclusion, the dry electrode EEG cap has shown a great potential for deciphering ERP outputs to a small scale of analysis that are relatable to other workload measures. For instance, the differences in NASA-TLX results due to phase of flight simulation were in the same direction as those of flight-phase differences in ERP positive waveform analysis. This opens up multiple avenues for future research in this, and other, disciplines.

References

1. Sanei, S., Chambers, J.A.: EEG source localization. In: EEG Signal Processing, pp. 197–218 (2013)
2. Luck, S.J.: Event-related potentials. In: Cooper, H., Camic, P.M., Long, D.L., Panter, A.T., Rindskopf, D., Sher, K.J. (eds.) APA Handbook of Research Methods in Psychology: Foundations, Planning, Measures, and Psychometrics, vol. 1, pp. 523–546. American Psychological Association, Washington, DC, US (2012)
3. Polich, J.: Updating P300: an integrative theory of P3a and P3b. Clin. Neurophysiol. **118** (10), 2128–2148 (2007)
4. Waryasz, S.A.: The Clinical Utility of P300 Evoked Responses in Post-Sport-Related Concussion Evaluation, Pathways On, Neuroanatomy at the Neuroaudiology Lab at the University of Arizona (2017)
5. Kramer, A.F., Wickens, C.D., Donchin, E.: Processing of stimulus properties: evidence for dual-task integrality. J. Exp. Psychol. Hum. Percept. Perform. **11**(4), 393–408 (1985)
6. Chi, Y.M., Jung, T.P., Cauwenberghs, G.: Dry-contact and noncontact biopotential electrodes: methodological review. IEEE Rev. Biomed. Eng. **3**, 106–119 (2010)
7. Lopez-Gordo, M.A., Sanchez-Morillo, D., Valle, F.P.: Dry EEG electrodes. Sensors **14**(7), 12847–12870 (2014)
8. Callan, D.E., Durantin, G., Terzibas, C.: Classification of single-trial auditory events using dry-wireless EEG during real and motion simulated flight. Front. Syst. Neurosci. **9**, 11 (2015)
9. Cecotti, H., Rivet, B., Congedo, M., Jutten, C., Bertrand, O., Maby, E., Mattout, J.: A robust sensor-selection method for P300 brain–computer interfaces. J. Neural Eng. **8**(1), 016001 (2011)

10. Squires, N.K., Squires, K.C., Hillyard, S.A.: Two varieties of long-latency positive waves evoked by unpredictable auditory stimuli in man. Electroencephalogr. Clin. Neurophysiol. **38**(4), 387–401 (1975)
11. Nowak, K., Oron, A., Szymaszek, A., Leminen, M., Näätänen, R., Szelag, E.: Electrophysiological indicators of the age-related deterioration in the sensitivity to auditory duration deviance. Front. Aging Neurosci. **8**, 2 (2016)
12. Rizzo, L., Dondio, P., Delany, S.J., Longo, L.: Modeling mental workload via rule-based expert system: a comparison with NASA-TLX and workload profile. In: IFIP International Conference on Artificial Intelligence Applications and Innovations, pp. 215–229. Springer, Cham (2016)
13. Delorme, A., Makeig, S.: EEGLAB: an open source toolbox for analysis of single-trial EEG dynamics including independent component analysis. J. Neurosci. Methods **134**(1), 9–21 (2004)
14. Lopez-Calderon, J., Luck, S.J.: ERPLAB: an open-source toolbox for the analysis of event-related potentials. Front. Hum. Neurosci. **8**, 213 (2014). https://doi.org/10.3389/fnhum.2014.00213
15. Yogev-Seligmann, G., Rotem-Galili, Y., Mirelman, A., Dickstein, R., Giladi, N., Hausdorff, J.M.: How does explicit prioritization alter walking during dual-task performance? Effects of age and sex on gait speed and variability. Phys. Ther. **90**(2), 177–186 (2010)
16. Ewolds, H.E., Bröker, L., De Oliveira, R.F., Raab, M., Künzell, S.: Implicit and explicit knowledge both improve dual task performance in a continuous pursuit tracking task. Front. Psychol. **8**, 2241 (2017)
17. Clark, D.J.: Automaticity of walking: functional significance, mechanisms, measurement and rehabilitation strategies. Front. Hum. Neurosci. **9**, 246 (2015)
18. Carson, H.J., Collins, D.: The fourth dimension: a motoric perspective on the anxiety–performance relationship. Int. Rev. Sport Exerc. Psychol. **9**(1), 1–21 (2016)

Analysis of Alternatives for Neural Network Training Techniques in Assessing Cognitive Workload

Colin Elkin(✉) and Vijay Devabhaktuni

EECS Department, The University of Toledo, Toledo, OH, USA
Colin.Elkin@rockets.utoledo.edu,
Vijay.Devabhaktuni@utoledo.edu

Abstract. Cognitive workload serves as a vital component in many human factors applications. Furthermore, the ability to make assessments, classifications, and predictions of mental load is a well-established yet ongoing research challenge. A wide arsenal of machine learning mechanisms has become available that address cognitive workload assessment, such as support vector machines and artificial neural networks. Due to the longevity of and continuing interest in the latter technique, this paper focuses on neural networks, diving into the many intricate variables and parameters that can make or break an effective neural network model in this area. To evaluate and compare these approaches, we obtain two distinct physiological datasets. Overall, the results indicate that under both datasets, the quasi-Newton optimizer contains a slight edge in accuracy, while stochastic gradient descent is more computationally efficient. Under the second and larger dataset, however, an unsupervised model boasts significantly lower computational runtime while maintaining similar levels of accuracy.

Keywords: Cognitive workload · Machine learning
Artificial Neural Networks · Restricted Boltzmann Machines

1 Introduction

The ability to make classifications, predictions, and assessments of cognitive workload in essential applications remains an ongoing research challenge. More recently, a variety of machine learning algorithms have become available that address cognitive workload assessment quite well. These range from support vector machines and decision trees to k-nearest neighbors and artificial neural networks. As a result, recent research efforts have aimed to compare and analyze many of these different mechanisms [1], aiming to determine the costs and benefits of one over the other in different contexts of cognitive workload applications. Due to the longevity and current relevance of the latter technique, however, this paper places specific focus on neural networks, diving into the many experimental parameters that can make or break an accurate neural network model in vital human factors applications.

H. Ayaz and L. Mazur (Eds.): AHFE 2018, AISC 775, pp. 27–37, 2019.
https://doi.org/10.1007/978-3-319-94866-9_3

1.1 Cognitive Workload

In general, cognitive workload refers to the level of mental resources being stored in one's working memory. In most essential activities, the underlying goal is to balance such a load so that it is neither too high nor too low. Based on prior research efforts, detection and prevention of mental overload appear to be the more established objectives, particularly in applications such as general cognitive tests [2], piloting of remotely operated vehicles [3], and science education [4]. Overload generally occurs when one must undergo numerous parallel tasks and/or difficult tasks, which results in increased human error. Depending on the application, consequences can range from decreased productivity to mission failure or even imminent danger. With that said, underload is another extreme that often warrants a significant amount of attention. For instance, a student in a classroom may be susceptible to cognitive overload, such as taking a difficult exam with too short of a time limit and encountering countless distractions such as muffled conversations outside the room, a marching band practice outside, and a classmate's cell phone ringing. However, he or she may at other times experience underload when sitting through a lecture that is uninteresting or difficult to understand, functioning on very little sleep, and periodically looking at the clock to find that the class in nowhere near over. Thus, when considering cognitive load, the optimal approach is to treat it as a balancing act, keeping individuals at a peak amount of mental focus to increase productivity and reduce human error.

Ideally, one can achieve this optimal balance by early detection of overload or underload through prediction, classification, or a more subjective form of assessment. Through the wide realm of machine learning, there exist many different mechanisms for achieving such assessments.

1.2 Machine Learning

Over the years, machine learning has become an invaluable interdisciplinary research tool that has aided in many different theoretical and applied fields. Thus, cognitive workload assessment is no exception, as established by many well-known machine learning mechanisms. Artificial neural networks (ANNs) remain one of the most established methods, spanning general classification of workload [5, 6] as well as specific applications such as air traffic control [7]. Support vector machines (SVMs) have been successful in assessing a driver's cognitive state [8–11] as well as addressing mental load from a more general context [12, 13]. Classification of driver workload has also succeeded under decision trees (DTs) and k-nearest neighbors (KNNs) [14]. Additional kNN applications include cognitive stress recognition [15] and brain-computer interfaces [16].

Regardless of application, successful implementation of machine learning begins with proper identification of desired inputs and outputs. The former typically consists of physiological data, such as electroencephalogram (EEG) [2] and Electrodermal activity (EDA) [17], while the latter is often a classification regarding a cognitive task [1]. This paper incorporates all these inputs and outputs in order to form a comprehensive cost-benefit analysis of different machine learning parameters. Building upon the work conducted in [1], all comparisons here are based solely on ANNs, evaluating different optimization techniques of both supervised and unsupervised neural networks.

1.3 Artificial Neural Networks

In supervised ANNs, two of the most established optimization methods include quasi-Newton and stochastic gradient descent. Regarding cognitive load, the latter method has modeled temporal dependencies using memory structures [18] and workload model performance [19], while the former aided in inferring mental load based on pupillary dilation [20] and both methods served as classifiers in decision support systems [21]. It is worth noting that only some of these applications have used these optimizers specifically for ANNs while others have utilized these as standalone numerical methods. For unsupervised ANNs, restricted Boltzmann machines have found success in assessing cognitive workload across multiple tasks [12].

1.4 Overview of Subsequent Sections

The rest of this paper is organized as follows. Section 2 provides more technical background regarding supervised and unsupervised ANNs. Section 3 gives an overview of how the research was conducted, and Sect. 4 presents and discusses the results. Finally, conclusions are drawn in Sect. 5.

2 Technical Background

An artificial neural network (ANN) is a common machine learning algorithm based on biological neural networks. Many types of supervised ANNs, such as Multilayer Perceptron (MLP), consist of multiple layers of nodes. In the case of MLP, these neural networks consist of input, output, and hidden nodes [22]. The latter can have one or more layers while inputs and outputs get one each. MLP employs a supervised learning technique that updates training weights across multiple iterations until forming an accurate model. Many types of training techniques are available for such an ANN, including stochastic gradient descent, which is a stochastic approximation technique and iterative method of minimizing an error function, and quasi-Newton, which is a low-cost alternative to Newton's method that approximates matrix inversion.

Unsupervised ANNs, such as restricted Boltzmann machines (RBMs), are a variation in which knowledge of the outputs is not required. While supervised ANNs such as MLP have neurons that connect to all other neurons in a neighboring layer, RBM neurons must form a bipartite graph between hidden and visible layers and have no connections between neurons within the same layer [23]. As a result, this type of mapping allows for greater training efficiency.

3 Simulation Setup

To address the costs and benefits of all three ANN optimizers, we conducted a variety of simulation scenarios using the scikit-learn machine learning library in Python [24]. An overview of the default simulation parameters in the experimental setup is as follows:

- Results presented as the averages of 30 independent trials for redundancy
- Data samples randomized during each trial with 80% for training and 20% for validation
- Supervised ANN classifier tested from 1 to 10 hidden layers of neurons and from 1 to 10 hidden neurons per layer
- Supervised ANN classifier trained with quasi-Newton and stochastic gradient descent optimizers with α value of 0.1 and a maximum of 1000 iterations
- Unsupervised ANN classifier trained with Bernoulli restricted Boltzmann machine under learning rates of 10^{-4}, 10^{-3}, 10^{-2}, 10^{-1}, and 1 and from 1 to 5 hidden binary units.

The simulation was performed on a Windows desktop with a quad core processor and 16 GB of RAM, with relevant conditions executed consistently among all training methods, parameter changes, and datasets. Figure 1 provides a flowchart detailing the overall machine learning process.

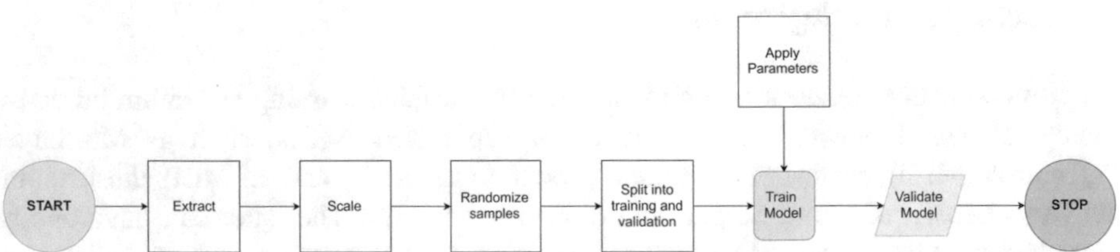

Fig. 1. Flowchart of machine learning process.

3.1 Data Acquisition

To provide a more comprehensive comparison of the different ANNs, two distinct datasets were obtained and formatted consistently. The first and smaller set stems from an experiment by [17] used to evaluate a combination of cognitive load and fear conditioning based on EDA input. Because our focus is on the former, we extracted the data samples that focused only on cognitive tasks, which consisted of human subjects repeatedly adding 2, repeatedly subtracting 7, or resting for 30 s intervals. Ultimately, our ANNs were modeled to address the question: *Given a time step, a subject number, and that subject's EDA, what task was he or she performing?* Over the course of 851 time steps and 24 subjects, there were two periods of adding, two periods of subtracting, and five periods of rest surrounding the beginning and end of each cognitive task. Due to the increased complexity of subsequent arithmetic operations, each period between rests was treated as a separate classification while all rest periods were considered as the same classification, hence a total of five possible output choices.

We obtained the second and larger data set from [25], which examines genetic predisposition among alcoholic and control subjects based on EEG correlation. Over the course of 122 subjects and 120 trials, the cognitive activity consisted of being shown either a single image, two matching images, or two non-matching images. We treated each of these three activities as separate classifications between alcoholic subjects and control subjects, hence a total of six possible output choices. Thus, our ANNs in this circumstance resulted in the modified question: *Given a time step, a subject number, a trial number, and the subject's EEG across 64 channels, what task was he or she performing?* Because the original dataset was too large to run in memory on the previously described hardware, we used only 10% of the total available data samples, which were selected randomly.

4 Results and Discussion

This section presents the comprehensive results for the Python simulations in terms of computational runtime and accuracy. This is followed by a discussion of the pros and cons of the two supervised training techniques as well as the unsupervised restricted Boltzmann machine.

4.1 Fear Conditioning Data

Figure 2 provides a surface plot of the training accuracy for both of the chosen supervised ANN training methods when modeling the smaller dataset.

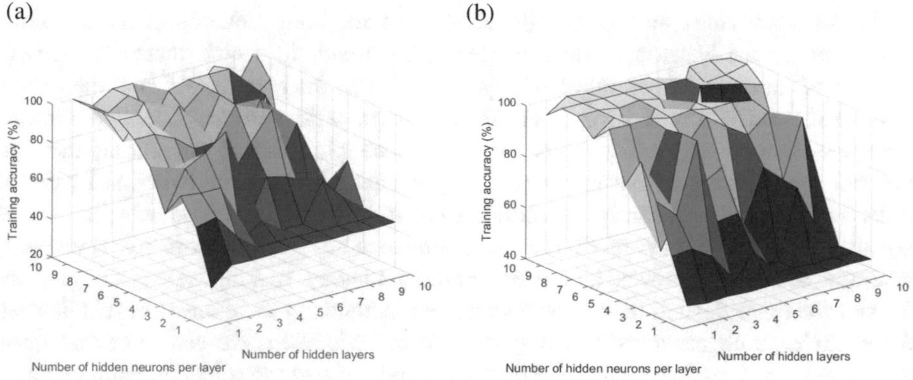

Fig. 2. Surface plots of training accuracy for supervised ANN on fear conditioning data trained by (a) quasi-Newton and (b) stochastic gradient descent optimizers, in terms of number of hidden layers and number of hidden neurons per layer.

As each of the subplots indicate, and as can be expected, an increase in the number of hidden neurons results in a significantly higher accuracy. In gradient descent, no significant correlation seems to exist when adjusting for number of hidden layers, although in quasi-Newton, accuracy appears to plummet after seven or more hidden layers. Ultimately, the latter method is capable of reaching training accuracy up to 99.9%, while the former reaches up to 99.7%.

In Fig. 3, the validation accuracy can be observed under the same parameters and constraints, following similar trends but with a slight reduction in top accuracy, down to 99.8% and 99.6%, respectively.

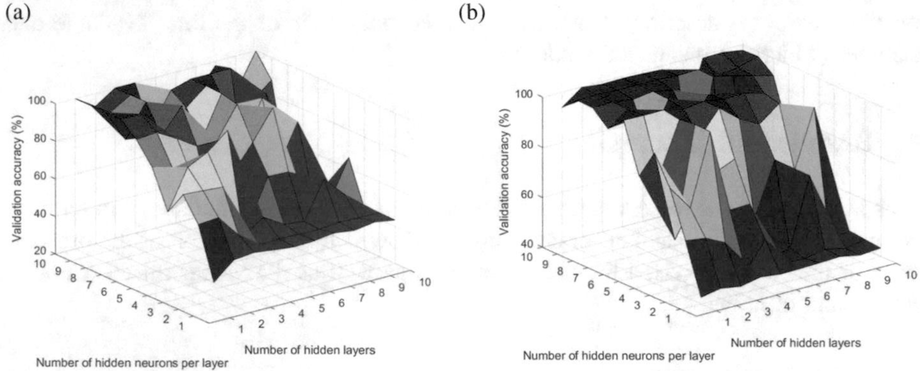

Fig. 3. Surface plots of validation accuracy for supervised ANN on fear conditioning data trained by (a) quasi-Newton and (b) stochastic gradient descent optimizers, in terms of number of hidden layers and number of hidden neurons per layer.

Figure 4 provides the computational runtime for both methods under the same conditions. Quasi-Newton follows a predictable trend, in which increases in both number of neurons and number of layers results in an increase of runtime, while gradient descent follows a more arbitrary pattern, with various spikes of runtime increases throughout various parts of the surface plot. When comparing the two methods, gradient descent seems to have a significant edge in speed over quasi-Newton under most conditions, rarely exceeding 10 s. In most cases, this is achieved at the expense of only a slight reduction of maximum accuracy, and under many parameters, accuracy actually improves alongside runtime, thereby making stochastic gradient descent the more desired of the two supervised methods in the context of this dataset. When taking both accuracy and runtime into consideration, we can infer that these supervised ANN models work most optimally under six or more hidden neurons and a small number of hidden layers. In the case of quasi-Newton, having only one hidden layer achieves the optimal balance of speed and accuracy, while gradient descent does best with multiple, but few, hidden layers.

Figure 5 presents all three of the aforementioned evaluation metrics for unsupervised ANNs via restricted Boltzmann machines. From here, both sets of accuracy indicate that this ANN type is ultimately unsuccessful at modeling this dataset, given

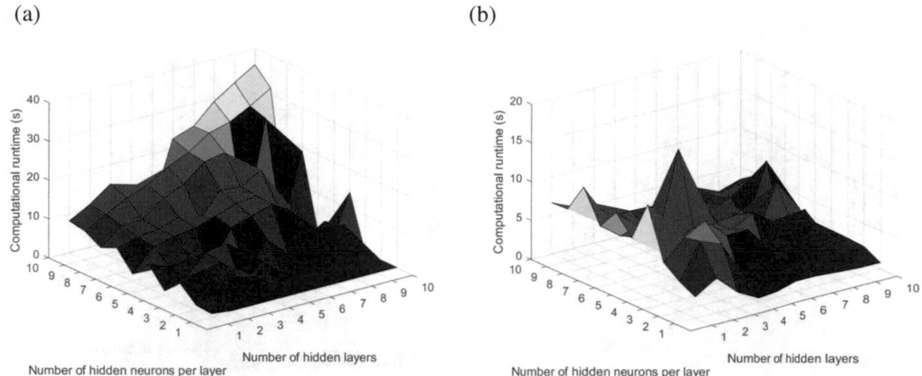

Fig. 4. Surface plots of computational runtime for supervised ANN on fear conditioning data trained by (a) quasi-Newton and (b) stochastic gradient descent optimizers, in terms of number of hidden layers and number of hidden neurons per layer.

that accuracy never surpasses 45%. Runtime appears to be significantly faster than the supervised methods by never exceeding 1.3 s, but this is inherently unhelpful when accuracy is too low.

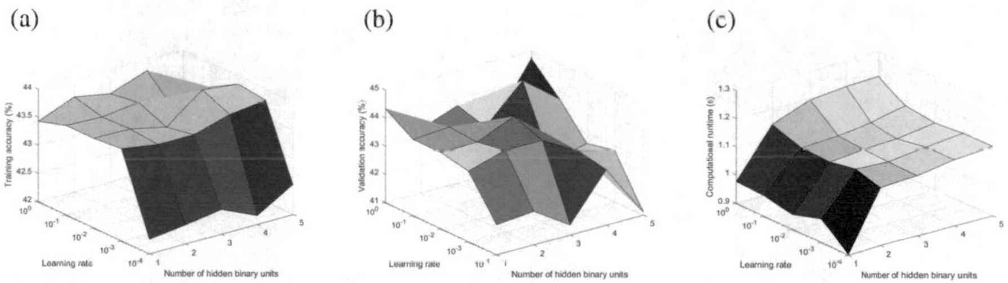

Fig. 5. Surface plots of (a) training accuracy, (b) validation accuracy, and (c) computational runtime for unsupervised ANN on fear conditioning data in terms of learning rate and number of hidden binary units.

4.2 Visual Matching Data

In a manner similar to the previous subsection, the next four figures present the results for the larger dataset. Beginning with Fig. 6, validation accuracy follows a similar trend to before in terms of number of hidden neurons. However, there appears to no meaningful correlation on number of hidden layers. Figure 7 provides the validation accuracy with results more oscillatory between parameters but otherwise exhibiting similar trends to those of the training accuracy. Overall observations include the fact that accuracy is expectedly much lower under the larger data, never exceeding 75%. However, there is also less fluctuation in accuracy, as all values consistently appear to be within 2% of one another. This is a stark comparison to a 60–80% accuracy difference among parameters under the smaller dataset.

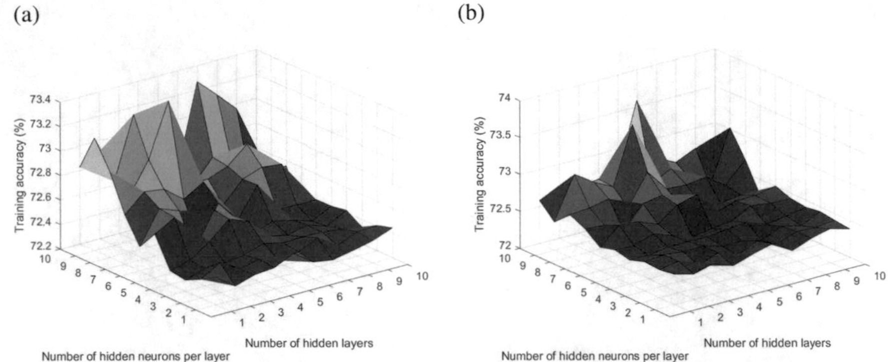

Fig. 6. Surface plots of training accuracy for supervised ANN on visual matching data trained by (a) quasi-Newton and (b) stochastic gradient descent optimizers, in terms of number of hidden layers and number of hidden neurons per layer.

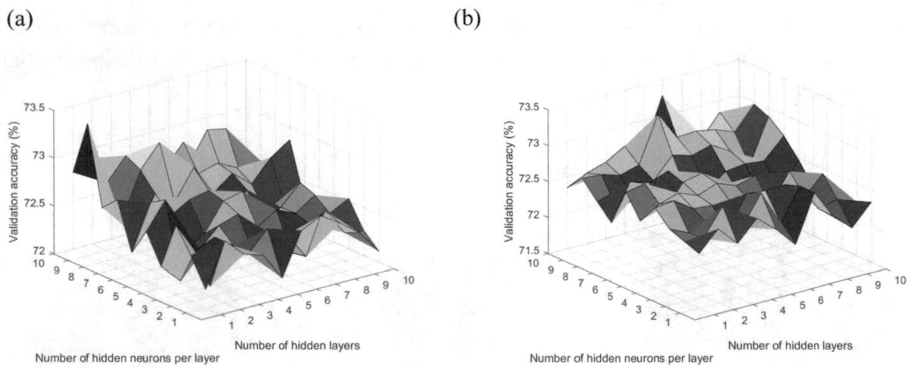

Fig. 7. Surface plots of validation accuracy for supervised ANN on visual matching data trained by (a) quasi-Newton and (b) stochastic gradient descent optimizers, in terms of number of hidden layers and number of hidden neurons per layer.

The computational runtime in Fig. 8 provides similar patterns to those of the other data but with higher runtime values, which is expected given the significantly higher number of samples. Quasi-Newton still shows an increase when either parameter increases, but unlike before, gradient descent now also follows these trends, albeit with some arbitrary spikes like before. When comparing the two, the latter method performs better in computational speed, while the differences in accuracy are relatively negligible, thereby giving gradient descent a clear advantage on both datasets when accounting for both speed and accuracy.

Figure 9 presents the results for the unsupervised ANN. For this data, the top accuracy is only slightly lower than that of the supervised methods. In addition, there is less than 1% variation in accuracy across all tested parameters, resulting in a higher

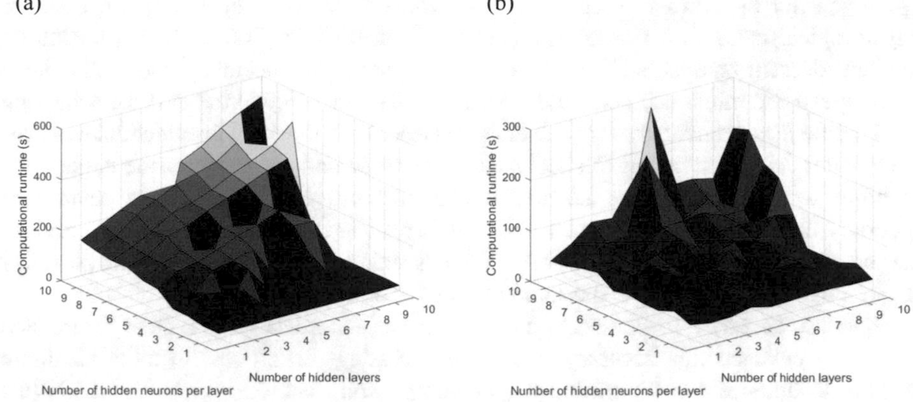

Fig. 8. Surface plots of computational runtime for supervised ANN on visual matching data trained by (a) quasi-Newton and (b) stochastic gradient descent optimizers, in terms of number of hidden layers and number of hidden neurons per layer.

minimum accuracy. Due to these small differences, correlations regarding learning rate and number of hidden units with accuracy are mostly negligible.

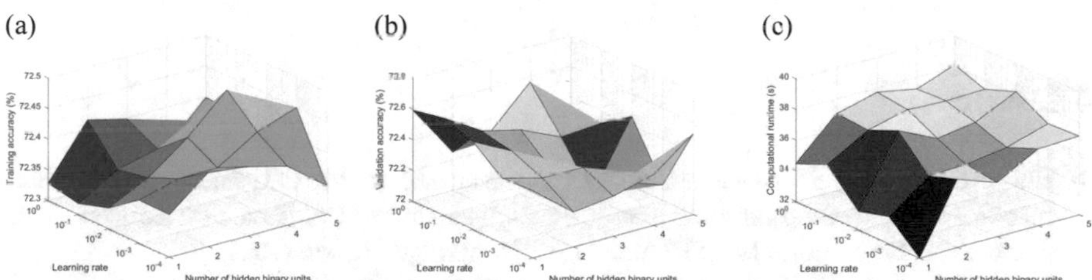

Fig. 9. Surface plots of (a) training accuracy, (b) validation accuracy, and (c) computational runtime for unsupervised ANN on visual matching data in terms of learning rate and number of hidden binary units.

When observing runtime, there exists a clear advantage in this unsupervised method, as the amount of runtime is significantly less than either of the supervised ANNs. Coupling this with the minor differences in accuracy, the results are apparent that restricted Boltzmann machines are highly superior for modeling this large dataset when both speed and accuracy are under consideration.

5 Conclusions and Future Work

This paper provided a comprehensive analysis of the pros and cons of supervised and unsupervised artificial neural networks, with particular focus on restricted Boltzmann machines for the latter as well as quasi-Newton and stochastic gradient descent optimizers for the former. These neural networks were tested on two distinct sets of physiological data. The results ultimately indicate that on both datasets, stochastic gradient descent possesses the best balance of runtime and accuracy and is also ideal when speed and efficiency take sole priority, while quasi-Newton is best for achieving an absolute maximum accuracy. The unsupervised ANN does not provide much use for the smaller data but contains the best balance of speed and accuracy for the larger data. In other words, for the first dataset, gradient descent provides the best balance of accuracy and computational speed, while the restricted Boltzmann machine provides that best balance for the second dataset. When seeking the maximum possible accuracy at any cost, quasi-Newton is the ideal choice for both datasets.

Because accuracy for the second dataset never exceeded 75%, a clear future step can be to enhance this accuracy using preprocessing techniques. In addition, more training techniques can be introduced for comparison, including global methods like genetic algorithm or gray wolf optimization. Acquisition and testing of additional physiological datasets form another viable direction for additional future work.

Acknowledgments. This research is supported by the Dayton Area Graduate Studies Institute (DAGSI) fellowship program for a project titled "Assessment of Team Dynamics Using Adaptive Modeling of Biometric Data." The authors wish to thank their DAGSI sponsor Dr. Gregory Funke for his continued guidance and support throughout the project. The authors also thank the EECS Department at the University of Toledo for partial support through assistantships and tuition waivers.

References

1. Elkin, C., Nittala, S., Devabhaktuni, V.: Fundamental cognitive workload assessment: a machine learning comparative approach. In: 8th International Conference on Applied Human Factors and Ergonomics (AHFE), pp. 275–284. Springer, Cham (2017)
2. Berka, C., et al.: EEG correlates of task engagement and mental workload in vigilance, learning, and memory tasks. Aviat. Space Env. Med. **78**(5), B231–B234 (2007)
3. Durantin, G., Gagnon, J.-F., Tremblay, S., Dehais, F.: Using near infrared spectroscopy and heart rate variability to detect mental overload. Behav. Brian Res. **259**, 16–23 (2014)
4. Niaz, M., Logie, R.H.: Working memory, mental capacity and science education: towards an understanding of the 'working memory overload hypothesis'. Oxford Rev. Educ. **19**(4), 511–525 (1993)
5. Wilson, G.F., Russell, C.A.: Real-time assessment of mental workload using psychophysiological measures and artificial neural networks. Hum. Factors **45**(4), 635–644 (2016)
6. Baldwin, C.L., Penaranda, B.N.: Adaptive training using an artificial neural network and EEG metrics for within- and cross-task workload classification. NeuroImage **59**(1), 48–56 (2012)
7. Chatterji, G.B., Sridhar, B.: Neural network based air traffic controller workload prediction. In: Proceedings of the 1999 American Control Conference, pp. 2620–2624 (1999)
8. Jin, L., et al.: Driver cognitive distraction detection using driving performance measures. Discrete Dyn. Nat. Soc. **2012**, 12 (2012)
9. Son, J., Oh, H., Park, M.: Identification of driver cognitive workload using support vector machines with driving performance, physiology and eye movement in a driving simulator. Int. J. Precis. Eng. Manuf. **14**(8), 1321–1327 (2013)

10. Putze, F., Jarvis, J., Schultz, T.: Multimodal recognition of cognitive workload for multitasking in the car. In: 2010 20th International Conference on Pattern Recognition (ICPR), pp. 3748–3751. IEEE Press, New York (2010)
11. Liang, Y., Reyes, M., Lee, J.: Real-time detection of driver cognitive distraction using support vector machines. IEEE Trans. Intel. Transp. Syst. **8**(2), 340–350 (2007)
12. Ziegler, M. et al.: Sensing and assessing cognitive workload across multiple tasks. In: Foundations of Augmented Cognition: Neuroergonomics and Operational Neuroscience, pp. 440–450. Springer, Cham (2016)
13. Yin, Z., Zhang, J.: Identification of temporal variations in mental workload using locally-linear-embedding-based EEG feature reduction and support-vector-machine-based clustering and classification techniques. Comput. Methods Prog. Biomed. **115**(3), 119–134 (2014)
14. Solovey, E. et al.: Classifying driver workload using physiological and driving performance data: two field studies. In: Proceedings of the SIGCHI Conference on Human Factors in Computing Systems, CHI 2014, pp. 4057–4066. ACM, New York (2014)
15. Calibo, T.K., Blanco, J.A., Firebaugh, S.L.: Cognitive stress recognition. In: 2013 IEEE International Instrumentation and Measurement Technology Conference (I2MTC), pp. 1471–1475. IEEE Press, New York (2013)
16. Girouard, A.: Distinguishing difficulty levels with non-invasive brain activity measurements. In: Human-Computer Interaction – INTERACT 2009, pp. 440–452. Springer, Berlin (2009)
17. Natarajan, A., Xu, K.S., Eriksson, B.: Detecting divisions of the autonomic nervous system using wearables. In: IEEE 38th Annual International Conference of the Engineering in Medicine and Biology Society (EMBC), pp. 5761–5764. IEEE Press, New York (2016)
18. Hefron, R.G., et al.: Deep long short-term memory structures model temporal dependencies improving cognitive workload estimation. Pattern Recogn. Lett. **94**, 96–104 (2017)
19. Gianazza, D.: Analysis of a workload model learned from past sector operations. In: SID 2017, 7th SESAR Innovation Days, pp. 1–9 (2017)
20. Juhaniak, T., et al.: Pupillary response: removing screen luminosity effects for clearer implicit feedback. In: UMAP (Extended Proceedings) (2016)
21. Tran, C., Abraham, A., Jain, L.: Decision support systems using hybrid neurocomputing. Neurocomputing **61**, 85–97 (2004)
22. Malsburg, C.: Frank rosenblatt: principles of neurodynamics: perceptrons and the theory of brain mechanisms. In: Brain Theory, pp. 245–248. Springer, Berlin (1986)
23. Salakhutdinov, R., Mnih, A., Hinton, G.: Restricted Boltzmann machines for collaborative filtering. In: Proceedings of the 24th International Conference on Machine Learning, ICML 2007, pp. 791–798. ACM, New York (2007)
24. Pedregosa, F., et al.: Scikit-learn: machine learning in python. J. Mach. Learn. Res. **12**, 2825–2830 (2011)
25. Lichman, M.: UCI Machine Learning Repository. University of California, Irvine, School of Information and Computer Sciences. http://archive.ics.uci.edu/ml

Changes in Physiological Condition in Open Versus Closed Eyes

Kazuya Onishi[1(✉)] and Hiroshi Hagiwara[2]

[1] Graduate School of Information Science and Engineering, Ritsumeikan University, 1-1-1 Noji Higashi, Kusatsu, Shiga 525-8577, Japan
is0250ip@ed.ritsumei.ac.jp

[2] College of Information Science and Engineering, Ritsumeikan University, 1-1-1 Noji Higashi, Kusatsu, Shiga 525-8577, Japan
hagiwara@ci.ritsumei.ac.jp

Abstract. With the development of information technology, modern society has become a 24-h society, and the burden on employees regarding longer working hours and the need to process greater amounts of information has increased. This excessive burden on employees will likely result in more errors. Employee workloads need to be monitored and adjusted to prevent these types of overwork-related errors. This study aimed to investigate the changes in physiological conditions that take place during specific tasks conducted with the eyes open and closed. Oxygenated hemoglobin concentrations increased and alpha waves in electroencephalograms recorded from the frontal and parietal regions decreased during the high workload task. This trend appeared in both the closed and open eyes states. Theta waves showed different trends between these brain regions. In the open eyes state, theta waves decreased in the occipital region but increased in the frontal region during the high workload task. The theta waves also showed a different trend between the closed and open eyes states. Theta waves in the frontal region increased during the high workload task in the open eyes state, but did not change in the closed eyes state during any task. Based on these results, we consider that it is better to use the theta waves in the occipital region and the alpha waves in the parietal region to assess mental workload in the closed eyes state.

Keywords: Open eyes · Closed eyes · Electroencephalogram
Auditory n-back task

1 Introduction

A large amount of information can be obtained instantaneously with today's information technology, and this informatization has substantially changed our work culture and lifestyle. Workers are required to not only have skills for gathering, organizing and analyzing this information, but also for developing new ways to process and manage it. Unfortunately, the modern 24-h society brought about by advances in information technology has increased the potential for overwork. This excessive burden on workers will likely result in more human errors, which could ultimately lead to serious mistakes and accidents. Performance decreases when the workload is too large; however,

H. Ayaz and L. Mazur (Eds.): AHFE 2018, AISC 775, pp. 38–48, 2019.
https://doi.org/10.1007/978-3-319-94866-9_4

performance may also be degraded if the workload is too small [1]. In particular, in an overburdened situation, there is a high possibility of human error.

In light of this, research to estimate the workload and mental state of workers is garnering attention. If the workload and mental condition of workers could be estimated, the workload could be appropriately adjusted to achieve optimal performance. Many researchers have studied the physiological and psychological states during task execution. It is well established that mental workload is positively correlated with cerebral activity of dedicated brain areas, such as the prefrontal cortex [2]. Some researchers postulate that oxygenated hemoglobin (oxyHb) concentrations reflect the mental workload [3]. However, few studies have investigated the effect of open and closed eyes on the physiological state by simultaneous measurement of electroencephalograms (EEGs) and oxygenated hemoglobin (oxyHb) concentrations. Therefore, in this study, we investigated the effect of open and closed eyes on the psychological and physiological states of participants as they conducted short-term memory tasks with different levels of difficulty.

2 Experimental Method

2.1 Subjects

The subjects were 10 healthy male university students (age range: 19–22 years). All subjects provided written informed consent prior to participation in this study. Subjects refrained from excessive eating and drinking the night before the experiment. In addition, subjects refrained from engaging in prolonged or strenuous exercise on the morning of the experiment.

2.2 Experimental Protocol

The experimental protocol is shown in Fig. 1 and represents one complete cycle of activities that subjects performed for each of no task, and the 1-back and 2-back tasks. The order of the tasks was randomized. As a subjective evaluation, subjects rated their fatigue and arousal using the Roken Arousal Scale (RAS) [4] before and after each task. At the end of one cycle, they completed the NASA Task Load Index (NASA-TLX) [5].

2.3 Tasks

Subjects performed an auditory n-back task, a representative task used to evaluate working memory. In this task, subjects heard numbers between 0 and 4 delivered in a random order for 3 min. The stimulus interval was 2 s, and there were 85 stimuli in one task. There were three experimental conditions: no task, 1-back, and 2-back. Subjects were instructed to press a key as soon as they heard a number that was identical to a number heard n steps earlier. For instance, during the 1-back task, subjects pressed a key when they heard the same number twice in a row. Figure 2 shows the schematic of the 1-back and 2-back tasks. During each auditory n-back task, subjects simultaneously

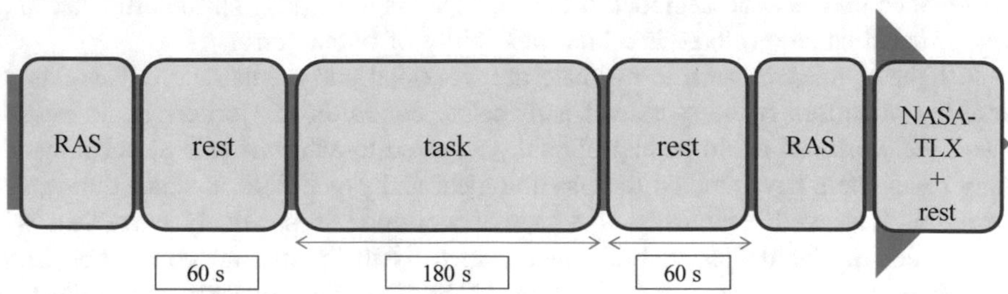

Fig. 1. The experimental protocol. This schematic shows the complete cycle of activities that subjects performed for each of the tasks. RAS: Roken Arousal Scale; NASA-TLX: NASA Task Load Index.

performed an alpha attenuation test (AAT) [6] specific activity (eyes closed [30 s], eyes open [30 s] × 3) for 3 min. During the AAT, subjects closed their eyes for the first 30 s, and then opened their eyes for the subsequent 30 s (Fig. 3). Subjects repeated this pattern three times for one task, so one task lasted 180 s.

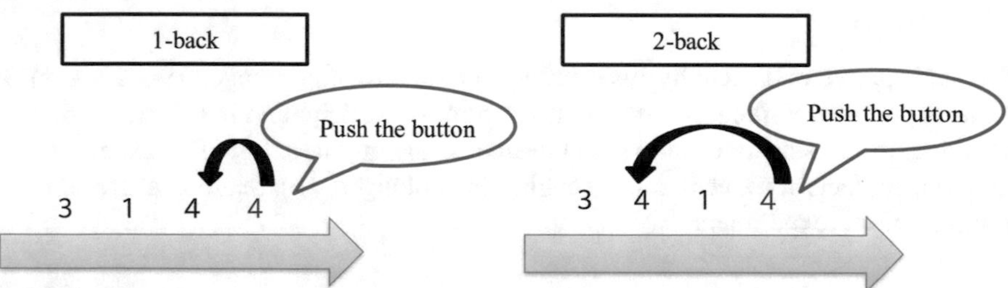

Fig. 2. Schematic of the auditory n-back task. The subject was required to press a button when the presented number was identical to the number presented n numbers previously.

2.4 Experimental Equipment and Electrode Fixation Points

OxyHb concentrations in the brain were measured using a near-infrared imaging device (OMM3000; Shimadzu, Kyoto, Japan). The sites of the near-infrared spectroscopy (NIRS) measurements are shown in Fig. 4. In this study, we analyzed data from the frontal cortex (channels 1, 10, and 19), which are involved in cognition or judgment, and the dorsolateral prefrontal cortex (channel 25), which is involved in attention and control. An EEG1100 (Nihon Kohden, Tokyo, Japan) was used to record EEGs. The sites of the EEG measurements are shown in Fig. 5. We obtained the data from electrodes C3, C4, Fz, Pz, O1, O2, A1 and A2. The subjects placed their jaw on a stand during the experiment to keep their head stationary, because NIRS data are affected by noise from head movements.

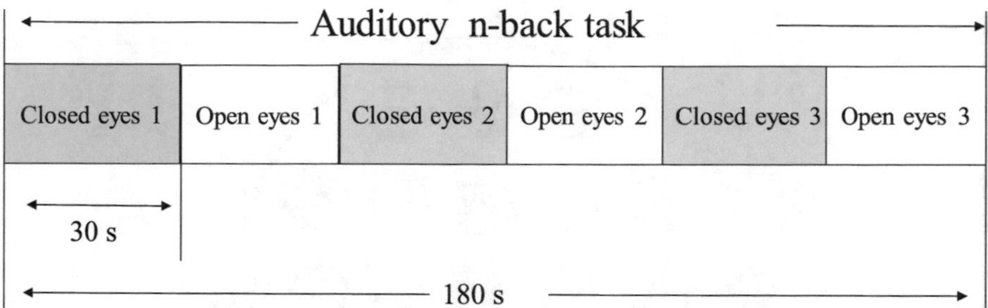

Fig. 3. Schematic showing the pattern of closed and open eyes during the task. During each auditory n-back task, subjects kept their eyes closed for 30 s and then open for 30 s, and repeated this sequence three times.

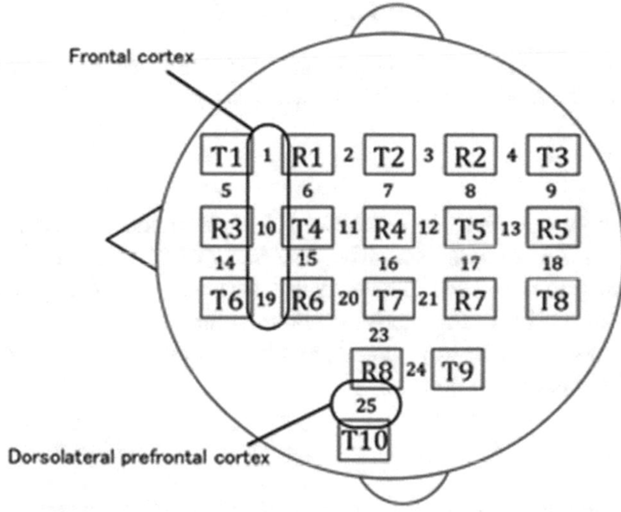

Fig. 4. Attachment sites of the NIRS probes. The NIRS probes were positioned on the head according to the international 10–20 system.

3 Analysis

3.1 Nasa-TLX

The NASA-TLX was used to evaluate the workload. The NASA-TLX evaluates six aspects of perceived workload: mental demand, physical demand, temporal demand, overall performance, frustration level and effort. A weighted score for these six aspects was calculated and used as the weighted workload (WWL) score.

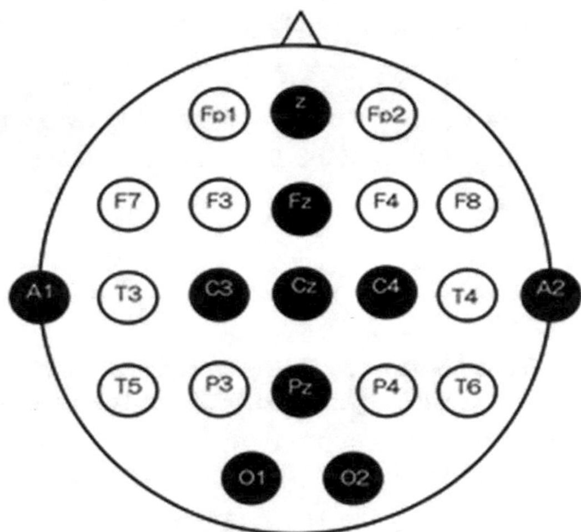

Fig. 5. Schematic diagram indicating the positions of the EEG electrodes. The EEG electrodes were positioned on the head according to the international 10–20 system.

3.2 EEGs

The sampling rate for the EEG data was 500 Hz. Noise in the EEGs recorded from O2-A1, Fz-A1 and Pz-A1 was removed using low-pass (120 Hz), high-pass (0.1 Hz), and band-stop filtering (57–60 Hz). Frequency analysis was then performed using a fast Fourier transform (FFT). In this study, we defined the power spectrum at frequencies from 8 to 13 Hz as the alpha wave power, and the power spectrum at frequencies from 4 to 8 Hz as the theta wave power.

3.3 NIRS

NIRS can be used to measure the oxygenated hemoglobin (oxyHb) concentration, deoxygenated hemoglobin concentration, and total hemoglobin concentration. In this study, we focused on the oxyHb concentration because it correlates with changes in regional cerebral blood flow [7]. We used the mean and standard deviation of the oxyHb concentration to assess the NIRS waveforms. Hasan et al. confirmed the validity of using the mean oxyHb concentration to evaluate workload when subjects performed single tasks [3]. The sampling rate for the oxyHb data was 10 Hz, and the oxyHb signals were bandpass filtered (0.001–01 Hz).

3.4 Statistical Analyses

All statistical analyses were performed using the SPSS statistical software package (version 20; IBM SPSS Statistics, Tokyo, Japan) and MATLAB software (MathWorks Inc., Natick, MA, USA).

4 Results

4.1 Performance

Figure 6 shows the average number of errors in the auditory n-back tasks. We defined the number of errors as the number of times subjects pressed a key when they did not have to or vice versa. Comparing the differences between the 1-back and 2-back tasks, the average number of errors in the 2-back task was significantly higher than that in the 1-back task ($p < 0.05$). The average reaction time also increased with increasing task workload (data not shown).

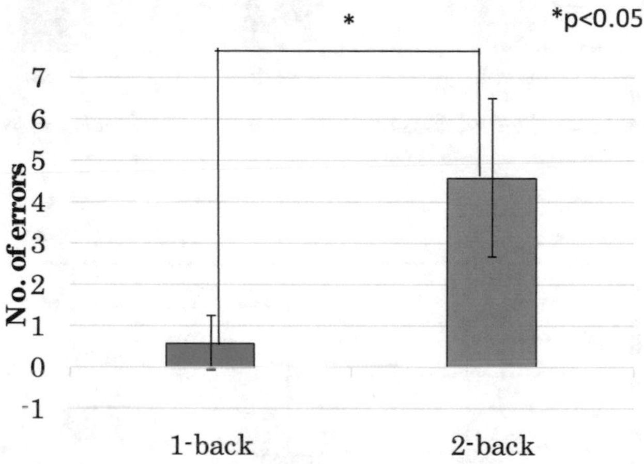

Fig. 6. Average number of errors in the auditory n-back tasks.

4.2 WWL Score

Figure 7 shows the average WWL scores for 10 subjects for no task, and the 1-back and 2-back tasks. The WWL score indicates the mental workload during the task and is obtained from the NASA-TLX. WWL scores showed a significant increase of 1% ($p < 0.01$) as the workload of the task increased.

4.3 NIRS

Figure 8 shows the average oxyHb concentrations at channel 25 for no task, and the 1-back and 2-back tasks. We measured the oxyHb concentrations during tasks for 180 s, summed these values and then normalized them for each subject. Compared to the concentrations measured during no task and the 1-back task, oxyHb concentrations during the 2-back task were significantly increased ($p < 0.01$). There was no significant difference in oxyHb concentrations between no task and the 1-back task. The results for channels 1, 19, and 10 were similar to those for channel 25 (data not shown).

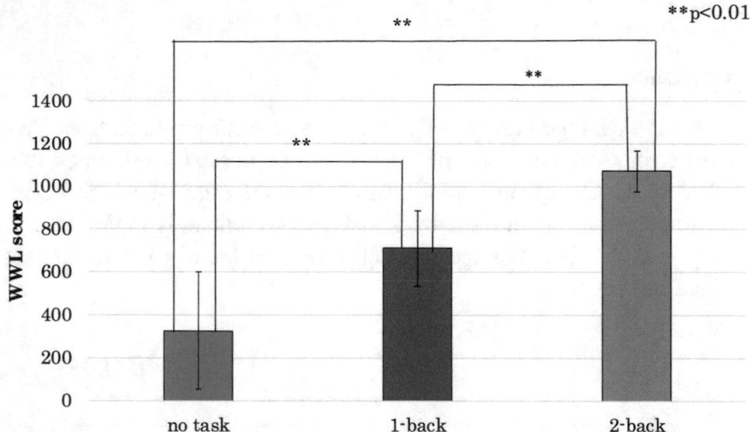

Fig. 7. The average weighted workload (WWL) scores for 10 subjects for no task and the n-back tasks.

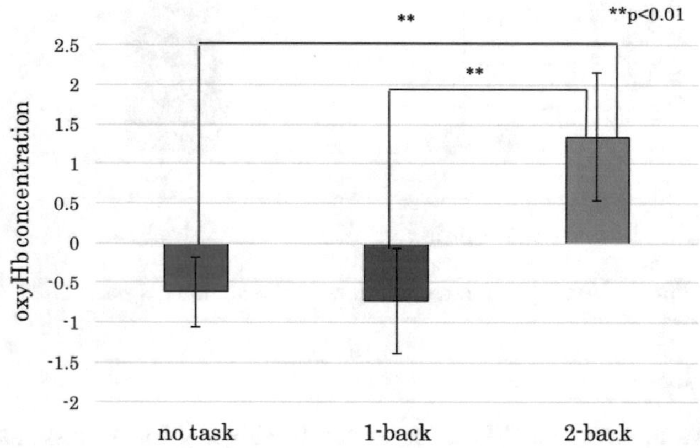

Fig. 8. The average oxyHb concentrations at channel 25 across all subjects.

4.4 EEGs

We analyzed the alpha waves and theta waves in EEG signals recorded from O2-A1, Fz-A1 and Pz-A1. We separated each task into the closed and open eyes states and calculated the power of the alpha waves and theta waves in each part. The figures that do not show significant differences and tendencies are not shown here. Only the data showing significant differences or tendencies are presented in the following figures. The values were normalized for each subject.

Figure 9 shows the average power of the theta waves at O2-A1 in the closed and open eyes states. In the closed eyes state, the average power of the theta waves during the 2-back task was significantly lower than that during no task ($p < 0.05$). In the open eyes state, the average power of the theta waves during the 2-back task tended to be lower than that during no task ($p < 0.1$).

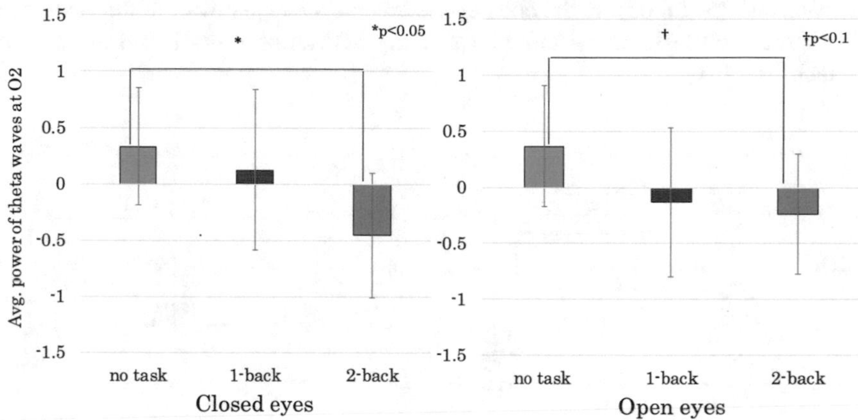

Fig. 9. Average power of theta waves at O2-A1 for no task, 1-back task, and 2-back task across all subjects. Left: closed eyes; Right: open eyes.

Figure 10 shows the average power of the theta waves at Fz-A1 in the closed and open eyes states. In the closed eyes state, there were no significant differences between tasks. In the open eyes state, however, the average power of the theta waves during the 2-back task was significantly higher than that during no task ($p < 0.05$) and during the 1-back task ($p < 0.01$). Thus, the power of the theta waves shows an opposite tendency between O2 and Fz in the open eyes state. There were no significant differences in the average power of the theta waves at Pz-A1 between tasks in the closed and open eyes states.

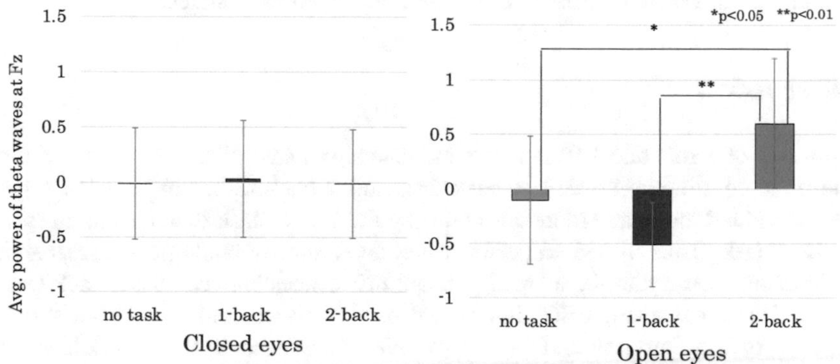

Fig. 10. Average power of theta waves at Fz-A1 for no task, 1-back task, and 2-back task across all subjects. Left: closed eyes; Right: open eyes.

Figure 11 shows the average power of the alpha waves at Fz-A1 in the closed and open eyes states. In the closed eyes state, the average power of the alpha waves during the 2-back task was significantly lower than that during no task ($p < 0.01$) and during the 1-back task ($p < 0.05$). In the open eyes state, the average power of the alpha waves during the 2-back task tended to be lower than that during no task and during the 1-back task ($p < 0.1$).

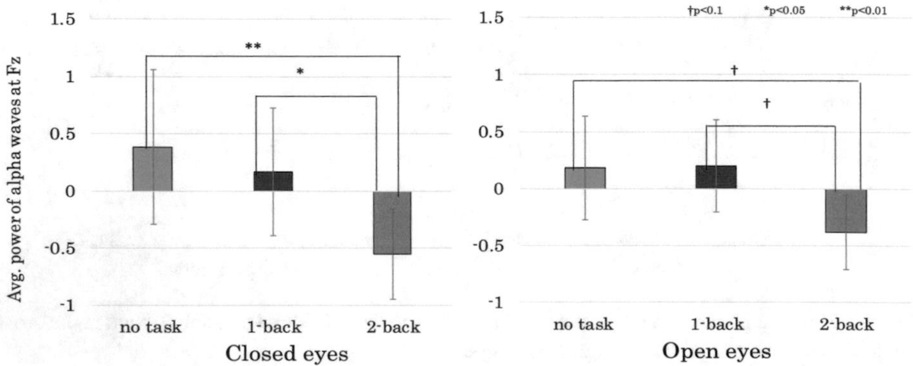

Fig. 11. Average power of alpha waves at Fz-A1 for no task, 1-back task, and 2-back task across all subjects. Left: closed eyes; Right: open eyes.

Figure 12 shows the average power of the alpha waves at Pz-A1 in the closed and open eyes states. In the closed eyes state, the average power of the alpha waves during the 2-back task tended to be lower than that during no task ($p < 0.1$) and was significantly lower than that during the 1-back task ($p < 0.05$). In the open eyes state, the average power of the alpha waves during the 2-back task tended to be lower than that during no task ($p < 0.1$), and it was significantly lower than that during the 1-back task ($p < 0.05$). There were no significant differences in the average power of the alpha waves at O2-A1 between tasks in the closed and open eyes states.

5 Discussion

The number of errors and WWL scores indicate significant differences in the difficulty of the three conditions: no task, 1-back task and 2-back task. Subjects felt a greater mental workload and made more errors during the 2-back task than during no task and the 1-back task. Thus, based on performance level and psychological assessment, we were able to confirm that there are significant differences between the 2-back task and the other tasks. However, EEG signals and oxyHb concentrations did not show any differences between no task and the 1-back task, likely because the 1-back task is very easy. Therefore, we hereafter refer to no task and the 1-back task collectively as the "low workload task" and the 2-back task as the "high workload task".

In the closed and open eyes states, the power of the alpha waves was lower during the high workload task than during the low workload task. This indicates that the power

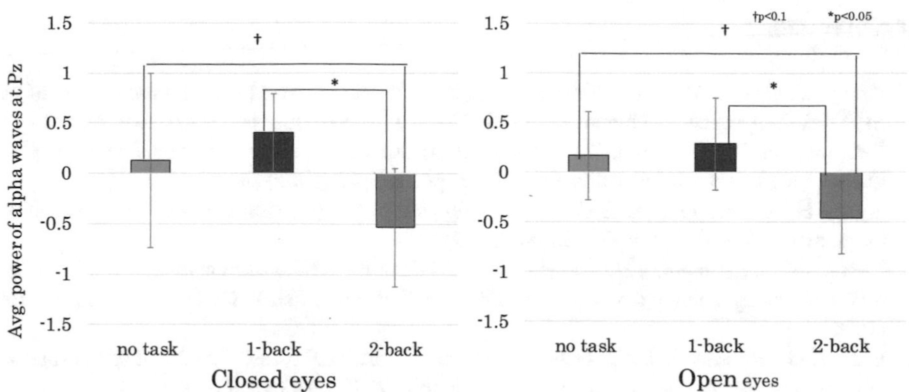

Fig. 12. Average power of alpha waves at Pz-A1 for no task, 1-back task, and 2-back task across all subjects. Left: closed eyes; Right: open eyes.

of the alpha waves decreases as the burden of the mental workload increases in the closed and open eyes states. This tendency was shown at both Fz-A1 and Pz-A1, but not at O2-A1. It is believed that the alpha waves in the occipital region are related to wakefulness and sleepiness. Regarding before and after the tasks, there were no significant differences in the values for "activation" and "sleepiness" from the RAS for all difficulties of task. This is likely why the power of the alpha waves at O2-A1 did not differ between the tasks.

In the open eyes state, the power of the theta waves at O2-A1 was lower during the high workload task than during the low workload task. In contrast, the power of the theta waves at Fz-A1 was higher during the high workload task than during the low workload task. Based on these results, we surmise that the power of the theta waves decreased in the occipital region but increased in the frontal region when subjects felt the burden of the mental workload in the open eyes state. In thc closed eyes state, the power of the theta waves at O2-A1 showed the same tendency as in the open eyes state. However, there were no differences in the power of the theta waves at Fz-A1 between tasks. The authors reported that $\theta Fz/\alpha Pz$ indicates the degree of the workload only when the eyes are open, but does not have utility in the closed eyes state [8]. This is because the power of the theta waves at Fz-A1 does not change with the workload. Therefore, it is better to use the theta waves at O2 and the alpha waves at Pz when we measuring mental workload in the closed eyes state. In the present study, there were negligible differences between no task and the 1-back task, and there was a high percentage of correct answers in the 2-back task. Therefore, we plan to conduct additional studies using a 3-back task to investigate the changes in physiological conditions during overwork.

Acknowledgements. This study was supported by the Ministry of Education, Culture, Sports, Science and Technology (MEXT)-Supported Program for the Strategic Research Foundation at Private Universities, 2013–2017.

References

1. Afergan, D., et al.: Dynamic difficulty using brain metrics of workload. In: Proceedings of the SIGCHI Conference on Human Factors in Computing Systems, pp. 3797–3806 (2014)
2. Parasuraman, R., Caggiano, D.: Neural and genetic assays of human mental workload. In: Quantifying Human Information Processing, pp. 123–49 (2005)
3. Ayaz, H., et al.: Optical brain monitoring for operator training and mental workload assessment. Neuroimage **59**(1), 36–47 (2012)
4. Takahashi, M., Kitagima, H., Honjo, Y.: Analysis of the relationship between sleepiness and relaxation using a newly developed mental work strain checklist. J. Sci. Labour **72**(3), 89–100 (1996)
5. Hart, S.G., Staveland, L.E.: Development of NASA-TLX (Task Load Index): results of empirical and theoretical research. Adv. Psychol. **52**, 139–183 (1988)
6. Hagiwara, H., Araki, K., Michimori, A., Saito, M.: A study on the quantitative evaluation method of human alertness and its application. Psychiatr. Neurol. Jpn. **99**(1), 23–34 (1997)
7. Pylyshyn, Z.W., Storm, R.W.: Tracking multiple independent targets: evidence for parallel tracking mechanism. Spat. Vis. **3**, 179–197 (1988)
8. Onishi, K., Hagiwara, H.: Effects of open versus closed eyes on physiological conditions during a working memory task. In: Proceedings of the IEEE 17th International Conference on Bioinformatics and Bioengineering (IEEE BIBE 2017), pp. 171–174 (2017)

The Effect of Mental Fatigue on Response Processes: An ERP Study in Go/NoGo Task

Xiaoli Fan[1(✉)], Chaoyi Zhao[1], Hong Luo[1], and Wei Zhang[2]

[1] AQSIQ Key Laboratory of Human Factors and Ergonomics, China National Institute of Standardization, Beijing 100191, China
fanfan19851414@163.com, {zhaochy, luohong}@cnis.gov.cn

[2] Tsinghua University, Beijing 100084, China
zhangwei@tsinghua.edu.cn

Abstract. The effects of mental fatigue on response processes were examined using event-related brain potentials in this study. Twenty subjects were requested to perform a 2-back task for 100 min continuously which was used as the inducement of mental fatigue, meanwhile they were required to perform a 10-min Go/NoGo task (pretest) before and after the 2-back task (posttest) respectively. The subjective scores of fatigue scales, behavioral performance and electroencephalogram were recorded during the whole experiment. Increased fatigue scale scores and impaired performance demonstrated mental fatigue was indeed induced. In ERP analysis, paired T-test was used to analyze Go/NoGo-P3, Go/NoGO-N2 and ERN/Pe between the pretest and posttest and the results were as follows. The amplitude of NOGO-P3 decreased significantly ($P < 0.05$), while latencies of NOGO-P3/N2 and GO-P3 increased (all $P < 0.05$). Larger ERN amplitude was evoked by erroneous response to NoGo trials ($P < 0.05$). These results suggested that mental fatigue attenuated the response processes, and the metrics above are possible indices for measuring mental fatigue.

Keywords: ERP · Mental fatigue · Go/NoGo · P3 · N2 · ERN/pe

1 Introduction

With the development of science technology, mechanized and intelligent machines being updated and improved constantly, work for operators turns from physical labor to mental labor gradually. The work model is mainly in the form of information processing task including monitoring and controlling, which leads to higher workload for average person. Mental fatigue can be easily induced during the long-term mental stress. There is no uniform definition for mental fatigue, but the symptoms are common: tiredness, weakness, decrease of alertness, decrements in performance, which would lead human error in operations or decrease productivity in workplace [1]. Mental fatigue may lead to temporary deterioration of attentional functioning and response readiness, which is common in daily life such as driving. Brown [2] demonstrated that a progressive withdrawal of attention from the road and traffic demands was induced with the time on task, and it might impair vehicle control capabilities and the ability to avoid collision.

H. Ayaz and L. Mazur (Eds.): AHFE 2018, AISC 775, pp. 49–60, 2019.
https://doi.org/10.1007/978-3-319-94866-9_5

Therefore, substantial efforts should be undertaken to clarify the mechanisms underlying mental fatigue and develop efficient methods for overcoming it.

There have been many measurements developed for the study of mental fatigue, including subjective measurements and objective measurements. The subjective evaluation is mainly in the forms of subjects' self-reports according to various single-item or multi-item questionnaires [3, 4], and the mostly used questionnaires are the Karolinska Sleepiness Scale (KSS), the Epworth Sleepiness Scale (ESS), Visual Analogue Scale (VAS), the Stanford Sleepiness Scale (SSS) and so on [5]. However, subjective evaluation is easily affected by individual difference. Therefore, objective evaluations obtain more attention, among which, reaction time is usually used to study the information processing. It has been proved that the reaction time increased with the time on cognitive-motor tasks, including switching, visual attention and flanker compatibility [6–8]. These studies indicate that mental fatigue impairs the performance along with the deterioration of information processing functions, such as attention and cognitive control. People have to regulate their cognitive resources and try to maintain their task performance [9]. It is considered that the mental fatigue occurs along with insufficient resource allocation, and to attenuated motivation for maintaining the original task performance [10–12].

Physiological measurements such as electroencephalogram (EEG), event-related potential (ERP), electrooculogram (EOG), electromyogram (EMG), and electrocardiogram (ECG) are proved effective for the mental fatigue evaluation. Because of the high temporal resolution, ERP is considered as an ideal indicator of structure and timing of information processing that occurs between stimulus onset and response. P3, a positive ERP component, its amplitude is considered to reflect the allocation of cognitive resources to a task, while its latency reflects the stimulus evaluation time [9, 13–16]. It was reported that the amplitude of P3 decreased, while the latency increased with the time on task [17]. Another negative ERP component, N2, is thought to be related to cognitive control, including behavioral inhibition and conflict monitoring. Boksem [18] demonstrated that cognitive control ability was impaired when the mental fatigue occurred, accompanied by the decrease of N2 amplitude.

The current study investigated the effects of mental fatigue on resource, behavioral inhibition and conflict monitoring, and error monitoring by using indicators, P3, N2 and ERN/Pe respectively.

2 Method

2.1 Participants

Twenty healthy males between 24 and 32 years of age took part in the experiment. They were all right-handed and had normal or corrected-to-normal vision. None of them had a history of mental illness. They all had normal sleep-wake habit and had more than 7 h sleep before the formal experiment. In addition, all of them were required to sign informed consent prior to the experiment and paid $30 for their participation.

2.2 Task Design

2-back task was used to induce mental fatigue in the current study. The stimulus were ten numbers, 0–9, and each of them appeared at the center of black screen randomly in white, as shown in Fig. 1 Participants were required to judge whether or not the number presented currently at the screen was the same with the last number but one, and press 'F' if it is, otherwise press 'J'. The whole task lasted for 100 min.

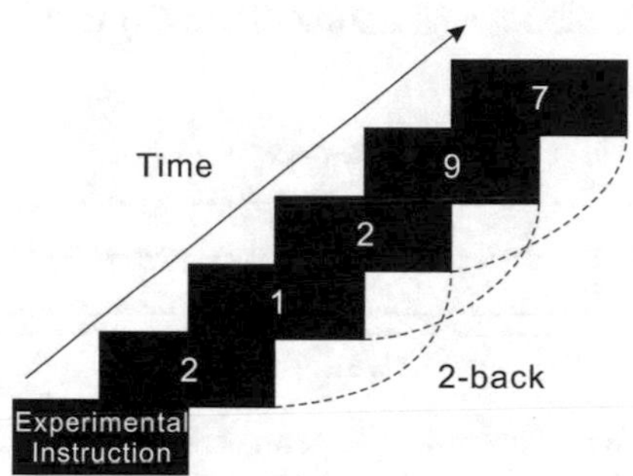

Fig. 1. Experiment paradigm for 2-back. The stimulus are ten numbers, 0–9, and each of them appear at the centre of black screen randomly in white. The inter-stimulus interval (ISI) is 1000 ms. Participants are required to judge whether or not the number presented currently at the screen is the same with the last number but one.

A Go/NoGo task was designed for the ERP study of response processes. Figure 2 showed the presentation pattern for one trial. Firstly, fixation point '+' appeared at the centre of screen, after an inter-stimulus interval (ISI) of 50 ms, cue stimulus, '1' or '−1', was presented for 300 ms. After an inter-stimulus interval (ISI) of 1000 ms, a probe stimulus was displayed for 1000 ms. When the cue stimulus was '1', the participants were required to keep a watchful eye on probe stimulus: if the probe stimulus was an even number, it was the target (Go) and they had to press 'F' as soon as possible or else didn't need to respond to it (NoGo). When the cue stimulus was '−1', the participants didn't need to make any responses whether probe stimulus was an even number or an odd number (NoGo).

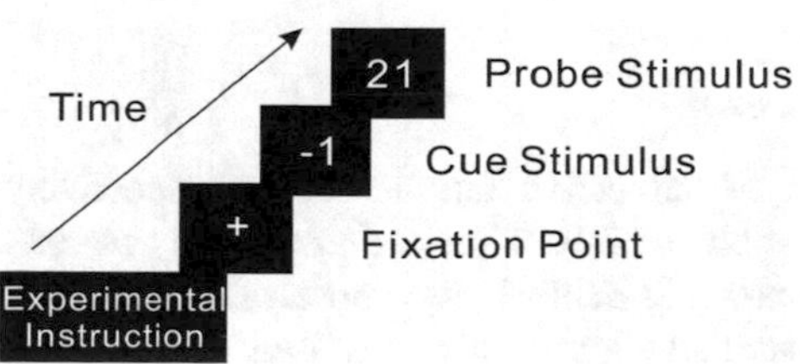

Fig. 2. Experiment paradigm for Go/NoGo. Fixation point '+' appears at the centre of screen, after an inter-stimulus interval (ISI) of 50 ms, cue stimulus, '1' or '−1', is presented for 300 ms. After an inter-stimulus interval (ISI) of 1000 ms, a probe stimulus is displayed for 1000 ms. Participants are required to judge whether or not probe stimulus is the target.

2.3 Procedure

The experiment was conducted in an electrically shielded and quiet room. The whole experimental procedure was shown in Fig. 3, which consisted of three sections, including 100-min 2-back task and 10-min Go/NoGo task for pretest and posttest respectively. Before and after the 2-back task, the participates were required to rate their scores according to mental fatigue scale from 0 to 10 in three aspects: thinking, attention and sleepy. The performance during the Go/NoGo and 2-back tasks were recorded. Meanwhile the EEG signal during Go/NoGo tasks were recorded.

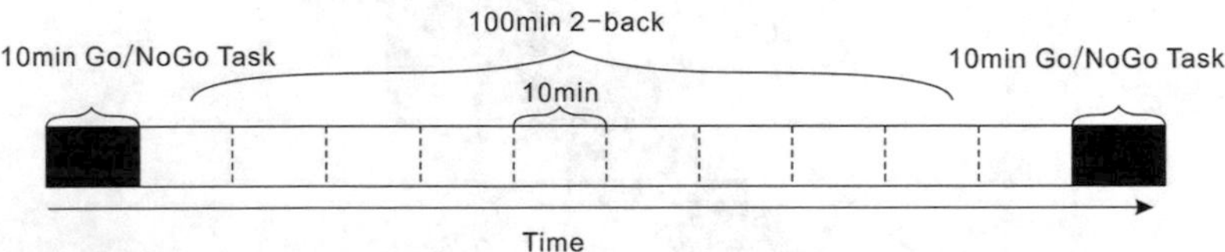

Fig. 3. Experiment procedure. The whole experimental consists of three sections, including 100-min 2-back task and 10-min Go/NoGo task for pretest and posttest respectively.

2.4 Acquisition and Processing of EEG Data

The EEG signal was acquired using a Brain Products 32-channel system. The Ag/AgCl electrodes were placed with the international 10–20 electrodes placement system. Two mastoid electrodes (M1 and M2) were used as reference, and the ground electrode was paced at Fpz. Moreover, electrooculograms (VEOG and HEOG) were also recorded. The EEG signals were sampled at 500 Hz with lowering all electrode resistances below 10 KΩ.

The raw EEG signals were off-line processed by the software of BrainVision Analyser, and the procedures are: down-sampling, reference reset, ocular artifacts removing, other artifacts removing, filtering, segmentation, baseline correction, overlapping averaging, peak detection and so on. P3, N2 and ERN/Pe were assured as the EEG characteristic parameters studied in the current study.

3 Result

3.1 Subjective Scores

There were two sets of subjective fatigue scores respectively from the beginning (pretest) and end (posttest) of 2-back task. Paired-t test was used to analyze the data, and the results were shown in Table 1. It demonstrated that the subjects' fatigue scores of the three aspects were all significantly increased after finishing the 100 min 2-back task (all $P < 0.05$). Most subjects felt that they had difficulty in concentrating on the task, in confusion and drowse at the end of the task. Thus, mental fatigue was induced by 100 min of performing a 2-back task.

Table 1. Paired-t test results of subjective fatigue scores

Aspect	Pretest	Post test	t	P
Thinking	1.267 (1.387)	8.400 (2.640)	−1.730	0.020
Attention	1.867 (1.356)	9.467 (2.326)	−2.502	0.000
Sleepy	1.267 (1.870)	8.867 (2.295)	−2.793	0.039

3.2 Behavior Performance

The 2-back task was divided into ten 10-min blocks, and the reaction time (RT) and accuracy of each block was computed. Then ten sets of performance data were obtained. The changes of accuracy and RT with time on task were shown in Fig. 4. One-way repeated-measures ANOVA was used to analyze the time main effect on accuracy and RT. It was obvious that the accuracy decreased significantly with time on task ($F = 2.3$, $P < 0.05$), meanwhile the reaction time increased obviously ($F = 9.3$, $P < 0.05$). These findings demonstrated that the mental fatigue was induced and impaired the behavior performance after subjects' continuously performing 2-back task for 100 min (Table 2).

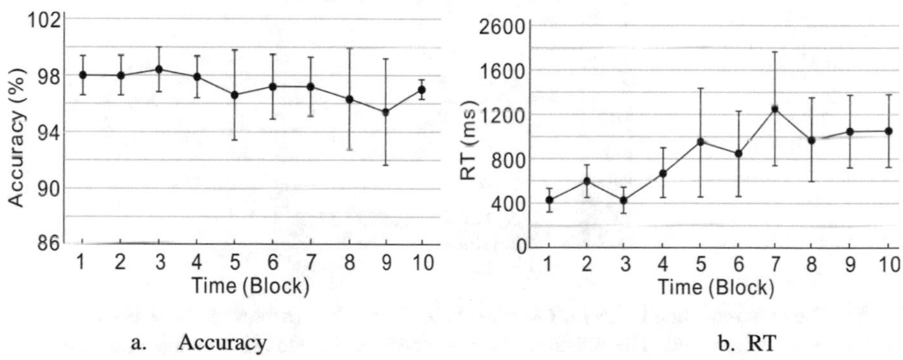

a. Accuracy b. RT

Fig. 4. Performance changes over time during the whole experiment. The curves show the accuracy decreases, meanwhile the reaction time increases with time on task.

Table 2. One-way repeated-measures ANOVA results of performance

Name	Parameter	Accuracy	RT
Time	(F, P)	(2.3, 0.016)	(9.3, 0.0)
Adjustment coefficient	H-F	0.567	0.672

3.3 ERPs

P3 and N2. The EEG signals obtained during Go/NoGo tasks before and after 100 min 2-back task were processed by ERP technology. The epoch for analysis of P3 and N2 components was 1000 ms (200 ms preceding mark onset to 800 ms following mark onset), and GO- and NoGo- were defined as the marks.

The grand-average ERPs for Go trial of pretest and posttest were shown in Fig. 5, which demonstrated that obvious Go-N2 and Go-P3 were stimulated. Paired-t test was used to analyze the amplitudes and latencies for each ERP components, and the results were shown in Tables 3 and 4. It demonstrated that the latencies of Go-P3 increased significantly in Fz, Cz and Pz after fatigue ($t = -2.359$, $P < 0.05$; $t = -4.865$, $P < 0.05$; $t = -2.719$, $P < 0.05$), while there were no obvious changes in other components (all $P > 0.05$).

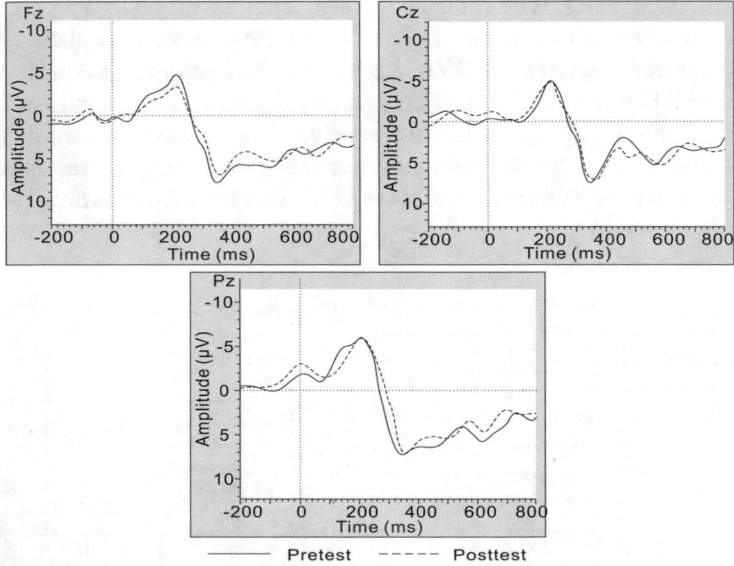

Fig. 5. The grand-average ERPs for Go trial at Fz, Cz and Pz. The curves show obvious Go-N2 and Go-P3 are stimulated. The full-line denotes pretest, and dotted line denotes posttest.

Table 3. Descriptive statistics and paired-t test results for Go-N2

Go		Pretest	Post test	t	df	P
Amplitude (μV)	Fz	−4.77 (2.19)	−3.29 (3.98)	−1.041	19	0.052
	Cz	−4.98 (5.43)	−4.78 (1.78)	−0.125	19	0.903
	Pz	−6.02 (5.29)	−5.97 (2.22)	−0.03	19	0.977
Latency (ms)	Fz	212.6 (12.92)	214.4 (12.02)	−0.307	19	0.766
	Cz	214.2 (18.36)	214.0 (12.68)	0.025	19	0.981
	Pz	207.2 (7.55)	209.8 (7.78)	−0.374	19	0.717

Table 4. Descriptive statistics and paired-t test results for Go-P3

Go		Pretest	Post test	t	df	P
Amplitude (μV)	Fz	7.88 (1.19)	6.96 (1.62)	1.36	19	0.207
	Cz	7.51(1.05)	7.12 (0.82)	1.362	19	0.206
	Pz	7.21 (1.16)	7.07 (1.05)	0.453	19	0.661
Latency (ms)	Fz	343.8 (12.27)	355.8 (9.11)	−2.359	19	0.043
	Cz	343.8 (4.56)	365.4 (8.54)	−4.865	19	0.001
	Pz	341.2 (11.93)	356.0 (14.24)	−2.719	19	0.024

The grand-average ERPs for NoGo trial of pretest and posttest were shown in Fig. 6, which demonstrated that obvious NoGo-N2 and NoGo-P3 were stimulated. Paired-t test was used and the results were shown in Tables 5 and 6, and it demonstrated that the latencies of NoGo-N2 increased significantly in Fz, Cz and Pz after fatigue ($t = -17.71$, $P < 0.05$; $t = -4.751$, $P < 0.05$; $t = -4.325$, $P < 0.05$). Moreover, the amplitudes of NoGo-P3 decreased significantly ($t = -3.68$, $P < 0.05$; $t = -4.27$, $P < 0.05$; $t = -4.39$, $P < 0.05$), and the latencies of NoGo-N2 increased significantly in all three elcctrodes ($t = -3.96$, $P < 0.05$; $t = -3.32$, $P < 0.05$; $t = -2.31$, $P < 0.05$).

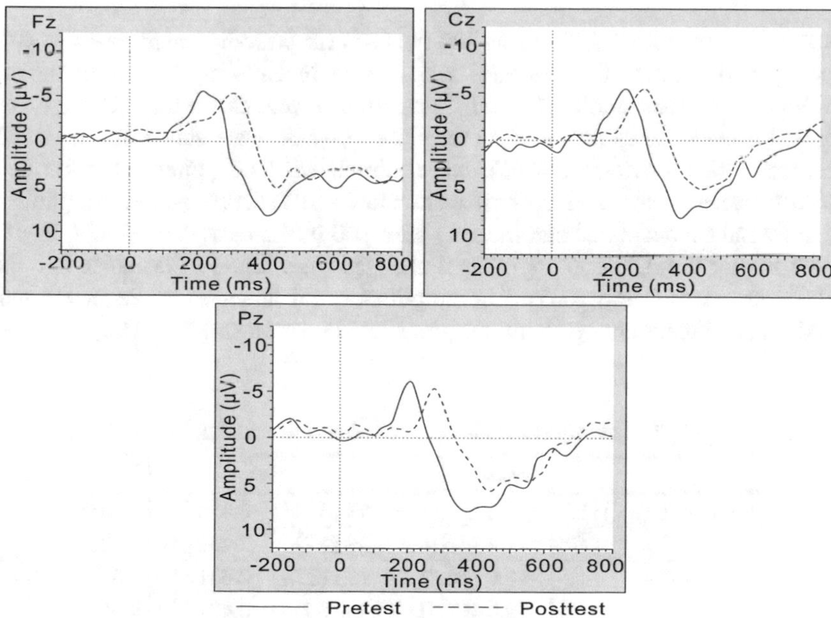

Fig. 6. The grand-average ERPs for NoGo trial at Fz, Cz and Pz.. The curves show obvious NoGo-N2 and NoGo-P3 are stimulated. The full-line denotes pretest, and dotted line denotes posttest.

Table 5. Descriptive statistics and paired-t test results for NoGo-N2

NoGo		Pretest	Post test	t	df	P
Amplitude (μV)	Fz	−5.55 (5.40)	−5.33 (2.94)	−0.167	19	0.871
	Cz	−5.39 (6.74)	−3.40 (2.29)	−1.132	19	0.287
	Pz	−5.37 (2.76)	−6.10 (3.89)	0.62	19	0.551
Latency (ms)	Fz	214.8 (19.87)	307.8 (18.46)	−17.71	19	0.00
	Cz	218.6 (19.67)	274.2 (43.59)	−4.751	19	0.001
	Pz	207.6 (9.79)	280.0 (54.82)	−4.325	19	0.002

Table 6. Descriptive statistics and paired-t test results for NoGo-P3

NoGo		Pretest	Post test	t	df	P
Amplitude (μV)	Fz	8.20 (1.18)	6.42 (0.96)	4.61	19	0.001
	Cz	8.20 (1.49)	5.11 (1.30)	5.32	19	0.00
	Pz	7.98 (1.23)	5.81 (0.73)	5.18	19	0.001
Latency (ms)	Fz	405.6 (23.51)	431.4 (19.96)	2.40	19	0.04
	Cz	380.0 (25.54)	441.6 (40.68)	3.667	19	0.005
	Pz	375.6 (26.86)	420.0 (35.28)	2.724	19	0.023

ERN/Pe The epoch for ERN/Pe was 800 ms (400 ms preceding mark onset to 400 ms following mark onset), and the false response to NoGo was defined as the mark. According to previous studies, Fz and Cz electrodes were chose for ERN/Pe analysis. The grand-average ERPs for the false response to NoGo were shown in Fig. 7, which demonstrated that obvious ERN/Pe were stimulated. The paired-t test results for ERN/Pe between pretest and posttest were shown in Tables 7 and 8. Amplitudes of ERN in Fz and Cz decreased significantly after 100 min 2-back ($t = -2.434$, $P < 0.05$; $t = -2.406$, $P < 0.05$), but no significant changes were found in latencies ($t = 0.631$, $P > 0.05$; $t = -0.451$, $P > 0.05$). The amplitudes and latencies of Pe in Fz and Cz showed no significant change between pretest and posttest (all $P > 0.05$).

Table 7. Descriptive statistics and paired-t test results for ERN

		Pretest	Posttest	t	df	P
Amplitude (μV)	Fz	−12.6 (3.6)	−9.57 (3.7)	−2.434	19	0.025
	Cz	−11.4 (4.5)	−7.34 (4.2)	−2.406	19	0.026
Latency (ms)	Fz	78.6 (16.4)	75.45 (13.6)	0.631	19	0.535
	Cz	74.5 (17.5)	76.4 (17.2)	−0.451	19	0.661

Table 8. Descriptive statistics and paired-t test results for Pe

		Pretest	Posttest	t	df	P
Amplitude (μV)	Fz	9.76 (3.90)	8.83 (1.24)	0.856	19	0.403
	Cz	10.39 (2.13)	10.81 (1.61)	−0.75	19	0.462
Latency(ms)	Fz	254.6 (29.5)	254.9 (33.7)	−0.35	19	0.973
	Cz	251.0 (34.7)	246.1 (35.72)	0.45	19	0.656

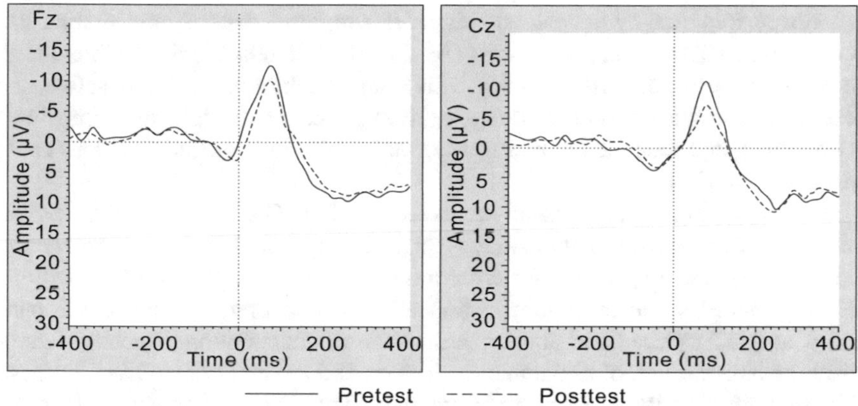

Fig. 7. The grand-average ERPs for false response to NoGo at Fz and Cz. The curves show obvious ERN and Pe are stimulated. The full-line denotes pretest, and dotted line denotes posttest.

4 Discussion

In this study, twenty subjects were requested to perform a 100 min 2-back task continuously to induce mental fatigue. Increased fatigue scale scores in different aspects demonstrated that the subjects felt fatigue in the end of the task. Moreover, impaired performance was found with the time on task, which demonstrated that mental fatigue resulted in deteriorations of cognitive-motor performance. Therefore, after performing 100 min 2-back task, mental fatigue was indeed induced successfully and effectively.

It had been proved that the latencies of Go-P3 and NoGo-N2, both amplitudes and latencies of NoGo-P3 had obvious fatigue effect. Go trials and NoGo trials consist of conflict task, in which Go trial reflects execution to reaction, while the NoGo trial reflects restraint to reaction. It was obvious Go-P3 was produced because of sufficient attention was concentrated on target stimulus (Go). The P3 amplitude is proved to be in positive correlation with attention resource allocation, and it is easily affected by stimulus probability, stimulus meaning, motivation and so on. The latency of P3 is in positive correlation with reaction time (RT), but shorter than RT. RT reflects the process of 'stimulus-cognitive processing-reaction', while the latency of P3 reflects the process of 'stimulus-cognitive processing'. The increase of Go-P3 latency in this study demonstrated that the speed of information processing declined because of mental fatigue.

More obvious N2 and P3 were produced by NoGo trial than Go trial, which was consistent with the results of the previous studies. NoGo-N2/P3 reflect the suppressions to the fixed tendency to respond to stimuli [19]. Moreover, NoGo-N2 and NoGo-P3 were proved to reflect different processes of response inhibition function. NoGo-N2 mainly reflects the pre-preparation of inhibiting motion, and is related to conflict monitor and error detection [20, 21]. NoGo-P3 mainly reflects the inhibition process of action [22]. In the current study, the NoGo-P3 amplitude decreased significantly, meanwhile the latencies of NoGo-N2/P3 were prolonged obviously after fatigue. These results demonstrated that the subjects' abilities of conflict monitor and inhibition reaction were impaired by the mental fatigue. It is reported that the intracerebral source was located in anterior cingulate cortex (ACC), which is related with conflict monitor. That is to say, when the task referring with reaction inhibition, ACC is activated and realize the adjustment of conflict. Therefore, it might be the mental fatigue impaired the ACC function, and then the conflict monitor and inhibition reaction functions can't be put to good use [23].

Error monitoring ability is usually reflected by ERN. The ERN amplitude decreased significantly after fatigue in this study, which demonstrated that the behavior monitoring function was impaired. The error monitoring is located in basal ganglia, and ERN is produced when error information affected dopamine system, the dopamine transmit to ACC reduced and some neuron were in disinhibition. Mental fatigue can destroy the transmission of dopamine in striatum and callosal convolution, because of which the error monitoring function is impaired. That is to say, mental fatigue might attribute to the change of ACC function. Moreover, there was no significant variation in Pe before and after fatigue. This is because ERN and Pe reflect different phases of error monitoring process, and ERN is the elementary automatic reaction to fault, while Pe reflects the adjustment based on behavioral consequences when fault occurred. Mental fatigue didn't make any difference to the behavioral adjustment after fault obviously.

5 Conclusion

The effect of mental fatigue on response processes was studied by ERP technology in Go/NoGo Task. Increased fatigue scale scores and impaired performance proved 100 min 2-back task indeed induced effective mental fatigue. In the ERP study, the increase of Go-P3 latency demonstrated the speed of information processing declined because of mental fatigue. NoGo-N2/P3 data demonstrated that the subjects' abilities of conflict monitor and inhibition reaction were impaired by the mental fatigue. ERN data indicated that mental fatigue reduces the efficiency of error monitoring. The metrics above are possible indices for measuring mental fatigue.

Acknowledgements. This research was funded by Presidential Fund of CNIS (522018Y-5943) and National Key R&D Program of China (2017YFF0206604).

References

1. Liu, J., Zhang, C., Zheng, C.: EEG-based estimation of mental fatigue by using KPCA–HMM and complexity parameters. Biomed. Signal Process. Control **5**, 124–130 (2010)
2. Brown, I.D.: Driver fatigue. Hum. Factors **36**, 298–314 (1994)
3. Li, H.C.O., Seo, J., Kham, K., Lee, S.: Method of measuring subjective 3D visual fatigue: a five-factor model. In: Digital Holography and Three-Dimensional Imaging, Florida, America (2008)
4. Gutierrez, J., Perez, P., Jaureguizar, F., Cabrera, J.: Subjective assessment of the impact of the transmission errors in 3DTV compared to HDTV. In: 3DTV Conference: The True Vision Capture, Transmission and Display of 3D Video (3DTV-CON), Madrid, Spain, pp. 1–4. IEEE (2011)
5. Yang, C., Wu, C.H.: The situational fatigue scale: a different approach to measuring fatigue. Qual. Life Res. **14**(5), 1357–1362 (2005)
6. Jong, D.: An intention–activation account of residual switch costs. In: Attention and Performance XVIII: Control of Cognitive Process, pp. 357–376. MIT Press, Cambridge (2000)
7. Lorist, M.M., Boksem, M.A.S., Ridderinkhof, K.R.: Impaired cognitive control and reduced cingulate activity during mental fatigue. Cogn. Brain. Res. **24**, 199–205 (2005)
8. Boksem, M.A.S., Meijman, T.F., Lorist, M.M.: Effects of mental fatigue on attention: an ERP study. Cogn. Brain. Res. **25**, 107–116 (2005)
9. Kok, A.: Event-related-potential (ERP) reflections of mental resources: a review and synthesis. Biol. Psychol. **45**, 19–56 (1997)
10. Fairclough, S.H.: Mental effort regulation and functional impairment of the driver, Hancock. In: Hancock, P.A., Desmond, P.A. (eds.) Workload and Fatigue, pp. 211–231. Mahwah (2001)
11. Warburton, D.M.: A state model for mental effort. In: Hockey, G.R.J., Gaillard, A.W.K., Coles, M.G.H. (eds.) Energetics and Human Information Processing, pp. 217–232. Martinus Nijhoff, Dordrecht (1986)
12. Hockey, G.R.: Compensatory control in the regulation of human performance under stress and high workload: a cognitive-energetical framework. Biol. Psychol. **45**, 73–93 (1997)
13. Kramer, A., Spinks, J.: Handbook of cognitive psychophysiology. In: Jennings, J.R., Coles, M.G.H. (eds.) Central and Autonomic Nervous System Approaches, pp. 179–242. Wiley, New York (1991)
14. Duncan-Johnson, C.C.: P300 latency: a new metric of information processing. Psychophysiology **18**, 207 (1981)
15. McCarthy, G., Donchin, E.: A metric for thought: a comparison of P300 latency and reaction time. Science **211**, 77–80 (1981)
16. Verleger, R.: On the utility of P3 latency as an index of mental chronometry. Psychophysiology **34**, 131 (1997)
17. Humphrey, D.G., Kramer, A.F., Stanny, R.R.: Influence of extended wakefulness on automatic and nonautomatic processing. Hum. Factors **36**, 652–669 (1994)
18. Boksem, M.A.S., Meijman, T.F., Lorist, M.M.: Mental fatigue, motivation and action monitoring. Biol. Psychol. **72**(2), 123 (2006)
19. Nieuwenhuis, S., Yeung, N., Widenberg, W.: Electrophysiological correlates of anterior cingulated function in Go/NoGo task: effect of response conflict and trial type frequency. Cogn. Affect. Behav. Neurosci. **3**(1), 17–26 (2003)
20. Botvinick, M.M., Braver, T.S., Barch, D.M.: Conflict monitoring and cognitive control. Psychol. Rev. **108**(3), 624–652 (2001)

21. Folstein, J.R., Van Petten, C.: Influence of cognitive control and mismatch on the N2 component of the ERP: a review. Psychophysiology **45**(1), 152–170 (2008)
22. Smith, J.L., Johnstone, S.J., Barry, R.J.: Movement-related potentials in the Go/NoGo task: the P3 reflects both cognitive and motor inhibition. Clin. Neurophysiol. **119**(3), 704–714 (2008)
23. Bekker, E.M., Kenemans, J.L., Verbaten, M.N.: Source analysis of the N2 in a cued Go/NoGo task. Cogn. Brain. Res. **22**(2), 221–231 (2005)

Cognitive Function Evaluation of Dementia Patients Using P300 Speller

Studies on Spelling Error and Counting Task

Ryo Morooka[1(✉)], Hisaya Tanaka[2], Takahiko Umahara[3], Akito Tsugawa[3], and Haruo Hanyu[3]

[1] Informatics Major, Graduate School, Kogakuin University, Tokyo, Japan
j214119@ns.kogakuin.ac.jp
[2] Kogakuin University, 1-24-2 Nishi-Shinjuku, Tokyo 163-8677, Japan
hisaya@cc.kogakuin.ac.jp
[3] Department of Geriatric Medicine, Tokyo Medical University, Tokyo, Japan
{takahiko,h-hanyu}@tokyo-med.ac.jp,
tsugawa-l@hotmail.co.jp

Abstract. The character-input BCI has been studying as a communication aid for neuromuscular illness patients such as ALS. We focused on input-error in the character-input BCI. It is necessary that the subject has to concentrate high attention to input correct characters in the character-input BCI. On the other hand, it causes an input-error of far character when the subject is distractedness. Hence, we considered that the number of input error of far character may increase when the dementia patient uses the character-input BCI. Therefore, we conducted a character gazing experiment using the character-input BCI to the elderly. Then, we quantified the result of the character estimation of gazed character by calculating the SEDV (Spelling-Error Distance Value). As a result, in the Modarate1 group, the SEDV was 0.41 characters farther than the Questionably significant group. Hence, there is possibility that the character-input BCI may be able to screening dementia.

Keywords: BCI · ERP · P300 · Dementia · Attention

1 Introduction

Recently, with the aging of the population in Japan, dementia patients are on the rise and it is becoming a big social problem. According to reports by the Ministry of Health, Labor and Welfare in Japan, the number of patients is expected to reach about 7.3 milion in 2025 [1]. With the progress of medicine, if we can find dementia early, we can delay treatment of progression. Therefore, early diagnosis of dementia is becoming more important from the viewpoint of maintaining patient's QoL (Quality of Life) [2].

Dementia is mainly screened by performing brain image examination using MRI and SPECT, and neuropsychological examination such as MMSE (Mini-Mental State Examination), and comprehensively summarizing these results. However, the MRI examination has high running cost, and the neuropsychological examination has a

H. Ayaz and L. Mazur (Eds.): AHFE 2018, AISC 775, pp. 61–72, 2019.
https://doi.org/10.1007/978-3-319-94866-9_6

problem that it is difficult to apply to people who understand Japanese insufficiently. Therefore, it is necessary to develop an inexpensive screening tool that can diagnose in a short time and research on technology.

So far research on event related potentials for dementia patients has been conducted in Japan. Morita showed that both the amplitude of P300 is significantly decreased and latency of P300 is significantly prolonged between the cognitive group and the dementia low risk group and the dementia group and healthy group at the visual event-related potential [3]. Therefore, it is considered that Event-Related Potential (ERP) measurement using character input BCI (Brain-Computer Interface) using P300 is possible even if dementia patients. Many papers that have studied dementia from the perspective of brain science have been posted. However, technological research to combine with brain information technology and diagnose at the same time with measurement is not conducted worldwide. Therefore, we aim to develop a screening tool for dementia using character input type BCI. If dementia can be screened using character input BCI, We can conduct inspections at low cost for introduction cost and running cost. We can also expect reduction in human cost by automating explanation of inspection. Also, while neuropsychological examination, which is a conventional method, takes about 1 h to perform, screening by character-input BCI can be performed in about 30 min.

Therefore, it is expected that the medical time can be shortened greatly. So far, character-input BCI was used for practical realization as a communication aid device for amyotrophic lateral sclerosis (ALS) patients, and researches on reducing the number of electrodes and reducing the time for inputting characters have been conducted based on patient needs [4, 5]. We focused on erroneous input in BCI. It is necessary that the subject has to concentrate high attention to input correct characters in the character-input BCI. On the other hand, it causes an input-error of far character when the subject is distractedness. Hence, we considered that the number of input error of far character may increase when the dementia patient uses the character-input BCI. In this research, in order to develop a dementia screening tool using character input type BCI, we calculated the weighted mean distance SEDV (Spelling - Error Distance Value) in consideration of the character estimation principle in character input type BCI. And we visualized and quantified distraction in the elderly.

2 The Character-Input BCI

2.1 Event-Related Potential P300

Event-related potential (ERP) is electric activity of the brain that occurs in response to stimuli such as light and sound and movement such as bending and stretching of a finger [6]. P300 is a positive potential associated with attention. Generally P300 is observed in the oddball task, but it is obtained with most tasks seeking some kind of mental judgment such as selection and understanding [7].

2.2 P300 Speller

P300 Speller is a character input type BCI proposed by Farwell, Donchin in 1988 [8]. The P300 Speller is a BCI that discriminates P300 at online and inputs characters by converting the discrimination result into switch information. Displays a dial on the screen and gives visual stimulus by blinking characters in rows or columns at random. The subject searches for the character intended for input and gazes at it. P300 is triggered when the gazing character (target character) flashes, characters are input by detecting P300 on-line (Fig. 1). In this study, P300 Speller do not use as a character input device, but it uses for visual stimulus presentation and ERP recording. Or later, when we mention BCI in this paper, we indicate the P300 Speller.

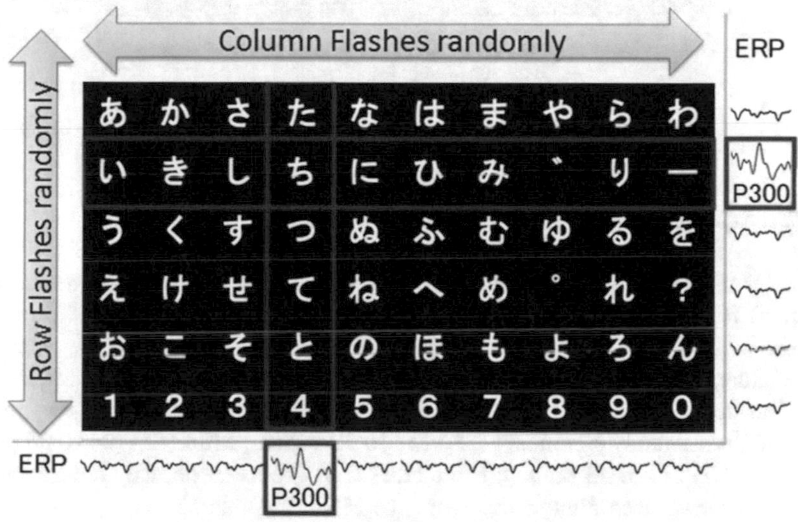

Fig. 1. Character estimate principle in BCI

2.3 Measurement Principle of Distraction Using BCI

In this study, I thought that the degree of distraction of distraction can be known by the position of incorrect input of BCI. In character input BCI, if attention to characters is sufficiently high, P300 appears strongly in rows and columns including target characters, and P300 does not appear in other rows and columns (Fig. 1). As an erroneous input that may occur when attention to characters is high, there is a neighbor error. This is caused by erroneous reaction when characters adjacent to the target character blink [9]. On the other hand, if the attention is distracted, P300 of the row and column including the target character become weak, and P300 appears extensively depending on distraction. Therefore, BCI can not correctly detect P300, making input-error of far character more likely to occur (Fig. 2). Therefore, it is considered that the average distance of input-error represents the degree of distraction.

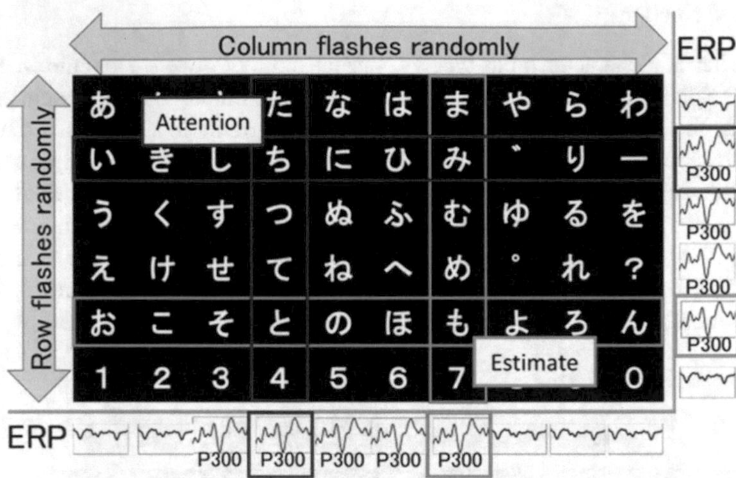

Fig. 2. Measurement principle of distraction

3 Neuropsychological Examination MMSE

MMSE (Mini-Mental State Examination) is a screening test for dementia that can be performed from 5 min to 10 min [10]. It can examine the functions of time, place orientation, immediate and delayed reproduction of 3 words, attention, calculation, nomenclature, repetition, language understanding, reading, writing and drawing functions. The upper limit is 30 points. In this study, subjects were classified into five groups of Questionably significant (L0) (27 to 30 points), Mild (L1) (24 to 26 points), Moderate1 (L2) (20 to 23 points), Moderate2 (L3) (16 to 19 points), and Severe (L4) (15 points or less) according to the scores of MMSE (Table 1).

Table 1. Subject grouping using MMSE scores

Score of MMSE	Cognitive function decline level
27 to 30 points	L0 (Questionably significant)
24 to 26 points	L1 (Mild)
20 to 23 points	L2 (Moderate1)
16 to 19 points	L3 (Moderate2)
15 points or less	L4 (Severe)

4 Evaluation Method of BCI Character Estimation Result

4.1 Relative Position Plot of False Input

In evaluating character estimation results by BCI, we visualized the relative positional relationship between the target character and the estimated character of BCI (erroneous input plot). Let the position of the target character in the dial used be (0, 0). When the position of the character estimated by BCI is regarded as (Ca, Cb) that the position when "a" character to horizontal direction and "b" character to vertical direction from target character shifted, add 1 to the coordinates (*Ca*, *Rb*) of the plot of Fig. 3. The erroneous input plot of Fig. 2 is an example in the case where the target character is "Annkomochi", and the estimated character of BCI is "Ifukamomo". By performing addition for the total number of input characters N_{char} times, the relative positional relationship of erroneous input becomes clear. By dividing each coordinate of the erroneous input plot by "N_{char}", the relative frequency distribution of each erroneous input can be obtained [11].

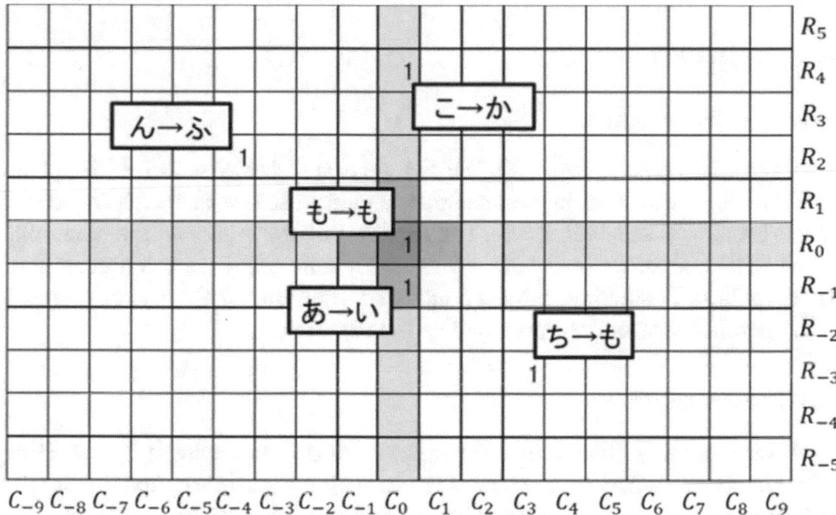

Fig. 3. The input-error plot

4.2 Spelling-Error Distance Value

SEDV (Spelling-Error Distance Value) is the weighted average distance taking into account the character estimation principle of BCI. Upon calculation of SEDV, weight of incorrect input of location of "a" character to horizontal direction and "b" character to vertical direction from target character was defined by Eq. (1). Here, D_1 indicates a case where one of the line and the column of the target character matches the character estimated by the BCI. And D_2 indicates a case where character estimated by the BCI and the row/column of the target character are different. In character estimation in BCI, it is only possible to enter correct characters only after the target character and the row/column of characters estimated by BCI match. Thus, matching one of the rows and columns means that the character estimation of BCI is half correct [9]. Therefore, we add processing that halves the weight. SEDV asked by multiplying the erroneous input plot (l) obtained in 4.1 by the weight (W) and dividing by the total number of characters input (N_{char}) (Eq. (2)) [11]. In the Eq. (2), r represents the number of rows of the dial and c represents the number of columns of the dial. SEDV is statistically a value corresponding to the standard deviation. When it was plotted as shown in Fig. 3, the SEDV is 2.24 character.

$$W = \begin{cases} D_1 \frac{1}{2} \sqrt{(a^2 + b^2)} (a = 0 \cup b = 0) \\ D_2 \sqrt{(a^2 + b^2)} (a \neq 0 \cap b \neq 0) \end{cases} \quad (1)$$

$$\mathrm{SEDV} = \frac{1}{N_{char}} \sum_{i=-(r-1)}^{r-1} \sum_{j=-(c-1)}^{c-1} (l \times W) \quad (2)$$

5 Experiment

5.1 Subject Information

Subjects were 24 persons (aged 80.5 ± 5.3 years old, MMSE 22.5 ± 5.0 points) in their 70's to 90's who were judged healthy or mild to severe dementia by specialists. The experiment was conducted with the approval of the ethics review committee of Tokyo Medical University (early diagnosis of dementia using brain computer interface (BCI), 2016-083). In addition, subject conducted a neuropsychological examination by a clinical psychologist on the same day as the experiment.

5.2 BCI Parameters

The dial was randomly blinked in units of rows and columns using a 6 × 10 Hiragana dial. Five or six characters of hiragana were gazed per experiment, and the experiment was conducted three or four times. The number of blinks (*F*) in each row and each column was set to 5 times (10 blinks per character). SOA (Stimulus Onset Asynchrony) was based on 210 ms (character lighting time 120 ms, character light off time 90 ms).

When subjects were difficult to recognize blinking, they were adjusted appropriately, and the maximum was 300 ms (character lighting time 170 ms, character lighting time 130 ms). Also, the target character search time *S* [*s*] is 6 s. The time required for a single experiment *ET* [*s*] can be obtained by the Eq. (3). In this study, the time required for one experiment was up to 198 s (SOA = 210 [ms], r = 6 [row], c = 10 [column], F = 10 [Times], S = 6 [s], N_{char} = 5 [characters]).

$$\mathrm{ET[s]} = \left(\frac{SOA}{1000} \times (r + c) \times F + S \right) \times N_{char} \quad (3)$$

5.3 BCI System

The system diagram of the BCI used in this study is shown in Fig. 4. Electroencephalograms were measured using an active electrode (LADY bird electrode manufactured by g.tec), collected using an electrode box (g.SAHARAbox manufactured by g.tec), and amplified the signal using a bio-amplifier (g.USBamp manufactured by g.

tec) and then incorporated into a PC. MATLAB 2012a was used for brain wave recording, stimulus presentation, and analysis processing. The electrodes were placed at eight location, Fz, Cz, P3, P4, Pz, O1, O2 and Oz, as defined by the international 10–20 system. Also, the reference electrode was placed on the back of the right earlobe and the ground electrode was placed on the forehead. The brain wave was derived using the monopolar induction method.

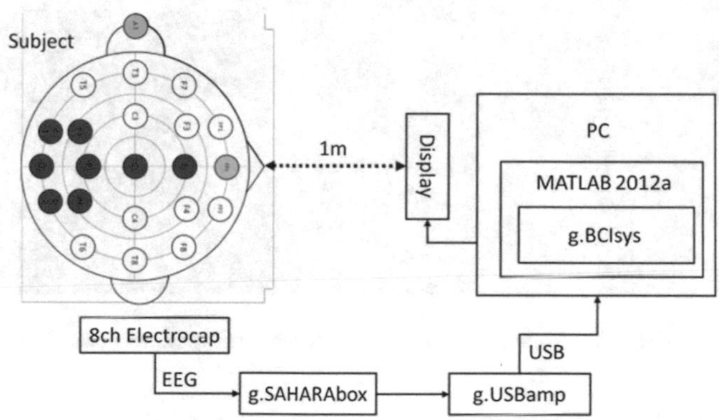

Fig. 4. BCI system configuration

5.4 Teaching

Before starting the experiment, we made the subject check the characters that are displayed thin on the dial and whether the characters are blinking. After confirming that subjects can see the blinking of characters, we taught the subjects using the actual experiment screen as follows. "Look at the shining green characters. Next, various rows and columns will flash, but please only look at the characters that was shining green. Last, please tell me the number of the count that shined." When judging from the state of the subject that the task contents of BCI were not understood, we taught to subjects until they understood the BCI task. During the experiment, if the experimenter judged that the subject was not able to search for the target character, it pointed to the target character before the character flashing began. We made the subjects report the number of blinks after the end of presentation of the blinking stimulus. Here, the number of times the subjects were not counted or counted unnecessarily was recorded as the number of erroneous reactions. It shows appearance of the character gaze experiment using BCI at Fig. 5. As shown in Fig. 5, blinking of characters was given to subjects as visual stimuli, and the subjects aim attention to blinking of characters. At this time, P300 appears when the gazing character blinks.

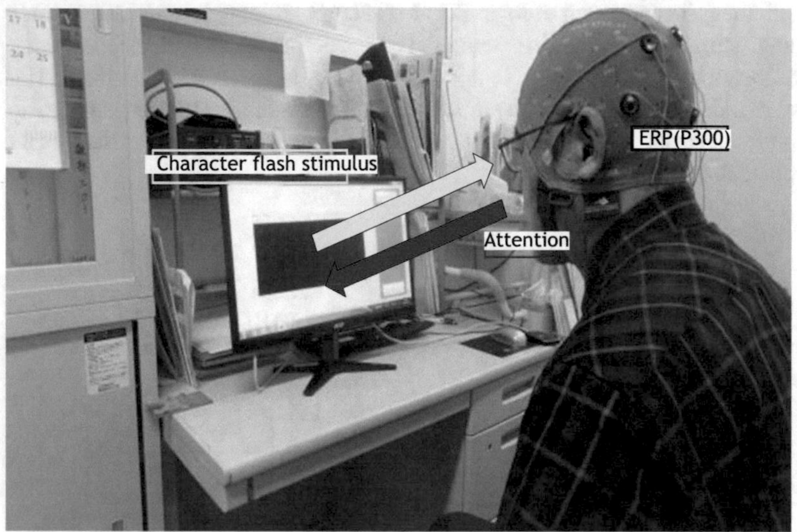

Fig. 5. Appearance of the character gazing experiment.

5.5 P300 Discrimination Rate

In this study, we examined whether it is possible to diagnose patients with dementia who can not accurately perform the counting task by calculating the correlation between the P300 discrimination rate and the number of erroneous reactions in the counting task. When the target character flashes, the number of times P300 could be properly discriminated is divided by the total number of stimuli and the value expressed in percentage is set as P300 discrimination rate (Eq. (4)).

$$\text{P300 discrimination rate} = \frac{collect\ classification}{all\ data} \times 100[\%] \tag{4}$$

6 Result

6.1 BCI Character Gaze Test

Figure 6 is a graph of SEDV in each subject group. From Fig. 6, there was a difference of 0.56 characters in SEDV between L1 group and L2 group, 1.81 characters in SEDV between L3 group and L4 group. As a result of analysis of variance, there was no significant difference, but as a result of multiple comparison, SEDV between L0 group and L4 group, SEDV between L1 group and L4 group showed a significant tendency difference.

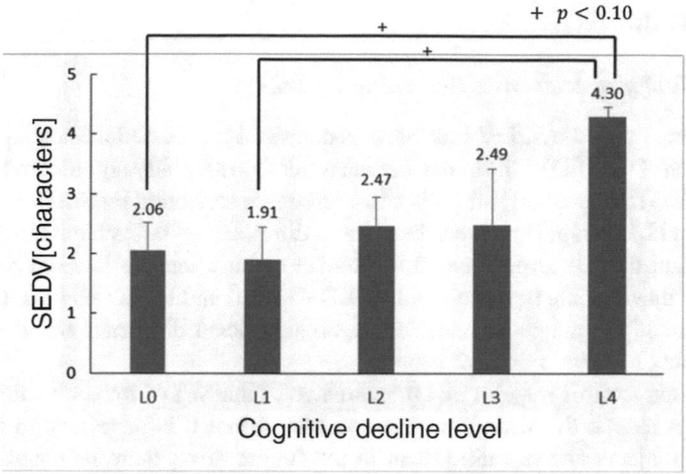

Fig. 6. SEDV between each level of cognitive decline

6.2 Relationship Between Erroneous Reaction Frequency and P300 Discrimination Rate

Correlation between the number of erroneous responses in the counting task and the P300 discrimination rate was determined, and the relationship between the two was examined. From Fig. 7, a significant weak negative correlation was obtained with a correlation coefficient r of −0.38 ($p < 0.01$). Therefore, it is indicated that the P300 discrimination rate tended to decrease as the counting task could not be performed.

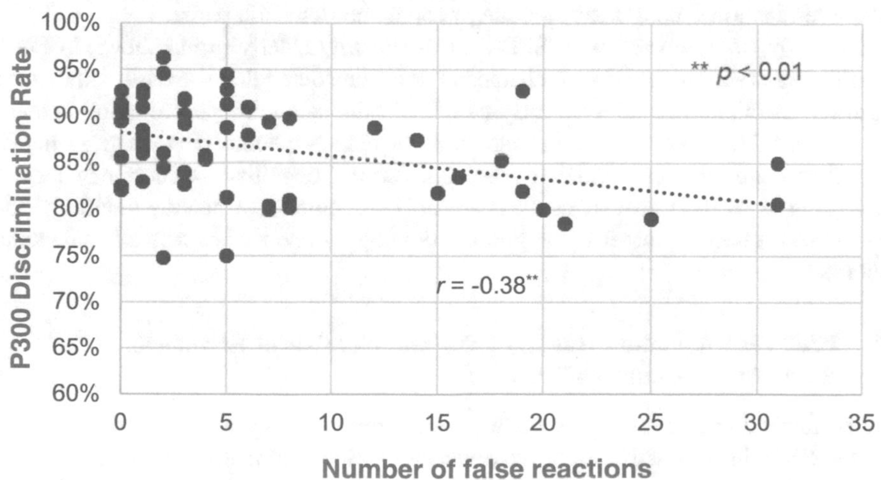

Fig. 7. Erroneous reaction frequency and P300 discrimination rate

7 Discussion

7.1 Possibility of Dementia Screening by BCI

As shown in Fig. 6, the SEDV increased as dementia progressed, indicating a tendency to distraction. The SEDV difference between the Questionably significant (L0) and the Mild (L1) is 0.15 characters, and it is considered that screening by SEDV is difficult for L0 group and L1 group. However, there was a difference of 0.41 characters between the L0 group and the L2 group, and 0.56 characters between the L1 group and the L2 group, so It may be able to screening by SEDV for L0 and L2, L1 and L2. However, as a result of multiple comparison, there was no significant difference between L0 group and L2 group, L1 group and L2 group.

The reason for this may be that the variance of the SEDV for each subject is large, therefore for reduce the variation, it is considered that it is necessary to increase the number of subjects and examine them in the future. Also, there is a significant trend difference of 2.24 characters for SEDV between L0 and L4 and 2.39 characters for SEDV between L1 and L4, therefore screening is considered possible.

7.2 Comparison Between Young Healthy Subjects and Dementia Patients by SEDV

As a result, experiments similar to this study in 12 young healthy people in their 20s to 30s, SEDV was 0.90 characters [11]. Therefore, as a result of comparison with SEDV of mild dementia group, dementia group, moderate dementia group, and severe dementia group shown in Fig. 6, the mild dementia group was 1.11 characters, the dementia group was 2.09 characters, We obtained 1.63 characters for the moderate dementia group and 3.14 characters for the severe dementia group, with SEDV differing by more than 1.00 characters in each group. From this, it can be said that there is a big difference in cognitive function among healthy young people and dementia patients.

Similarly, as compared with SEDV of the healthy elderly group shown in Fig. 6, since there is a difference of 1.06 characters, it is considered that there is a difference in cognitive functions in healthy elderly people. At the same time, since aging is one of the causes of the decrease in the attention of elderly people diagnosed as healthy according to the score of MMSE, it was suggested that cognitive function may decrease with aging as well. However, as there are only 5 members of healthy elderly people, there is also a need to increase the number of subjects and review as well as dementia patients.

7.3 P300 Discrimination Rate by LDA and Number of Erroneous Responses in Counting Task

The correlation coefficient r was −0.36 in the erroneous response in the counting task and the P300 discrimination rate, and a weak negative correlation was observed. From this fact, even if the degree of achievement of the counting task is low, there is a possibility that accurate diagnosis cannot be made but the result of character input is not greatly adversely affected. “Double flash” can be cited as a cause of poor performance of counting tasks. Double flash is a phenomenon in which target characters shine continuously. It is expected that even young healthy people will have difficulty reacting and the priority of solution will be high. As a solution to this problem, it is necessary to create an algorithm to present the stimulus so as not to cause double flash.

8 Conclusion

In this study, the erroneous input distance value SEDV was obtained for the development of dementia screening tool using character-input BCI, and the distraction of the elderly including dementia was quantified. And we examined the relationship between the number of incorrect responses in the counting task and the P300 discrimination rate and examined whether it is possible to diagnose dementia using the P300 Speller even for patients who cannot achieve the counting task.

As a result, SEDV increased as dementia progressed, indicating that the attention tended to be distracted.

Regarding the number of false responses in the counting task and the P300 discrimination rate, a significant weak negative correlation was observed with a correlation coefficient $r = -0.36$. It was also found that there was no major impact, so it was shown that diagnosis of dementia is possible even if the achievement rate of the counting task is bad. However, since a weak negative correlation appears, it is not always that the results of the counting task do not have a bad influence on the diagnosis of dementia, so we indicated that we improve the UI of the dial and develop an algorithm to suppress the double flash.

As a future subject, it is necessary to continue the experiment and to confirm the tendency of distraction that has been clarified in this research.

In addition, focusing on the character estimation result of each subject, it is necessary to consider whether screening is possible by comprehensive judgment using P 300 latency or SEDV.

Acknowledgments. I am deeply grateful to all the subjects who cooperated with the experiment pleas-antly when carrying out this study. We also deeply appreciate everyone Tokyo Med-ical University Hospital Elderly General Medicine Field who cooperated in securing and implementing the experiment space.

References

1. Ministry of Health, Labor and Welfare: Comprehensive Strategy for Promoting Dementia Measures (New Orange Plan). http://www.mhlw.go.jp/file/04-Houdouhappyou-12304500-Roukenkyoku-Ninchishougyakutaiboushitaisakusuishinshitsu/01_1.pdf
2. Wu, J.R., Tsumoto, S.: Neurosurgical engineering - fusion of neuroscience, engineering and information science - Ohmsha (2009)
3. Morita, K.: Characteristics of cognitive function of dementia: psychophysiological study. J. Senile Dement. Res. **19**(3), 70–72 (2012)

4. Ijichi, Y., Tanaka, H.: Study of the number and arrangement of BCI electrodes considering the attitude of ALS persons. Technical report of the Institute of Electronics, Information and Communication Engineers, vol. 115, no. 354, pp. 137–142 (2015)
5. Kitamura, S., Tanaka, H.: Examination of stimulus presentation time and stimulus count in P300 Speller. Technical report of the Institute of Electronics, Information and Communication Engineers, vol. 116, no. 104, pp. 11–16 (2016)
6. Nittono, H.: Event-Related Potential Guidebook for Psychology. Kitaoji Shobo, Kyoto (2005)
7. Hiraki, K., Kanayamai, N.: Using EEGLAB and SPM for EEG Analysis Introduction. Tokyo University Press, Tokyo (2016)
8. Farwell, L.A., Donchin, E.: Talking off the top of your head: Toward a mental prosthesis utilizing event-related brain potentials. Electroencephalogr. Neurophysiol. **70**, 510–523 (1988)
9. Townsend, G.: A novel P300-based brain-computer interface stimulus presentation paradigm: moving beyond rows and columns. Clin. Neurophysiol. **121**(7), 1109–1120 (2010)
10. Folstein, M.F., Folstein, S.E., McHugh, P.R.: "Minimental state": a practical method for grading the cognitive state of patients for the clinician. J. Psychiatr. Res. **12**(3), 189–198 (1975)
11. Kurihara, R., Morooka, R., Hamanaka, S., Tanaka, H., Umahara, T., Tsugawa, A., Hanyu, H.: BCI character input characteristics in patients with mild dementia. In: Human Interface Symposium 2017, DVD Papers Collection, 6C1-1 (2017)

Neurophysiological Sensing

The Relationship Between Aesthetic Choices, Ratings, and Eye-Movements

Elif Celikors[1](✉) and Chris R. Sims[2]

[1] Department of Design and Environmental Analysis, Cornell University, Ithaca, NY, USA
ec839@cornell.edu

[2] Department of Cognitive Science, Rensselaer Polytechnic Institute, Troy, NY, USA
simsc3@rpi.edu

Abstract. A previous study found that looking time is related to aesthetic choices, but not internally determined preferences [1]. The purpose of the current study is to expand on these findings by generalizing them to realistic environmental stimuli. In the choice task coupled with eye-tracking, participants chose the office design that was aesthetically more pleasant. In the preference task, participants rated the aesthetic value of each stimulus. Results showed that looking time predicts choice and aesthetic ratings; however, the magnitude of the latter relationship was small. We also found that rating had a significant main effect, but it didn't have a significant interaction effect with fixation duration on choice. Further research is necessary to understand the relationship between looking time, choices, and preference in the context of environmental stimuli.

Keywords: Aesthetic preferences · Eye-movements · Environmental aesthetics

1 Introduction

Aesthetic qualities have been shown to have cognitive benefits [2–4]. Consequently, environmental aesthetics have gained attention in the design of cognitively demanding environments like office spaces [5–7]. Despite the acknowledgement of the importance of environmental aesthetics, studying this topic remains challenging because defining and operationalizing aesthetic experiences is difficult. The goal of the current study then is to investigate how we can use eye-movement measures to predict certain aspects of aesthetic experiences. For this purpose, we have replicated the study by Isham and Geng [1] to generalize their findings about eye-movements and aesthetic experiences to environmental aesthetics.

Several models of attention propose different mechanisms by which eye-movements interact with various cognitive processes. Traditional models suggest that looking time represents internally determined preferences (i.e.: top-down processes), as measured by preference ratings [8]. Such models predict that during a choice task, subjects' looking time data represent their preference ratings for each stimulus. As a

H. Ayaz and L. Mazur (Eds.): AHFE 2018, AISC 775, pp. 75–82, 2019.
https://doi.org/10.1007/978-3-319-94866-9_7

more recent perspective on attention, the drift diffusion model highlights the importance of eye-movements as affecters rather than effectors of behavioral output [9]. Based on this model, fixations will not be a direct representation of internally determined preferences, but they will actively bias choices. Consistent with the drift diffusion model, Holmes and Zanker [10] found that looking time was a significant predictor of aesthetic choices. In another study, Isham and Geng [1] found that looking time predicted choices, but not internally determined preferences when subjects viewed abstract, black and white geometric compositions. While such findings are informative in understanding the relationship between looking time, aesthetic choices and preferences, we cannot generalize them to the study of environmental aesthetics.

To generalize the findings of Isham and Geng [1] to office room stimuli, we performed three alterations and additions in the original study design. First, we created stimuli that would induce aesthetic responses of different intensities (i.e.: very pleasing to very unpleasing). Choice tasks, where participants are forced to choose the image that has more aesthetic appeal, have the risk of not capturing a real aesthetic reaction. Through ensuring that we have extreme aesthetic responses, we wanted to be more confident that our stimuli induced genuine aesthetic experiences. Second, in the current study, we used real office designs as our stimuli to see whether previous findings with simple aesthetic images can be generalized to stimuli with more complex and realistic content. Lastly, we had separate sessions for the preference and choice tasks. In their study, Isham and Geng [1] established that when participants rated images after making a choice, ratings were biased by preceding choices. In order to prevent the effect of one task on the other, we introduced a one-week waiting period between the choice and the preference tasks.

The goal of this study is to understand the relationship between eye movements and aesthetic responses. In line with the findings from Isham and Geng [1], our hypothesis has two components: (1) Looking time predicts aesthetic choices and (2) looking time does not predict aesthetic preferences. We are expanding previous research by using real life stimuli rather than geometric shapes. Moreover, we were interested in seeing whether preferences moderated the relationship between looking time, and choice. Our findings have the potential to guide researchers and professionals about how they should interpret looking time while subjects evaluate aesthetic properties of office spaces.

2 Methods

2.1 Participants

Twenty-two students from Drexel University (17 females, age range 18–22) participated in the study after providing informed consent. Three of the participants were excluded because eye-tracking data was invalid.

2.2 Stimuli Selection

We did a preliminary survey to select the stimuli to be included in the study. After an extensive Google search for key words "office design," we gathered 207 stimuli with varying visual properties. These images were used in a Qualtrics survey to collect aesthetic responses. The survey asked respondents to evaluate the aesthetic value of each office room design on a scale of 1 to 9 (1 being not aesthetically appealing at all, 9 being as aesthetically appealing as it can be). We collected responses from ten graduate and undergraduate students from Drexel University. The purpose of the preliminary survey was to create a stimulus set with varying aesthetic values. To ensure that we had items of extreme aesthetic value (very low or very high), items that had a mean score of 4, 5 and 6 were excluded. In the end, a total of 64 items (32 highest and 32 lowest scoring) were selected to create three types of paired trials to replicate the study design of Isham and Geng [1]. The first type of trial included two items with similar ratings: two stimuli with high aesthetic rating or two stimuli with low aesthetic rating. This corresponded to the symmetrical design of Isham and Geng [1]. The second type of trial included two items with different aesthetic values: one item with high and one item with low aesthetic rating. This corresponded to the asymmetrical design of Isham and Geng [1]. For the choice task, the initial assignment of stimuli to same-rating and different rating trials was random. Each item appeared in the paired trials only once.

2.3 Preference Task

For the first part of the study, participants received a link to a Qualtrics survey. The survey contained all 64 of the office designs. Participants saw one item at a time and rated each office design based on their aesthetic value on a scale of 1 to 9, 1 being not aesthetically appealing at all and 9 being as aesthetically appealing as possible. The order of items was randomized for each participant.

2.4 Choice Task

A week after completing the online survey, participants were invited to the lab located at Drexel University for the eye-tracking portion of the experiment. Isham and Geng [1] showed that when the preference task comes immediately after the choice task, participants are more likely to give higher ratings to the items that they have chosen. To eliminate any possible effects of ratings on choice, we included a one-week gap between the two tasks. When participants came to the lab after one week, they were presented with pairs of stimuli and pressed a computer key to indicate which room design was aesthetically more appealing (Fig. 1). The order of paired trials was randomized for each participant. The stimuli remained on the screen until the participant made a decision. However, if participants took more than five seconds to respond, the trial was excluded from the final analysis. While participants were performing the choice task, we collected eye movement data using EyeLink 1000.

Fig. 1. Example stimulus pair from the choice task. Participants were given five seconds to press a key to select the stimulus with more aesthetic appeal.

2.5 Variables

We operationalized looking time as the fixation proportion to the left item. This is calculated by dividing the total fixation time on the left item by the total fixation time on the left and the right item. We created another variable called relative rating by subtracting the rating for the right item from the rating for the left item. This gave us sixteen relative ratings levels ranging from −8 to 8. Finally, choice was a binary variable with a value of 1 if left item is chosen, and 0 if right item is chosen.

3 Results

We first looked at the relationship between aesthetic ratings and choices. Replicating Isham and Geng [1], we calculated the probability of the left item being chosen as a function of relative rating levels (Fig. 2).

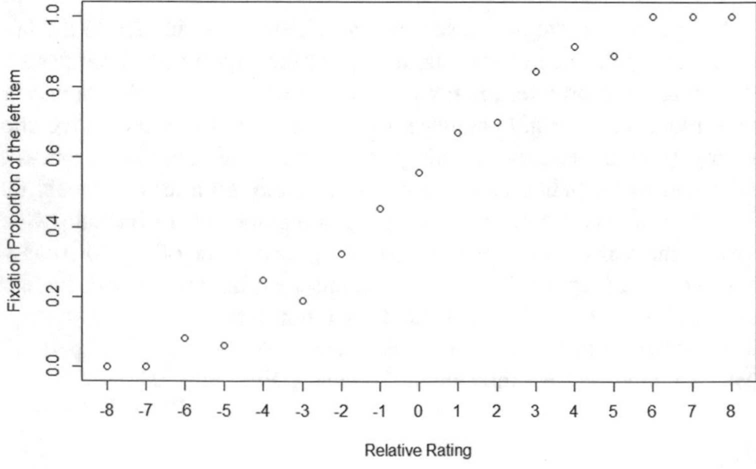

Fig. 2. Mean probability of choosing the left item for each relative rating level.

We then ran a generalized linear model using fixation proportion and relative rating as predictor variables of choice. As expected, relative rating was a strong predictor of choice $\beta = 1.907$, $z = 4.994$, $p < .001$. This means that the higher the left item was rated compared to the right item, the probability of choosing the left item increased significantly. To evaluate the first component of our hypothesis, we used the same generalized linear model look at the relationship between fixation duration and choice. We found that fixation duration significantly predicted choice ($\beta = .039$, $z = 6.983$, $p < .001$). This indicates that as fixation proportion to the left item increases by one percent, the probability of choosing the left item goes up by 0.039. Consistently, we observed that mean fixation proportion for the chosen item ($M = .543$, $SD = .157$) was higher than the mean fixation proportion for the unchosen item ($M = 0.457$, $SD = .157$) (Fig. 3). We did not observe a significant interaction effect of fixation duration and relative rating on choice ($\beta = -.009$, $z = -1.213$, $p = .225$).

To address the second component of our hypothesis that fixation duration does not predict relative rating and estimate the influence of fixation duration on relative rating, we ran a linear mixed-effect model. Fixation proportion significantly predicted relative rating, $\beta = .026$, $z = 4.823$, $p < .001$. However, this indicates a small increase of .026 in relative rating on a 16-point scale, per one percent increase in fixation proportion. Figure 4 shows the mean fixation proportion to the left item for each relative rating level.

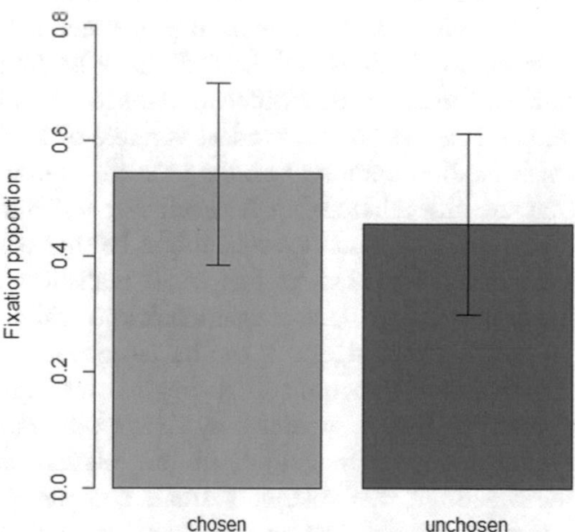

Fig. 3. Comparison of mean fixation proportion of the chosen ($M = .543$, $SD = .157$) and non-chosen ($M = .457$, $SD = .157$) items.

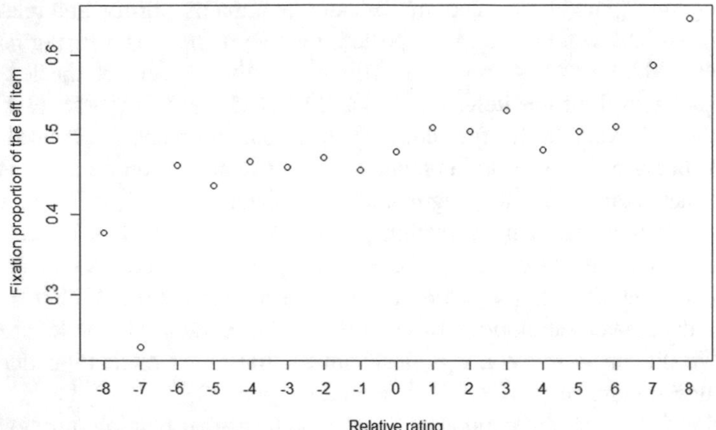

Fig. 4. The relationship between relative aesthetic ratings, as calculated by subtracting the right rating from the left rating, and mean fixation proportion for each relative rating category.

4 Discussion

Based on previous findings, we hypothesized that fixation duration would predict choices but not preference ratings. Our data supported the first component of our hypothesis that fixation durations strongly predict choices. However, we also found that fixation duration was a significant predictor of preference ratings.

The difference between our findings and the findings of Isham and Geng [1] might have several explanations. First, we used different statistical methods to analyze our data. We think that fixations act as an independent variable of choice. For this reason, we ran models in which fixation duration was the predictor variable. We believe that this approach may capture the relationship between our variables more accurately. Another reason for obtaining inconsistent results might be that we have used realistic stimuli instead of black and white abstract images. If realistic stimuli have, indeed, induced more genuine aesthetic experiences than black and white geometric ones, the responses gathered in the forced choice task can be more related to internally determined preferences. Finally, it is important to go beyond the significance levels and understand how well each variable is predicted by fixation duration. Even though we found a statistical significance, the magnitude of the relationship between fixation duration and preference ratings was relatively small compared to the relationship between fixation duration and choice. The findings that fixation duration is a significant predictor of choice might be caused by the large data set.

Moreover, we found that relative rating and fixation proportion did not have a significant interaction effect on choice, but both variables had significant main effects. The finding that fixations guided choices might explain the cases where subjects chose the item that they rated lower. Despite the strong correlation between ratings and choices, as the difference in rating of the left and the right item decreased, the probability of choosing the lower rated item went as high as one half (44%). This Again, this incongruency can be explained by fixations, as they have been showed by both previous and current study to bias choices. In future work, we plan to expand on these findings using structural equation modeling to see if ratings and fixations are two separate predictors of choice. Such an understanding would reinforce the idea that mechanisms other than internally determined preference guide eye-movements, which in turn bias choices.

Finding a clear answer to the question of whether looking time predicts choice and preferences will highlight the important difference between aesthetic preferences (i.e.: ratings) and comparative decision processes (i.e.: choices), as the latter might be biased by fixations. Establishing this relationship has several implications. First, eye-tracking is a promising tool to be used by design researchers because eye-movement measures overcome challenges of validity posed by self-report measures. Collecting looking time data to predict choices can be especially valuable in research that focuses on consumer behavior. In catalogues, design options are usually presented side by side, causing viewers to compare the options simultaneously. On the other hand, using eye-movements to predict behavior about three dimensional spaces in real life is more complex than using eye-movements to estimate choices on a computer screen. When we try to choose a study space, we rarely compare our options simultaneously. Rather, such spatial decisions are a product of mental comparison that does not involve fixations to two different options. In cases where we want to predict decisions about a three-dimensional space using a laboratory experiment with two-dimensional stimuli, looking time should be used with care because they might not successfully predict real-life choices. Future research should address this question of whether fixation durations, choices, or ratings would predict spatial behaviors based on environmental aesthetics.

The purpose of the current study was to generalize the findings of Isham and Geng [1] to complex and realistic environmental stimuli. We need further studies to confirm this hypothesis. The complex relationship between looking time, preferences, and choices indicates that multiple cognitive processes occur simultaneously during aesthetic experiences, and that measures used to predict aesthetic behaviors should be used with care.

Acknowledgments. We thank Cornell University Statistical Consulting Unit for their contribution in data analysis.

References

1. Lynnaw Twedt, E.: Restorative Environments: The Role of Visual and Observer Characteristics. University of Virginia, Charlottesville (2013)
2. Abkar, M., Mustafa Kamal, M.S., Maulan, S., Mariapan, M., Davoodi, S.R.: Relationship between the preference and perceived restorative potential of urban landscapes. HortTechnology **21**(5), 514–519 (2011)
3. Han, K.-T.: An exploration of relationships among the responses to natural scenes. Environ. Behav. **42**(2), 243–270 (2010)
4. Bringslimark, T., Hartig, T., Patil, G.G.: Psychological benefits of indoor plants in workplaces: putting experimental results into context. HortScience **42**(3), 581–587 (2007)

5. Lee, K.E., Williams, K.J.H., Sargent, L.D., Williams, N.S.G., Johnson, K.A.: 40-second green roof views sustain attention: the role of micro-breaks in attention restoration. J. Environ. Psychol. **42**, 182–189 (2015)
6. Raanaas, R.K., Evensen, K.H., Rich, D., Sjøstrøm, G., Patil, G.: Benefits of indoor plants on attention capacity in an office setting. J. Environ. Psychol. **31**(1), 99–105 (2011)
7. Isham, E.A., Geng, J.J.: Looking time predicts choice but not aesthetic value. PLoS ONE **8** (8), 1–7 (2013)
8. Orquin, J.L., Mueller Loose, S.: Attention and choice: a review on eye movements in decision making. Acta Physiol. **144**(1), 190–206 (2013)
9. Krajbich, I., Rangel, A.: Multialternative drift-diffusion model predicts the relationship between visual fixations and choice in value-based decisions. Proc. Natl. Acad. Sci. **108**(33), 1385–13857 (2011)
10. Holmes, T., Zanker, J.M.: Using an oculomotor signature as an indicator of aesthetic preference. I-Perception **7**, 42–439 (2012)

Does Comfort with Technology Affect Use of Wealth Management Platforms? Usability Testing with fNIRS and Eye-Tracking

Siddharth Bhatt[1(✉)], Atahan Agrali[2], Rajneesh Suri[1], and Hasan Ayaz[2]

[1] Lebow College of Business, Drexel University, Philadelphia, PA 19104, USA
{shb56, surir}@drexel.edu

[2] School of Biomedical Engineering, Science and Health Systems, Drexel University, Philadelphia, PA 19104, USA
{saa76, ayaz}@drexel.edu

Abstract. Most wealth management firms offer online platforms where investors with varied levels of comfort with technology manage their portfolios. Past research shows that comfort with technology is crucial for users' acceptance of new technologies. We investigated how users' comfort level with technology influences their use of a new wealth management online platform. We used a multi-modal approach that incorporates survey, behavioral, eye-tracking and neural measures to assess investors' comfort with technology on web-platform usability to provide a rigorous test of the effects of comfort with technology on usability experiences for a wealth management firm. Our findings suggest that traditional survey measures do not show any differences in users' evaluations. However, behavioral and neurophysiological measures reveal insights that traditional survey measures fail to reveal.

Keywords: Web usability · Fintech · Wealth management · Neuro-imaging Eye-tracking

1 Introduction

Digitization is changing the way business is conducted across many industries. Across industries, customers now demand online platforms that enable them to buy products, make payments, and provide feedback from the comfort of their homes [1]. To cater to these needs, companies from diverse industries are investing in improving consumers' experiences with their websites. The business logic behind such massive investment is sound. Websites have now become a major source of contact with customers. In many industries, a website serves as the primary channel of sales and distribution. Online channels enable a quicker and faster flow of information, product, services and word of mouth. Hence, billions of dollars are spent on web platforms. Per IBISWorld, a leading industry research company, the web design services market stood at $34.0 billion in 2017. The wealth management industry is no exception to this trend. In fact, given the nature of its services, wealth management industry is the prime beneficiary of the digitization wave. Players in the industry are investing a lot of money in creating

H. Ayaz and L. Mazur (Eds.): AHFE 2018, AISC 775, pp. 83–90, 2019.
https://doi.org/10.1007/978-3-319-94866-9_8

websites that provide a seamless experience to their clients. Such web platforms serve as the primary channel through which clients interact with the company. Clients buy and sell financial assets on such websites and actively manage their portfolios. Hence, it is important for wealth management companies to provide a good experience to their clients through such websites. Such a goal necessitates creation and maintenance of websites that are functionally up to date and aesthetically pleasing. Further, launching websites that do not perform as per the expectations of client leads to negative consequences [2]. Hence, it is necessary to have an accurate understanding of website quality and ease of use before its launch [3].

Furthermore, it is necessary to understand the effect of individual characteristics on web usage. Individuals with diverse demographic and psychographic characteristics have different levels of acceptance of technology and usage patterns. In this study, we investigated the effect of consumers' comfort with technology on their use and assessment of a web platform. We utilized a multi-modal approach that helps assess this effect beyond the limitations of traditional survey measures. The goal of this research is two-fold – (a) to demonstrate that a psychographic characteristic of consumers, level of comfort with technology, influences their assessment of a web-based wealth management platform, and (b) to demonstrate that traditional survey measures fail to uncover insights that behavioral and neurophysiological measures can.

2 Conceptual Background

While lots of investments are being made in creating websites, not enough money and rigor goes in testing the performance of such websites. Till date, most companies rely on traditional methods of website evaluation [4]. Such methods rely on asking users about their experience with the website. The questions used in such assessment studies are either open ended or close ended. Both these types of measures work on the assumption that users can articulate all aspect of their experience with the website. However, this is seldom the case. Due to natural and unavoidable difficulties in articulation of thought and feelings, such measures are subjective and prone to biases. Due to such biases, the company may obtain positive reviews about the platform and but when the platform is launched several issues surface. Lack of detail and specificity in users' responses during the testing phase lead to such an unfortunate situation. Hence, the need to create more objective and rigorous methods of assessing web usability cannot be overemphasized [5].

The issue is further complicated by the fact that the clientele of such companies is diverse. People of all age, income, gender, education and other demographic characteristics use the same website to manage their wealth. Hence, it is necessary to examine how such consumer characteristics affect their use and assessment of websites [6]. An important variable in examining the usability of websites is the level of comfort that users of the website possess with technology [7]. These authors show that users' level of comfort with technology is a key determinant of whether and how much they adopt online banking platforms. On similar lines, [8] finds that comfort with technology, or in other words, self-efficacy related to technology influences intention to use on-line shopping platforms. These findings suggest that users' level of comfort with technology influences their preference for certain types of media over other. Further, differing levels of comfort with technology will reflect in web platform usage. In this research, we enrolled participants with differing levels of comfort with technology who assessed a new web platform for a major wealth management firm in the US. To overcome the limitations of traditional self-reported measures, we employed a multi-modal approach. The usability testing methodology in this research comprised of four different metrics of assessing the web platform – survey measures, behavioral measures, eye-tracking measures and neuroimaging measures.

3 Methodology

3.1 Sample, Design and Procedure

We used a multi-method approach to assess the online platform for a large wealth management firm located in the US. The wealth management platform was designed by the firm for its clients who wish to manage their portfolio online. The web platform consisted of five major sections – (i) dashboard, (ii) goals, (iii) analytics and performance, (iv) investments, and (v) activities. The five sections encompass most of information that clients need to manage their portfolio.

Twelve participants (employees of the firm developing the fintech solution) representing differing levels of comfort with technology volunteered for the study. Comfort with technology was measured using a single item - "How comfortable are you with using technology?" (5-point scale; 1 = very uncomfortable, 5 = very comfortable). Participants were invited to a lab equipped with stimulus presentation computers, a research grade eye-tracker (Aurora, Smart Eye) and neuroimaging devices (Model 1100, fNIR Devices). After obtaining consent, participants were seated in front of a computer that presented them with the web platform. To measure neural responses, we mounted a 16-optode sensor that measured neural activity in the anterior pre-frontal cortex. The pre-frontal cortex is the judgement and decision-making area of the brain. Measures of activity in this area helps us understand the neural basis for the attitudinal and behavioral responses to the web platform. We measured neural responses through a non-invasive technique – Functional Near Infrared Spectroscopy (fNIRS). The fNIRS sensor measures the cortical oxygenation changes in the anterior prefrontal cortex. Relative changes in the levels of oxygenated and deoxygenated hemoglobin provide a measure of mental effort required to complete a task. This technique is described in detail by [9]. Past research has used the technique has been used to assess mental workload in a diverse array of applications [10, 11]. Additionally, we also measured eye movements of participants throughout the study. Participants were instructed that their eye movements will be tracked as they navigate through the web platform. Participants started using the web platform from the dashboard section. They then navigated to the other four sections of the website.

To assess usability, we created several tasks for participants to complete in each section. As participants completed the tasks on the web platform we measured (a) neural activity in the anterior pre-frontal cortex using fNIRS, (b) eye movements using a research grade eye-tracker, (c) accuracy in completing tasks across the five sections of the website, and (d) self-reported measures of website functioning, aesthetics, and liking and these four metrics were separately used in the analysis. Accuracy was coded as '0' if the participant failed to complete the task and '1' if the participant successfully completed the task. We used eleven items to measure participants self-reported experience with the website. The self-reported items which were measured toward the end of the study comprised of eleven items measured on a nine-point scale (1 = strongly disagree, 9 = strongly agree). The scale comprising of eleven items was found to have high reliability (Cronbach's alpha = 0.71. Eye tracking data was processed to calculate average saccade velocity on the screen during the study. Neural data was processed to calculate combined activation in the left and right hemispheres of the prefrontal cortex.

3.2 Results

Survey Measures – An ANOVA revealed no significant differences on the survey scale by level of comfort with technology (F (2, 9) = 3.48, p = 0.07). Mean ratings for the website by level of comfort with technology are reported in Table 1.

Table 1. Summary of results

Comfort with technology	Survey measures	Behavioral measures	Eye tracking measures	Neural measures
	Mean rating scale (1–9)	Mean accuracy per task (0 = inaccurate, 1 = accurate) (% total accuracy)	Mean gaze fixations	Mean oxygenated hemoglobin changes
Very uncomfortable (n = 0)	-	-	-	-
Somewhat uncomfortable (n = 0)	-	-	-	-
Somewhat comfortable (n = 2)	7.73	85.9	5.07	0.0141
Comfortable (n = 7)	6.92	79.4	4.94	0.0030
Very comfortable (n = 3)	7.48	83.0	4.65	0.0157

Behavioral Measures – Based on accuracy per task (0 = inaccurate response, 1 = accurate response), mean accuracy (in %) was calculated for all subjects. Next, an ANOVA was run to analyze differences in mean accuracy level (in %) by comfort with technology. The ANOVA revealed a significant difference in mean accuracy per task by level of comfort with technology (F (2, 9) = 0.89, p = 0.44). Mean accuracy per task and mean total accuracy (%) by level of comfort with technology are reported in Table 1.

Eye-tracking Measures – A significant difference was observed in average saccade velocity by level of comfort with technology (F (2, 467) = 3.26, p = 0.03). Mean gaze fixations are reported in Table 1.

Neural Measures – The ANOVA revealed no significant difference in oxygenated hemoglobin changes (HbO) in the pre-frontal cortex across the four levels of comfort with technology (F (2, 328) = 0.08, p = 0.91). Mean oxygenated hemoglobin (HbO) changes are reported in Table 1.

While survey measures, behavioral measures (accuracy), and neural measures (oxygenated hemoglobin changes) do not show any differences on participants' rating of the web platform, there are differences in visual patterns by which participants with different levels of comfort with technology navigate through the website. Our findings show that average saccade velocity decreased with increasing comfort with technology. Hence, participants with higher levels of comfort with technology were able to process the visual information on the website with greater ease and less visual hopping. On the other hand, participants with less comfort with technology made rapid visual scans of the website to locate the information they needed. Read in conjunction, these results suggest that level of comfort with technology is an important variable in examining usability of web platforms. These results also suggest a need to supplement survey measures with other objective measures such eye-tracking and brain-imaging (Figs. 1, 2, 3, and 4).

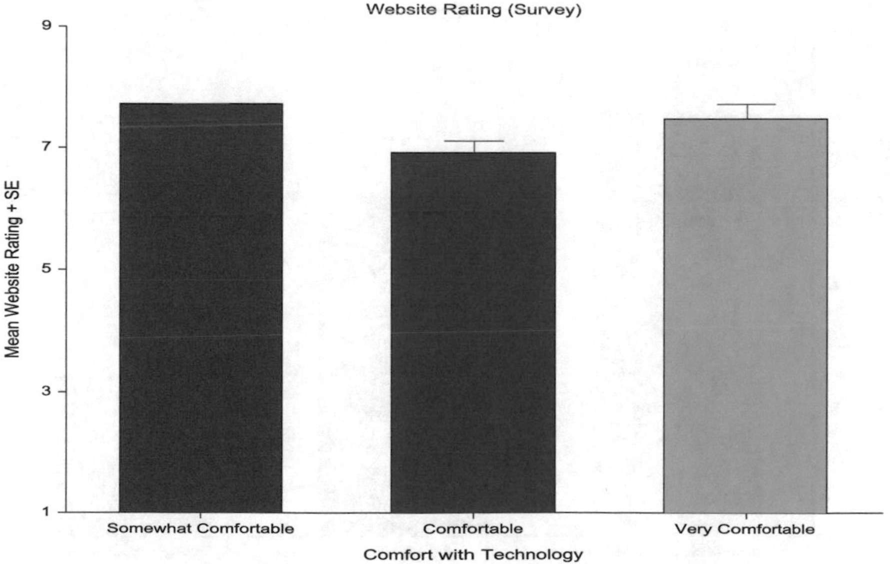

Fig. 1. Website rating (survey measures)

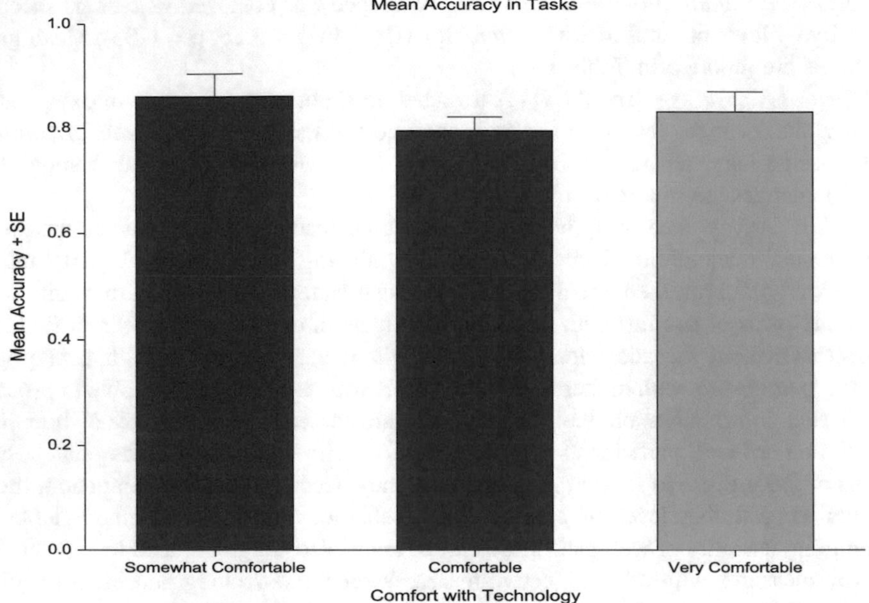

Fig. 2. Accuracy in tasks (behavioral measures)

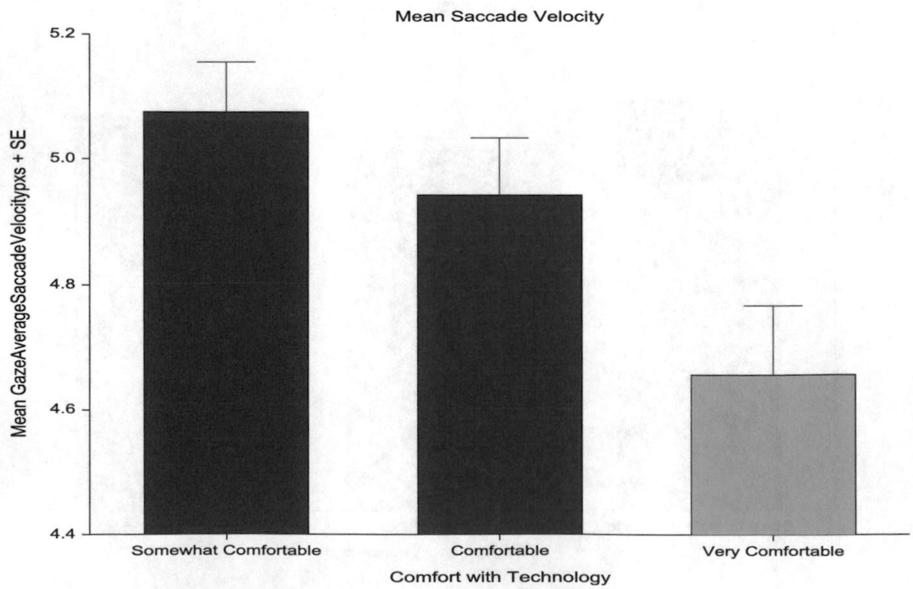

Fig. 3. Saccade velocity (eye-tracking measures)

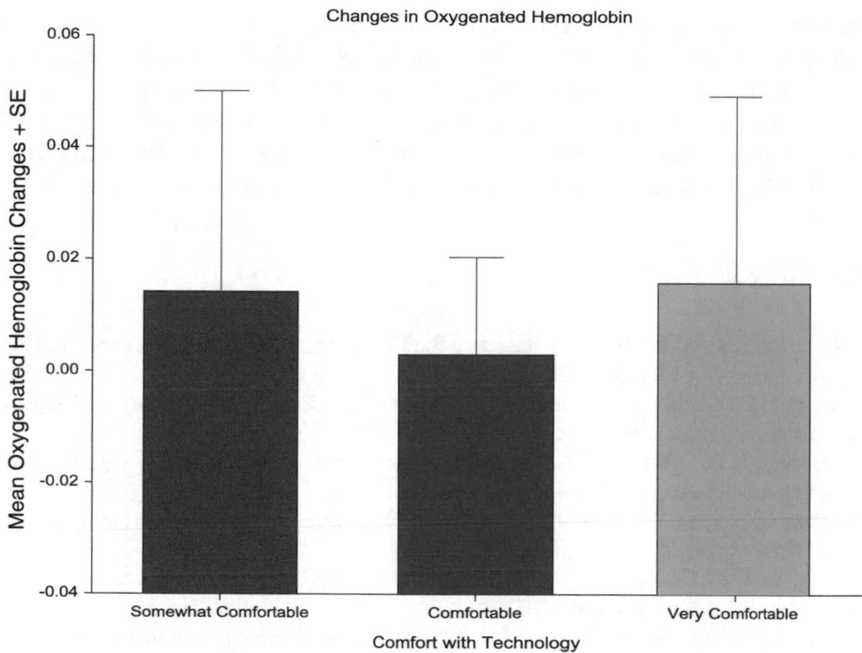

Fig. 4. Changes in oxygenated hemoglobin (neural measures)

4 Discussion

In this study, we empirically assessed the usability of a new web platform using a multi modal approach. The limitations of traditional survey measures have been acknowledged for a long time. However, little research has examined how these limitations can be overcome. To overcome the limitations, we used an approach that combines the strength of behavioral and neurophysiological measures alongside survey measures. As our findings reveal, survey measures did not show differences on users' assessment of the web portal. However, eye tracking measures (average saccade velocity) showed that participants with higher levels of comfort with technology had greater ease in visual processing of the website. This result suggests that participants with higher levels of comfort with technology were more adept in visually navigating the website and finding the information necessary to complete the tasks. Findings from brain activity suggest an inverse relationship between brain activity in the anterior prefrontal cortex (as measured through changes in oxygenated hemoglobin) and level of comfort with technology. Relying on traditional self-reported measures alone would not have helped discover such differences. The additional insights discovered through eye-tracking metrics reveal that the company may wish to visually simplify the website if majority of its clients have low to moderate levels of comfort with technology. Doing so could improve the search efficiency on the website and lead to a greater accuracy in completing the tasks. In sum, this research suggests two key findings about users' assessment of web platforms – (a) users' comfort with technology is an important variable in understanding website usability, and (b) traditional survey measures often fail to uncover facets of user experience that can be explored through objective neurophysiological measures. Additionally, this research provides a novel and more objective way of assessing online platforms such as wealth management websites. Future research should examine the role of other individual level characteristics that have a bearing upon the use and evaluation of web platforms.

References

1. Markovitch, S., Willmott, P.: Accelerating the Digitization of Business Processes. McKinsey & Company, San Francisco (2014)
2. Chang, H.H., Chen, S.W.: Consumer perception of interface quality, security, and loyalty in electronic commerce. Inf. Manag. **46**(7), 411–417 (2009)
3. Loiacono, E.T., Watson, R.T., Goodhue, D.L.: WebQual: a measure of website quality. Market. Theory Appl. **13**(3), 432–438 (2002)
4. Matera, M., Rizzo, F., Carughi, G.T.: Web usability: principles and evaluation methods. In: Web Engineering, Springer, Berlin (2006)
5. Lazar, J.: Web Usability: A User-Centered Design Approach. Addison-Wesley Longman Publishing Co., Inc., Boston (2005)
6. Lassar, W.M., Manolis, C., Lassar, S.S.: The relationship between consumer innovativeness, personal characteristics, and online banking adoption. Int. J. Bank Market. **23**(2), 176–199 (2005)
7. Goldfarb, A., Prince, J.: Internet adoption and usage patterns are different: implications for the digital divide. Inf. Econ. Policy **20**(1), 2–15 (2008)
8. Vijayasarathy, L.R.: Predicting consumer intentions to use on-line shopping: the case for an augmented technology acceptance model. Inf. Manag. **41**(6), 747–762 (2004)
9. Ayaz, H., Shewokis, P.A., Curtin, A., Izzetoglu, M., Izzetoglu, K., Onaral, B.: Using MazeSuite and functional near infrared spectroscopy to study learning in spatial navigation. J. Vis. Exp. **56**, e3443 (2011)
10. Ayaz, H., Shewokis, PA., Bunce, S., Schultheis, M., Onaral, B.: Assessment of cognitive neural correlates for a functional near infrared-based brain computer interface system. In: International Conference on Foundations of Augmented Cognition. Springer, Berlin (2009)
11. Ayaz, H., Onaral, B., Izzetoglu, K., Shewokis, P.A., McKendrick, R., Parasuraman, R.: Continuous monitoring of brain dynamics with functional near infrared spectroscopy as a tool for neuroergonomic research: empirical examples and a technological development. Front. Hum. Neurosci. **7**, 1–13 (2013)

Gaze Strategies Can Reveal the Impact of Source Code Features on the Cognitive Load of Novice Programmers

Andreas Wulff-Jensen(✉), Kevin Ruder, Evangelia Triantafyllou, and Luis Emilio Bruni

Department of Architecture Design and Media Technology, Aalborg University, A.C. Meyers Vaenge 15, 2450 Copenhagen, Denmark
{awj, evt, leb}@create.aau.dk, kevinruder@ymail.com

Abstract. As shown by several studies, programmers' readability of source code is influenced by its structural and the textual features. In order to assess the importance of these features, we conducted an eye-tracking experiment with programming students. To assess the readability and comprehensibility of code snippets, the test subjects were exposed to four different snippets containing or missing structural and/or textual elements. To assure that all subjects were at an equivalent level of expertise, their programming skills were also evaluated. During the eye-tracking experiment, the subjects were also asked to give a readability and comprehensibility score to each snippet. The absence of textual features showed to increase the average fixation duration. This indicates that for the test subjects the textual features were more essential for the comprehension of the code. Gaze pattern analysis revealed less ordered patterns in the absence of structural features compared to the absence of textual features.

Keywords: Eye-tracking · Code features · Readability · Programming Programmer

1 Introduction

The ubiquitousness of digital culture translates into the ubiquitousness of programming languages. The demand for programming literacy has widely increased in contemporary society, and it is not limited to specialized engineers but it has also become necessary for media workers, natural and social scientists and even humanity professionals and artists. As in natural language, not all readers are good writers. Still, being able to understand and troubleshoot a source code[1] has become a useful skill in many fields and situations. From the cognitive point of view, there may be many analogies but also specificities between reading and comprehending natural language and reading and comprehending a computer code. Both kinds of languages are symbolic and

[1] Source codes are written in a programming language, which a human can easily read. A large program may contain many different source code files within its architecture. Compilers translate source codes into machine language, which is processed faster by the digital machine but it is harder for a human to deal with.

H. Ayaz and L. Mazur (Eds.): AHFE 2018, AISC 775, pp. 91–100, 2019.
https://doi.org/10.1007/978-3-319-94866-9_9

encompass syntactic, semantic and even pragmatic levels [1–3], and both entail differential demands in terms of cognitive load, as it all depends on the readers expertise and the level of ambiguity used in the language [1, 4]. Source code readability, and thus comprehension, is a relative new area of research [5, 6], which seeks to understand the features that contribute to optimize code readability.

2 Readability and Comprehensibility of Source Code

Several studies have investigated the relationship between structural elements (spaces between lines, indentation of the code, etc.) and textual elements (words with single meaning, hypernyms, words, coherency between identifier and comments, etc.) of different code snippets and readability [5, 7–9]. In these studies this relationship was assessed by annotators' subjective evaluations. The remarks were typically assessed through machine learning techniques to classify which structural or textual elements had the highest influence on the readability of the individual code snippets. Through this process, Buse and Weimer [5] found a series of textual features that had high influence on the perceived readability score given by computer science students. These features were: (a) "average line length", (b) "average number of identifiers per line", (c) "average number of curly brackets and parentheses per line", (d) "maximum line length", (e) "average amount of dots per line", (f) "maximum number of identifiers", (g) "average number of identifiers per line", and (h) "average number of keywords per line". Posnett et al. [7] elucidated that size and entropy of the code had a relation to the readability. Dorn [9] further looked at the structure of the code and found "frequency of the lines", "line spacing" and "visual elements" to be important for the readability. Lastly, Scalbarino et al. [8] found the following textural features to be relevant for readability: "narrow meaning identifiers", "comments readability", and "identifier terms in dictionary".

While the previous studies determine the readability through algorithmically disentangling the features that have the biggest influence on the readability score, they limit themselves to subjective assessments and do not explore other available behavioral or psychophysiological tools for assessing readability. One such tool is eye-tracking, which has already been used to assess cognitive load and how people read code through analysis of fixation points, saccades and gaze patterns [10–12]. Even though eye-tracking technics have been used for assessing source code reading patterns and thus its comprehensibility, the effect of structural and textural features is yet to be understood. When a programmer reads source codes he creates an internal mental representation of it [13]. Thereby, the less comprehensible the code is the more mental workload is needed to read it. The eye-tracking parameters used as correlates of cognitive load are: pupillometry [14, 15], Saccade speed [14], fixation duration, and direction [14, 16]. However, for some of these measures there are different drawbacks. Pupillometry is highly sensitive to light changes [17], thus any change in the surrounding light settings will provoke a change in the pupil size. In the case of saccades, it is argued that there are no cognitive functions happening during a saccade [14]. However, this argument can be challenged if saccades are triangulated with other parameters. Despite these drawbacks, taking all the measurements into consideration might possibly reveal levels of cognitive load, and thus the reader's code comprehension while reading the source code and its relationship with different readability features.

In this study, eye-tracking together with subjective comprehension assessment was used to assess the effects of different readability features on source code comprehension. More specifically, this study aimed to answer the following research questions:

- How do structural and textual readability features affect program comprehension?
- How does the absence of textual features affect program comprehension?
- How does the absence of structural features affect program comprehension?

3 Experimental Setup

To understand the influence of textual and structural features on comprehensibility, four different code snippets were carefully chosen. These were written in Processing (builds on Java programming language) to accommodate the language taught on the early semesters at the Architecture, Design and Media Technology department at Aalborg University Copenhagen. Furthermore, each of the snippets were complete executable programs in order to avoid any bias on the comprehension and understandability of the code.

During the experiment, the test subjects were exposed to four different snippets containing: (1) only structural elements, (2) only textual elements, (3) none of these elements or (4) all of them. For later reference, we will refer to the first snippet type as SNT, to the second as NST, to the third as NSNT, and to the fourth as ST.

To subjectively evaluate the comprehensibility of the code, two different questionnaires were employed. One to assess the programmers experience [18] and another one for evaluating the readability and understandability of the code snippets. The questionnaire for measuring programming experience was designed by Feigenspan et al. based on questions from published comprehension experiments, and it has been validated as well. During validation, it was found that self-estimation seems to be a reliable way to measure programming experience. In the second questionnaire, the subjects had to rate the snippets on two Likert items one rating the readability and another the understandability of the snippets. Furthermore, the subjects had to write summaries of their understanding of the codes. The summaries were post hoc classified as either right or wrong.

The code snippets were presented in Notepad to avoid code color coding of the Processing environment. The questionnaires were developed and distributed in Google forms. Both of them were presented on a 15-in. laptop, which was situated in front of a bulletin board to avoid visual distractions. Furthermore, to be able to capture the gaze, an EyeTribe eye-tracker was connected to the computer. The computer had the following specifications: Processor. Intel®Core™ i7-4720HQ CPU @ 2.60 GHZ 2.59 GZ. 16.00 GB of RAM. The operation system was Windows 10 64-bit.

3.1 Participants and Experimental Procedure

Twenty-one programming students at Aalborg University Copenhagen participated in our study. Out of the 21, five were females and sixteen where males, while their average age was 26.42 (SD 4.37). Their average experience level was 2.35 (SD = 0.51) in a scale from 0–6, as evaluated by the programming experience assessment questionnaire. Thus, this score insinuates most of the subjects being novice programmers.

After a brief introduction to all the elements of the experiment, and filling in the experience assessment questionnaire, the eye-tracker was calibrated and the experiment began. The study was designed as a within group experiment in order to reduce the bias possibly introduced by the order of exposure to the four different code snippets (ST, NST, SNT, NSNT). The order of exposure was randomized (all four snippets can be seen in Figs. 1, 2, 3 and 4). When the subject was ready to see the first code snippet, the test conductor pressed the start button which also triggered the eye-tracking recording. The subjects were instructed to look at the code for as long as they needed, and then press a stop button. This action stopped the eye-tracking recording and made the *understandability and readability questionnaire* visible on the screen. After the subjects had been through the first code snippet and the questionnaire, the next random code snippet was shown. This continued until all four of them were shown. Upon completion of the test the subjects were debriefed.

```
//temperature in Kelvin
float Kelvin;

//Returns the temperature in celsius
float Celsius(float temp){
  return temp-273.15;
}

//Returns the temperature in fahrenheit
float Fahrenheit(float temp){
  return temp * 9/5 -459.67;
}

void setup(){

  Kelvin = 300;

  println("Temperature in Celsius: "+Celsius(Kelvin));
  println("Temperature in Fahrenheit: "+Fahrenheit(Kelvin));

  if( Celsius(Kelvin) < 0 ) {
    println("ITS FREEZING ");
  }
  else{
    println("IT'S NOT FREEZING");
  }

}
```

Fig. 1. Shows an example of a code with structural and textual features (ST)

```
//Calculates the histogram of an image. A histogram is the frequency distribution of the gray levels with the number
// of pure black values
//displayed on the left and number of pure white values on the right.
/Note that this sketch will behave differently on Android, since most
//images will no longer be full 24-bit color.
size(640, 360);
// Load an image from the data directory
// Load a different image by modifying the comments
PImage img = loadImage("frontier.jpg");
image(img, 0, 0);
int[] hist = new int[256];
// Calculate the histogram
for (int i = 0; i < img.width; i++) {for (int j = 0; j < img.height; j++) { int bright = int(brightness(get(i, j)));
hist[bright]++;}}
// Find the largest value in the histogram
int histMax = max(hist);
stroke(255);
// Draw half of the histogram (skip every second value)
for (int i = 0; i < img.width; i += 2) {
// Map i (from 0..img.width) to a location in the histogram (0..255)
int which = int(map(i, 0, img.width, 0, 255));
// Convert the histogram value to a location between
// the bottom and the top of the picture
int y = int(map(hist[which], 0, histMax, img.height, 0));
line(i, img.height, i, y);
}
```

Fig. 2. Shows an example of a code with no structural features (NST)

```
int[] A = { 30, 10, 45, 35, 60, 38, 75, 67 };

void setup() {
  size(640, 360);
  noStroke();
  noLoop();  // Run once and stop
}

void draw() {
  background(100);
  pC(300, A);
}

void pC(float d, int[] data) {
  float pA = 0;
  for (int i = 0; i < data.length; i++) {
    float color = map(i, 0, data.length, 0, 255);
    fill(color);
    arc(width/2, height/2, d, d, pA, pA+radians(A[i]));
    pA += radians(A[i]);
  }
}
```

Fig. 3. Shows an example of a code with no textual features (SNT)

```
// An array
int AoR[];
int rD(){int rNr = int(random(1,6));return rNr;}
//The array is filled using a function
void setup(){
  AoR = new int[10];for(int i = 0; i < AoR.length ; i++){AoR[i] = rD();}}
//Something happens here
void draw(){
  for(int i = 0; i < AoR.length ; i++){println(AoR[i]);
}
}
```

Fig. 4. Shows an example with neither structural nor textual features (NSNT)

4 Results

We ran a Wilcoxon signed-rank test on the questionnaire data, while we used a paired t-test for the eye-tracking parameters. The eye-tracking parameters were extracted using the Ogama [19] program, which was configured to work with the following assumption: a fixation is defined as the gaze staying within a radius of 20 pixels for a duration of 5 samples, which is equal to 83.33 ms. Through this configuration we could separate the fixation points from the saccades, thus understand the individual components of the subject's gaze behavior. This enabled us to elucidate the influence of textual and structural elements on comprehensibility of the code.

The Wilcoxon signed-rank test performed on the questionnaire data showed that the ST condition was significantly more comprehensible and readable compared than the three other conditions ($p < 0.05$), while SNT, NST, and NSNT were not significantly different from each other. This observation is further supported by the amount of correct and incorrect summaries that the subjects wrote based on the code snippets, as shown in the Table 1.

Table 1. Shows the amount of correct and incorrect summaries for the different snippets in all subjects

Features in the code	Correct summaries	Incorrect summaries
ST	19	2
SNT	8	13
NST	14	7
NSNT	14	7

Table 2 shows the results for the eye-tracking parameters. The values shown in this table are the average results from all subjects. The definition of the different eye-tracking parameters is as follows:

Mean fixation count: the amount of fixations per subject
Mean fixation count/second: the average amount of fixations per second
Mean fixation duration (ms): the average amount of milliseconds per fixation
Mean fixation to saccade ratio: relationship between number of fixations and number of saccades
Mean saccade length (px): average lengths of saccades
Mean saccade velocity (px/s): the average speed of saccades
Biased pupil size (px): pupil size influenced by subjective factors and distance to the eye-tracker.

All the above parameters were tested through 6 paired T-tests with an alpha value of 0.01 to avoid alpha inflation. The mean fixation count data showed to be significantly different in every pair ($p < 0.01$) apart from the pair ST vs NSNT. The tests for mean fixation count per second were only significantly different between NST and

Table 2. Shows the mean values for all the eye-tracking parameters with respect to the four different kinds of code snippets.

	ST	SNT	NST	NSNT
Mean fixation count	116.43	174.28	310.43	128.48
Mean fixation count/second	2.89	2.61	3.42	2.77
Mean fixation duration (ms)	241.31	306.44	251.51	302.43
Mean fixation to saccade ratio	698.96	792.82	844.31	823.97
Mean saccade length (px)	132.74	117.93	151.37	130.57
Mean saccade velocity (px/s)	1.91	1.85	2.14	2.01
Biased pupil size (px)	18.72	19.47	18.93	19.12

SNT, where NST was significantly higher than SNT. Moreover, the results for the mean fixation duration revealed that SNT and NSNT were both significantly higher than ST and NST.

The tests conducted on the mean fixation to saccade ratio data showed ST to be significantly lower than NSNT and NST. Furthermore, mean saccade length showed to be significantly higher for NST compared to the three other cases. However, saccade velocity was only significantly different between NST and SNT. The tests for the last parameter, biased pupil size, showed no statistically significant differences among the four cases.

Since all eye-tracking parameters apart from mean fixation duration and biased pupil size were significantly higher for NST compared to the other cases, we can conclude that the absence of structural elements in the source code increases the majority of the eye-tracking parameters. This could suggest that the absence of structural elements provokes the reader's gaze to be more rapid and chaotic, as both fixation count, and saccade length and velocity are higher compared to the other cases. This observation could suggest a vastly different visual strategy in the NST case, where the reader is furiously looking around to make her own structure and thus comprehend the code. This is an assumption that could be further investigated in further studies. On the other hand, the absence of textual features seems to provoke longer fixation times, which could be interpreted as the reader using more time to comprehend single elements of the code thus higher cognitive load. The focus on single elements could in turn provoke the subjects to forget about the overview and the meaning of the code. This assumption could be one of many possible reasons for the higher amount of incorrect summaries for SNT compared to the others, thus pointing out that the code is harder to comprehend.

5 Discussion

The study presented in this paper provides a behavioral and psychophysiological perspective to how the features of source code affect its comprehensibility. However, there are a few issues with the experiment material, which could have affected the results. Firstly, the character count of the four snippets of code varies, ranging from 273 characters for NSNT to 1186 characters for NST. This character variability could potentially have influenced the comprehensibility of the code, even though Buse and Weimer's readability model [5] did not find this feature to significantly impact the readability of the code. Secondly, the code snippets contain different syntactical units, which could have made some snippets natively easier to comprehend than others. For instance, a snippet containing arrays, nested loops or recursions would be more difficult to understand for novice programmers, such as the subjects of our study [20].

Apart from the experiment material, there are issues with the controllability of the experiment. Although pupil dilatation has been proved to be directly related to cognitive load [14], pupillometry can be become inadequate for this experiment due to difficulties in controlling lighting conditions and different distances from the subject to the eye-tracker among test subjects. Another factor in the experimental set-up, which could have affected the results, is the low sample size. However, the homogeneity of the sample population could eliminate the bias introduced by this factor.

The last issue, which could have influenced the results, is the fact that readability is not universal across programming languages and experience. In this study, the language used was Processing and the subjects were predominately novices. It would be interesting to consider different populations as well as different programming languages to see how this variability influences code comprehension and visual strategies. With respect to the current sample group, changing the programming language to C# could potentially elicit different results, as a Pearson correlation test on the programming experience questionnaire data showed a strong correlation between years of experience and perceived experience with C# programming ($r = 0.705$), while the correlation between years of experience and perceived experience with Processing was moderate ($r = 0.383$).

The results of this study provide a clear indication that novice programmers will have higher degree of cognitive load [16] while reading source code in the absence of textual features. While this paper was mostly focused on the effects that structural and textual features have on cognitive load, it would also be interesting to see how these features affect visual strategies for program comprehension. Tools such as the VET, code schemes and gaze transition flow charts could facilitate the investigation of differences in visual strategies.

6 Conclusion

This paper presented a study employing eye-tracking in order to assess the importance of structural and textual features of source code to its readability and comprehensibility. We contributed to current modelling and understanding of readability features that influence code comprehension by adding behavioral and psychophysiological parameters to the framework for its assessment in order to triangulate with existing self-report evaluations. For this we relied on methods for inferring visual strategies and cognitive load from eye-tracking data. The results of the experiment showed that the absence of textual features increase the average fixation duration and the amount summation mistakes. This indicates that, for novice programmers, textual features more so than structural features are essential towards the comprehension of source code. The eye metrics also showed indications of differing visual strategies between the experiment conditions. The absence of structural features showed a more chaotic pattern consisting of increased fixations and saccade length. The absence of textual features however showed a decrease in the gaze variability, which could be indicative of a slower, methodical process with less fixation transitions.

References

1. Kuhn, A., Ducasse, S., Gîrba, T.: Semantic clustering: identifying topics in source code. Inf. Softw. Technol. **49**, 230–243 (2007). https://doi.org/10.1016/j.infsof.2006.10.017
2. Busjahn, T., Bednarik, R., Begel, A., Crosby, M., Paterson, J.H., Schulte, C., Sharif, B., Tamm, S.: Eye movements in code reading: relaxing the linear order. In: IEEE International Conference on Program Comprehension August 2015, pp. 255–265 (2015). https://doi.org/10.1109/icpc.2015.36
3. Schulte, C., Clear, T., Taherkhani, A., Busjahn, T., Paterson, J.H.: An introduction to program comprehension for computer science educators. In: Proceedings of the 2010 ITiCSE Working Group Reports on Working Group Reports - ITiCSE-WGR 2010, p. 65 (2010). https://doi.org/10.1145/1971681.1971687
4. Macizo, P., Teresa Bajo, M.: When translation makes the difference: sentence processing in reading and translation. Psicológica **25**, 181–205 (2004)
5. Buse, R.P.L., Weimer, W.R.: Learning a metric for code readability. IEEE Trans. Softw. Eng. **36**, 546–558 (2010). https://doi.org/10.1109/TSE.2009.70
6. Boehm, B., Basili, V.R.: Software defect reduction top 10 list. Computer (2001). https://doi.org/10.1109/2.962984
7. Posnett, D., Hindle, A., Devanbu, P.: A simpler model of software readability. In: Proceeding of the 8th Working Conference on Mining Software Repositories - MSR 2011, p. 73 (2011). https://doi.org/10.1145/1985441.1985454
8. Scalabrino, S., Linares-Vasquez, M., Poshyvanyk, D., Oliveto, R.: Improving code readability models with textual features. In: IEEE International Conference on Program Comprehension, July 2016 (2016). https://doi.org/10.1109/icpc.2016.7503707
9. Dorn, J.: A general software readability model. MCS thesis (2012)
10. Sharif, B., Falcone, M., Maletic, J.I.: An eye-tracking study on the role of scan time in finding source code defects. In: Proceedings of the Symposium on Eye Tracking Research and Applications - ETRA 2012, pp. 381–384 (2012). https://doi.org/10.1145/2168556.2168642
11. Busjahn, T., Shchekotova, G., Antropova, M., Schulte, C., Sharif, B., Begel, A., Hansen, M., Bednarik, R., Orlov, P., Ihantola, P.: Eye tracking in computing education. In: Proceedings of the Tenth Annual Conference on International Computing Education Research - ICER 2014, pp. 3–10 (2014). https://doi.org/10.1145/2632320.2632344
12. Bednarik, R.: Expertise-dependent visual attention strategies develop over time during debugging with multiple code representations. Int. J. Hum. Compu. Stud. **70**, 143–155 (2012). https://doi.org/10.1016/j.ijhcs.2011.09.003
13. Gilmore, D.J.: Models of debugging. Acta Physiol. (Oxf) **78**, 151–172 (1991). https://doi.org/10.1016/0001-6918(91)90009-O
14. Rosch, J.L., Vogel-Walcutt, J.J.: A review of eye-tracking applications as tools for training. Cogn. Technol. Work (2013). https://doi.org/10.1007/s10111-012-0234-7
15. Hess, E.H.: Attitude and pupil size. Sci. Am. **212**, 46–54 (1965). https://doi.org/10.1038/scientificamerican0465-46

16. Meghanathan, R.N., van Leeuwen, C., Nikolaev, A.R.: Fixation duration surpasses pupil size as a measure of memory load in free viewing. Front. Hum. Neurosci. **8** (2015). https://doi.org/10.3389/fnhum.2014.01063
17. Pachymeningitis Abscess, Pain Control, and Pain Management: Pupillary response. In: Encyclopedia of Intensive Care Medicine, 1934–1938. Springer, Heidelberg (2012). https://doi.org/10.1007/978-3-642-00418-6
18. Feigenspan, J., Kastner, C., Liebig, J., Apel, S., Hanenberg, S.: Measuring programming experience. In: 2012 20th IEEE International Conference on Program Comprehension (ICPC), pp. 73–82 (2012). https://doi.org/10.1109/icpc.2012.6240511
19. Voßkühler, A., Nordmeier, V., Kuchinke, L., Jacobs, A.M.: OGAMA (Open Gaze and Mouse Analyzer): open-source software designed to analyze eye and mouse movements in slideshow study designs. Behav. Res. Methods **40**, 1150–1162 (2008). https://doi.org/10.3758/BRM.40.4.1150
20. Timmermann, D., Kautz, C.: Design of open educational resources for a programming course with a focus on conceptual understanding. In: Proceedings of the 44th SEFI Annual Conference on Design of Open, Tampere, Finland, pp. 12–15 (2016)

Computational Methods for Analyzing Functional and Effective Brain Network Connectivity Using fMRI

Farzad Vasheghani Farahani(✉) and Waldemar Karwowski

Department of Industrial Engineering and Management Systems,
University of Central Florida, Orlando, USA
farzad.vasheghani@knights.ucf.edu, wkar@ucf.edu

Abstract. Brain connectivity investigation using fMRI time series have begun since the mid-1990s and provided a new world for researchers, especially neuroscientists, to survey the human brain network with high precision. The present study seeks to provide an overview of the computational methods available for brain connectivity, which are divided into two general categories: functional connectivity and effective connectivity. The former examines the temporal correlation between spatially remote brain areas, and the latter is about the effects of brain regions on each other. Based on these two categories of connectivity, the computational methods presented in the literature along with their strengths and weaknesses are discussed.

Keywords: Brain connectivity · Functional connectivity
Effective connectivity · fMRI · Brain network

1 Introduction

Studies on brain connectivity are becoming incrementally widespread as neuroscientists look for revealing the comprehensive information underlying behavior, cognition, and perception [1]. Connectivity refers to the fact that how different brain regions are communicating with each other and how information is transferred among them, with the help of detailed analysis of the functional mechanism of the brain.

During the last two decades, the ability of functional magnetic resonance imaging (fMRI) to map the human brain functions has dramatically attracted researchers. Using blood oxygen level dependent (BOLD) contrast imaging [2–4], fMRI looks for different brain regions, which are active at any given time. Recently, rather than detecting the activation in different areas of the brain under specific conditions, attention has drawn on how distinct regions of the brain are correlated with each or how they affect each other when performing a particular kind of cognitive function [5].

One of the major applications of this field is to compare the connectivity characteristics between groups of subjects or among multiple sessions for one subject [6]. Such comparisons are commonly performed using resting-state data rather than task data. Further, another remarkable application is the study of mental disorders and brain diseases and, consequently, helping patients improve their health by analyzing

H. Ayaz and L. Mazur (Eds.): AHFE 2018, AISC 775, pp. 101–112, 2019.
https://doi.org/10.1007/978-3-319-94866-9_10

connectivity patterns and variations in the brain architectural features. Two types of connectivity approach termed functional connectivity and effective connectivity, have been explained in the fMRI literature [1, 7]. Briefly, functional connectivity provides information about the statistical dependencies or temporal correlation between spatially remote neurophysiological events, whereas effective connectivity is concerned with the directed influence of brain regions on each other [7].

In the following, we will review the computational methods that are presented in the literature for both types of connectivity in two different sections. In the first section, we explain conventional methods used for functional connectivity, including seed-based such as cross-correlation and coherence analysis, decomposition-based methods such as principal component analysis (PCA) and independent component analysis (ICA), clustering-based, and finally graph-based methods. Also, we discuss the advantages and possible pitfalls of each of these methods when performed on fMRI data. In the second section, we follow the same procedure for the methods proposed for effective connectivity. These methods include Granger casualty (GC), dynamic causal modeling (DCM), and Bayesian networks (BNs).

2 Functional Connectivity

Functional connectivity refers to the temporal correlations between BOLD signals from spatially remote brain regions [8, 9]. The objective of functional connectivity in comparison with other types of brain connectivity is to investigate the regional interactions in the brain from a macro perspective, while in effective connectivity researchers are seeking to analyze the effect that a neuronal system has on another [10].

Methods employed in the functional connectivity studies using fMRI data are mainly divided into model-based and model-free groups. The first group such as cross-correlation and coherence analysis has been extensively used in literature because of its easy implementation and interpretation although such methods require prior knowledge in their modeling. Therefore, model-based approaches can be challenging when applying on the resting-state fMRI data, since often there is no prior knowledge in these situations. On the other hand, methods used in the second group have a significant role in the fMRI studies when there are no available temporal or spatial patterns. These methods are based on decomposition (e.g., PCA and ICA), clustering or graph theory [10].

2.1 Cross-correlation Analysis

One of the traditional techniques for examining functional connectivity is the cross-correlation analysis introduced by Cao and Worsley in 1999 [11]. In this method, a correlation can be defined between each pair of brain regions that are functionally interconnected with the aid of their BOLD time series.

Many cross-correlation studies are limited to calculating correlation with zero lag, which may have a significant difference with the reality in that case. In general, the big drawback of using cross-correlation is when we seek to identify the correlation of two series regarding all the available lags since the computational complexity, in this case, is very high [12]. Although this problem is less evident in the fMRI studies, i.e., the short duration of the hemodynamic response function (a few seconds) does not require a calculation of correlation in all lags [13, 14], a large number of works consider correlations with zero lag.

2.2 Coherence Analysis

Despite the high use of cross-correlation technique in fMRI analysis, it has notable disadvantages, each of which can distort the results of the study compared to what is happening in reality. The first challenge is related to the complex calculations of this technique when examining all available lags of two given brain regions [12]. Secondly, the correlation is responsive to the form of HRF, which itself varies from one person to another and from one region to another [14–16]. The third issue refers to the cases where the two areas do not have any fluctuations in blood flow and the correlation of their BOLD time series is high, but we know that in practice there is no functional connectivity between such areas. Further, the physiological noise induced in the brain, such as cardiac and respiratory changes, would result in false correlation value [17].

To solve these problems, Sun et al. [18] developed a new technique called coherence by applying the correlation concepts in the frequency domain. They showed that the spectral demonstration of correlation in the frequency domain would help scientists to measure the communication of BOLD time series more realistically. For more information on the mechanism of these metrics and their calculations, refer to the original article [18].

2.3 Principal Component Analysis (PCA)

PCA is one of the most traditional and extensively used techniques for analysis of data. The core idea behind this method is to display the available signal, here the fMRI time series, using a linear combination of several orthogonal contributors. To maintain the maximum energy of the desired time courses, those contributors are selected which have the enormous impact on the data variance. PCA decomposes each contributor into a temporal pattern (a principal component) and a spatial pattern (an Eigen map) that determine the connectivity structure among brain regions.

The use of PCA in functional connectivity studies was first undertaken by Friston et al. [8] on PET images, followed by numerous studies using fMRI data mostly in resting state [19, 20]. The strengths of this technique can be observed in its relatively high ability to explore the whole-brain connectivity as well as the ability to reduce the dimensions of a complex system such as the brain, although some problems have caused the use of this method to be restricted in functional connectivity analysis. For instance, the PCA does not perform well when the contrast-to-noise ratio (CNR) is not sufficient due to the presence of physiological noise and other experimental factors that cause signal alteration [19]. Besides, selecting a proper number of components is another challenge to this method, which is often ignored by researchers [10]. Hence, one of the successful applications for PCA is helping to reduce the dimensions of brain connectivity as a pre-processing step for more efficient techniques, e.g., independent component analysis.

2.4 Independent Component Analysis (ICA)

ICA is another favorite way of examining the functional brain connectivity, especially for resting state fMRI data, which does not require prior knowledge (spatially or temporally) like PCA. The critical distinction between ICA and PCA is that here, the components should be as statistically independent as possible, whereas PCA looks for orthogonal contributors [21, 22]. Typically, ICA seeks to minimize the mutual information among the components, with two frequently used strategies known as Fixed-Points [23] and Infomax [24]. Esposito et al. made a comparison between them and showed that Infomax is superior to the other regarding overall estimation and coping with noise while the Fixed-Points has higher temporal and spatial accuracy [25].

The use of this technique for fMRI data, particularly in resting state, has been well received by researchers. However, it has some remarkable weaknesses. For example, if the underlying assumption of ICA (i.e., the independence of the signal source) is not met, the brain connectivity results will be in poor performance [26]. Further, how to select the number of independent components is utterly pivotal, e.g., the small number of components affects the results of ICA [27]. Also, since Gaussian distribution is generally used to convert the independent maps for thresholding while the nature of these maps is non-Gaussian, the false positive rate (FPR) is possibly overestimated [27, 28]. Most importantly, the presence of noise and unexplained signal variations is not considered in the initial ICA modeling, which leads to a variety of issues in analyzing fMRI data such as overfitting and inability to evaluate the statistical significance of the brain areas within spatial maps. To overcome these problems, Beckmann et al. presented a probabilistic ICA that allows for nonsquare mixing in the presence of Gaussian noise [29].

2.5 Clustering

Another technique used to examine the functional connectivity in fMRI data, especially those that are in resting-state, is clustering. Clustering algorithms look for grouping a given number of objects that are alike based on a set of similarity measures defined in the problem. When studying resting-state fMRI data, researchers attempt to group some voxels or regions of interest based on the similarities found in their BOLD time series using various distance metric, e.g., Pearson correlation.

The hierarchical clustering which creates a dendrogram of available objects, is one of the most popular clustering algorithms that also has been widely used in the brain connectivity studies [30, 31], Besides, k-means [32] and fuzzy clustering called fuzzy c-means (FCM) [33] are two other common types of clustering algorithms. These two methods are somewhat similar regarding updating memberships and cluster centers until convergence, which cause the latter to be known as soft k-means. However, their essential difference is that in FCM, each data has a membership probability value to each cluster, rather than entirely belonging to only one cluster as k-means [33]. Although FCM-based algorithms have been extensively studied in the clustering field, they use the Euclidean distance that is not suitable for non-Euclidean data such as MRI [34]. Finally, spectrum-based [35] and graph-based [36] techniques are among the other types of clustering algorithms used for the brain network exploration. For further information, see the referenced articles.

Heuvel et al. [37], in their study, compared the results of clustering algorithms and two types of traditional methods such as seed-based and decomposition-based (particularly ICA) ones. They concluded that although ICA has a certain superiority over the clustering and seed-based methods due to measuring directly the functional connectivity between subjects without any need for seed generation, the results of all three methods tend to present a high level of overlap [37].

2.6 Graph Theory

Graph theory provides a theoretical framework in which the pairwise relationships between elements of a system can be modeled through mathematical structures. Therefore, employing its concepts for the study of functional or effective connectivity supply researchers with a good understanding of a large-scale network such as the human brain [38]. In the brain functional networks, ROIs are displayed as nodes that are connected by edges, indicating functional architecture (i.e., correlations between time series). Several metrics are used to assess the topological patterns of networks and complex systems, including characteristic path length, clustering coefficient, centrality, node degree, and modularity [39, 40]. For instance, the *characteristic path length* is a measure to demonstrate the level of global connectedness within a graph by computing the average distance between all pairs of nodes in the network. As another example, the *clustering coefficient* reflects the connectedness among neighbors of a node in a graph, indicating the level of local clustering.

Many studies have revealed that the analysis of these metrics in the brain networks is associated with the small world topology, which was first expressed in social networks. The main feature of the networks based on this concept is that the number of edges incident to each node is relatively small, while there is a relationship between each pair of nodes with a short distance. This outstanding property only occurs in the presence of hubs in the network (that is, the critical nodes that have many connections and cause local clusters) [41, 42]. Therefore, firstly, the clustering coefficients of the small world networks are almost high, which results in cliques and local clusters within a graph, secondly, the characteristic path length in these networks is nearly low that helps network nodes interact with each other efficiently [31, 43, 44].

Although graph theory provides meaningful information about the development of functional neural networks, its weaknesses should be resolved. In research based on graph theory, inferences are agreed upon when nodes are well specified as voxels or areas of interest. Node specification is very tough in developmental studies since the nodes are probable to be similar across different subjects or sessions, so the graphs possibly contain a considerable amount of distortion. Hence, the employment of graph theory needs an attentive approach for node selection, and an understanding that created graphs are reliable only when nodes as are well defined [45].

3 Effective Connectivity

One of the areas that attracted the attention of researchers in the last decade is the study of directional interactions between different regions of the brain. In this domain which known as effective connectivity, the influences of neuronal systems on each other are evaluated [7]. Such studies play a pivotal role in our comprehension of functional integration of brain networks, neuronal dynamic, and physiological correlations of psychiatric and neurological disorders [46]. Granger Causality (GC) [47] is the most prevalently used technique for directional connectivity and analyzing statistical relationships over time. Despite the simple implementation, the existence of some assumptions in its modeling [48, 49] has led to severe challenges when using fMRI data, which will be discussed further. Dynamic causal modeling (DCM) [50] and structural equations modeling (SEM) [51] are two other traditional methods which can be used for detecting effective connectivity. Both methods are among hypothesis-driven approaches, which need prior connectivity knowledge to be implemented. Although these methods can supply hypothetically strong inferences, they are inconclusive in many cases, such as examining the resting state that prior assumptions are unavailable [52]. However, successful efforts have been made to improve DCM and, by enlarging the search space, researchers have enhanced the exploratory characteristic of DCM over conventional ones [53], while it has still some limitations in fMRI studies. On the contrary, probabilistic approaches such as Bayesian networks (BNs) [54] are exploratory in essence, but latency differences between fMRI time series is an absolute limitation for them [49]. We will discuss each of these methods in more detail along with their strengths and weaknesses.

3.1 Granger Casualty (GC)

One of the traditional methods that drive us toward the characterization of functional circuits that underlies many psychological issues such as behavior, cognition, and perception is Granger causality (GC) analysis, in which directional functional interactions are identified from time series data. This concept is defined in both areas of time and frequency [55].

We say that *X* Granger causes *Y* if *X* provides information that can estimate the future of Y more accurate than the past data of *Y*, more clearly, the past data of a given brain area can tell the current state of another. This technique was first proposed by Granger [47] for analyzing the temporally structured data. In a general sense, however, GC is based on this core concept that causes come before effects at all times, it does not require prior knowledge of a structural model. Due to the discrepancy between sampling intervals and neurodynamics events which are remarkably faster than the former, GC cannot directly be used for fMRI data, since in this case, instead of achieving interactions in neuronal responses that are the underlying goal, BOLD time series interactions are predicted [56, 57]. To overcome this problem, researchers implement this concept along with other models such as linear vector autoregressive (VAR) as their most common [58, 59]. Other time-varying, nonlinear and non-parametric models can also be used in this regard [60, 61].

Although its implementation of fMRI data is growing, various issues prevent its ability to interpret neural activity. First, latency discrepancies in HRF across several brain regions, second, low-sampling rates, which is extremely meaningful in time series modeling [62], and finally, the presence of noise. To address such issues, Wen et al. [48] examined whether there is a methodical relationship between GC at the BOLD fMRI times series and that at the neural responses. They first simulated a set of neural signals and afterward, generate simulated fMRI data using them. By comparing these two sets of simulated data, they demonstrated that BOLD data is a monotonically increasing function of neural responses. Also, using real fMRI data, they concluded that this monotonicity is a valid positive correlation. Eventually, they showed that after modifying the latency discrepancies, a significant improvement in detection occurs. Such findings indicate that Granger causality is a reliable method for examining fMRI data when the over-mentioned issues are adequately addressed [48].

3.2 Dynamic Causal Modeling (DCM)

One of the most widely used statistical methods for identifying causal architecture in distributed systems is dynamic causal modeling (DCM). DCM works based on the bilinear system of equations to see how neural activity variations in node x_1 are due to activation in node x_2 under external stimuli [50, 63, 64]. This framework contains information about the fixed connectivity between two regions, the variations in the coupling strength (resulting from the experimental manipulations), and the direct effects on a given node [50]. DCMs are designed in a way that they can examine the experimental modulation of self-connections or forwards and backwards connections among brain areas, which are active throughout a certain task. Then by performing a comparison based on a Bayesian framework the model that best matches the data, will be chosen. Moreover, this method estimates coefficients for the connection strength as well as the modulatory strength using an experimental condition. Accordingly, DCM provides a proper statistical platform that can assess mutual and hierarchical functional structures in response to external manipulations [38].

Since the use of DCM requires prior knowledge to formulate the hypotheses and specify a set of models for comparison and testing, it cannot be viewed as an exploratory tool for data analysis. Nevertheless, few methods have been trying to find themselves in an exploratory manner by examining a substantial number of models through post hoc analyzes wherein the inversion is performed only on the full model [53].

Friston et al. [59] showed that GC and DCM play a complementary role together and can reinforce each other to investigate the causal architecture. On the one hand, utilizing a generic inferential way provided by GC, one can analyze any given time series obtained from the experimentally selected brain regions to identify the directional relationships among them. Thus, a satisfactory perception of the brain's dynamical behavior is achieved, whether in rest or activity. On the other hand, in a more mechanistic framework using the concept of Bayesian, one can compare and test different models and hypotheses, in which case DCM will be helpful. However, this approach requires customized models for the system in question [65]. The therefore their difference is that DCM compares the evidence of various models and eventually chooses the best ones [66], but in GC the process is implicit in the form of checking the presence of GC followed by identifying the VAR model order [67].

3.3 Bayesian Network (BN)

In the last decade, a significant number of researchers have focused on Bayesian networks to extract effective connectivity across the brain regions. Bayesian network (BN) is a probabilistic graphical model that identify the conditional dependencies among a given set of random variables as well as a set of parameters representing the path coefficients [68, 69]. These networks are practical tools for machine learning and data mining, which can be used in many areas such as marketing, finance, robotics and health systems to identify and disclose valuable information [69, 70]. BN is a kind of directed acyclic graph (DAG) whose vertices indicate network variables and edges display conditional dependencies about the variables. The parameters represent the probabilities of the nodes given their parent nodes by conditional probability distributions (CPDs). For any two disconnected vertices, the corresponding variables are conditionally independent of each other.

Gaussian BN [71] and discrete dynamic BN (DBN) [72, 73] are two of the most widely studied techniques that have made a significant contribution to the studies in the field of BN. The primary assumption in the first one, i.e., Gaussian BN, is that Gaussian distribution is used to model the time series of each network node and a set of linear regression equations are formed to generate the network structure. Since such types of networks are static, they are only able to display a fixed set of brain connections and cannot extract the temporal dependencies of various brain areas among multiple processes [72]. The latter, i.e., discrete DBN, is also another technique derived from BN, which can model temporal dependencies among different regions of the brain by considering the first-order Markov chain as well as stationary assumptions [72]. Despite the dynamic nature of this technique, its main disadvantage is the use of binomial distribution in the production of time series of the brain nodes, which entails the data to be discretized, and subsequently, a substantial amount of information is lost.

To overcome the shortcomings of the two techniques as mentioned above, Wu and his colleagues [46] presented a method termed Gaussian DBN. Their designed network is a kind of first-order linear dynamic system whose brain activity in two consecutive instances indicates network nodes and edges indicate the conditional dependencies between brain region pairs. They used the Bayesian information criterion (BIC) [68] to identify both the network that best suited the probabilistic dependencies and the Gaussian DBN parameter that represents the connectivity strength. In this model, unlike the Gaussian BN, the existence of conditional dependencies among the variables is because the paths can be created both between and within the columns. Furthermore, Gaussian DBN assumes that the nodes time series follows a Gaussian distribution, hence protecting the network from the loss of information and prevailing over the main limitation of discrete DBN. Finally, using a set of synthetic data as well as real fMRI data, they compared their results with discrete DBN to evaluate the flexibility and robustness of Gaussian DBN [46].

4 Conclusion

Analysis of fMRI data resulted in remarkable advances in neuroscience and behavioral studies. This article provides an overview of computational methods for analyzing functional and effective connectivity in the brain network. In the first group, seed-based methods (e.g., cross-correlation and coherence analysis) and decomposition-based methods (e.g., PCA and ICA), as well as clustering-based methods and graph theory were discussed, and in each case, weaknesses and strengths were noted. In the second category, the computational methods available for effective connectivity such as Granger casualty, dynamics causal modeling, and Bayesian networks were reviewed. In addition to the methods described here, other approaches that are close to the nature of the brain architecture such as graph theory, system dynamics, and complex system can provide significant medical applications for mental disorders and brain diseases. Further, various brain imaging techniques such as EEG and diffusion-weighted MRI can be applied together to increase the efficiency of future studies on brain connectivity exploration.

References

1. Friston, K.J.: Functional and effective connectivity: a review. Brain Connect. **1**, 13–36 (2011)
2. Ogawa, S., Lee, T.M., Kay, A.R., Tank, D.W.: Brain magnetic resonance imaging with contrast dependent on blood oxygenation. Proc. Natl. Acad. Sci. USA **87**, 9868–9872 (1990)
3. Ogawa, S., Tank, D.W., Menon, R., Ellermann, J.M., Kim, S.G., Merkle, H., Ugurbil, K.: Intrinsic signal changes accompanying sensory stimulation: functional brain mapping with magnetic resonance imaging. Proc. Natl. Acad. Sci. USA **89**, 5951–5955 (1992)
4. Kwong, K.K., Belliveau, J.W., Chesler, D.A., Goldberg, I.E., Weisskoff, R.M., Poncelet, B. P., Kennedy, D.N., Hoppel, B.E., Cohen, M.S., Turner, R.: Dynamic magnetic resonance imaging of human brain activity during primary sensory stimulation. Proc. Natl. Acad. Sci. USA **89**, 5675–5679 (1992)
5. Greicius, M.D., Krasnow, B., Reiss, A.L., Menon, V.: Functional connectivity in the resting brain: a network analysis of the default mode hypothesis. Proc. Natl. Acad. Sci. USA **100**, 253–258 (2003)
6. Zhang, L., Guindani, M., Vannucci, M.: Bayesian models for functional magnetic resonance imaging data analysis. Wiley Interdiscip. Rev. Comput. Stat. **7**, 21–41 (2015)
7. Friston, K.J.: Functional and effective connectivity in neuroimaging: a synthesis. Hum. Brain Mapp. **2**, 56–78 (1994)
8. Friston, K.J., Frith, C.D., Liddle, P.F., Frackowiak, R.S.: Functional connectivity: the principal-component analysis of large (PET) data sets. J. Cereb. Blood Flow Metab. **13**, 5–14 (1993)
9. Lee, L., Harrison, L.M., Mechelli, A.: A report of the functional connectivity workshop, Dusseldorf 2002. Neuroimage **19**, 457–465 (2003)
10. Li, K., Guo, L., Nie, J., Li, G., Liu, T.: Review of methods for functional brain connectivity detection using fMRI. Comput. Med. Imaging Graph. **33**, 131–139 (2009)
11. Cao, J., Worsley, K.: The geometry of correlation fields with an application to functional connectivity of the brain. Ann. Appl. Probab. **9**(4), 1021–1057 (1999). (Ed. by, J. Cao, K. Worsley)

12. Cecchi, G.A., Rao, A.R., Centeno, M.V., Baliki, M., Apkarian, A.V., Chialvo, D.R.: Identifying directed links in large scale functional networks: application to brain fMRI. BMC Cell Biol. **8**, 1–10 (2007)
13. Friston, K.J., Jezzard, P., Turner, R.: Analysis of functional MRI time-series. Hum. Brain Mapp. **1**, 153–171 (1994)
14. Saad, Z.S., Ropella, K.M., Cox, R.W., DeYoe, E.A.: Analysis and use of fMRI response delays. Hum. Brain Mapp. **13**, 74–93 (2001)
15. Lee, S.P., Duong, T.Q., Yang, G., Iadecola, C., Kim, S.G.: Relative changes of cerebral arterial and venous blood volumes during increased cerebral blood flow: implications for bold fMRI. Magn. Reson. Med. **45**, 791–800 (2001)
16. Miezin, F.M., Maccotta, L., Ollinger, J.M., Petersen, S.E., Buckner, R.L.: Characterizing the hemodynamic response: effects of presentation rate, sampling procedure, and the possibility of ordering brain activity based on relative timing. Neuroimage **11**, 735–759 (2000)
17. Friston, K.J., Holmes, A.P., Worsley, K.J., Poline, J.B., Frith, C., Frackowiak, R.S.: Statistical parametric maps in functional imaging: a general linear approach. Hum. Brain Mapp. **2**, 189–210 (1995)
18. Sun, F.T., Miller, L.M., D'Esposito, M.: Measuring interregional functional connectivity using coherence and partial coherence analyses of fMRI data. Neuroimage **21**, 647–658 (2004)
19. Baumgartner, R., Ryner, L., Richter, W., Summers, R., Jarmasz, M., Somorjai, R.: Comparison of two exploratory data analysis methods for fMRI: fuzzy clustering vs. principal component analysis. Magn. Reson. Imaging **18**, 89–94 (2000)
20. Worsley, K.J., Chen, J.-I., Lerch, J., Evans, A.C.: Comparing functional connectivity via thresholding correlations and singular value decomposition. Philos. Trans. Roy. Soc. Lond. B: Biol. Sci. **360**, 913–920 (2005)
21. Hyvärinen, A., Oja, E.: Independent component analysis: algorithms and applications. Neural Netw. **13**(4–5), 411–430 (2000)
22. Comon, P.: Independent component analysis, a new concept? Sig. Process. **36**, 287–314 (1994)
23. Hyvärinen, A.: New approximations of differential entropy for independent component analysis and projection pursuit. In: Advances in Neural Information Processing Systems, pp. 273–279 (1998)
24. Bell, A.J., Sejnowski, T.J.: Information-maximization approach to blind separation and blind deconvolution. Neural Comput. **7**, 1129–1159 (1995)
25. Esposito, F., Formisano, E., Seifritz, E., Goebel, R., Morrone, R., Tedeschi, G., Di Salle, F.: Spatial independent component analysis of functional MRI time-series: to what extent do results depend on the algorithm used. Hum. Brain Mapp. **16**, 146–157 (2002)
26. Calhoun, V.D., Adali, T., Pearlson, G.D., Pekar, J.J.: Spatial and temporal independent component analysis of functional MRI data containing a pair of task-related waveforms. Hum. Brain Mapp. **13**, 43–53 (2001)
27. Ma, L., Wang, B., Chen, X., Xiong, J.: Detecting functional connectivity in the resting brain: a comparison between ICA and CCA. Magn. Reson. Imaging **25**, 47–56 (2007)
28. Zhao, X., Glahn, D., Tan, L.H., Li, N., Xiong, J., Gao, J.H.: Comparison of TCA and ICA techniques in fMRI data processing. J. Magn. Reson. Imaging **19**, 397–402 (2004)
29. Beckmann, C.F., Smith, S.M.: Probabilistic independent component analysis for functional magnetic resonance imaging. IEEE Trans. Med. Imaging **23**, 137–152 (2004)
30. Cordes, D., Haughton, V., Carew, J.D., Arfanakis, K., Maravilla, K.: Hierarchical clustering to measure connectivity in fMRI resting-state data. Magn. Reson. Imaging **20**, 305–317 (2002)

31. Salvador, R., Suckling, J., Coleman, M.R., Pickard, J.D., Menon, D., Bullmore, E.: Neurophysiological architecture of functional magnetic resonance images of human brain. Cereb. Cortex **15**, 1332–2342 (2005)
32. Golland, Y., Golland, P., Bentin, S., Malach, R.: Data-driven clustering reveals a fundamental subdivision of the human cortex into two global systems. Neuropsychologia **46**, 540–553 (2008)
33. Lee, M.H., Hacker, C.D., Snyder, A.Z., Corbetta, M., Zhang, D., Leuthardt, E.C., Shimony, J.S.: Clustering of resting state networks. PLoS ONE **7**, 1–12 (2012)
34. Farahani, F.V., Ahmadi, A., Zarandi, M.H.F.: Hybrid intelligent approach for diagnosis of the lung nodule from CT images using spatial kernelized fuzzy c-means and ensemble learning. Math. Comput. Simul. **149**, 46–68 (2018)
35. Bellec, P., Rosa-Neto, P., Lyttelton, O.C., Benali, H., Evans, A.C.: Multi-level bootstrap analysis of stable clusters in resting-state fMRI. Neuroimage **51**, 1126–1139 (2010)
36. van den Heuvel, M., Mandl, R., Pol, H.H.: Normalized cut group clustering of resting-state fMRI data. PLoS ONE **3**, e2001 (2008)
37. van den Heuvel, M.P., Hulshoff Pol, H.E.: Exploring the brain network: a review on resting-state fMRI functional connectivity. Eur. Neuropsychopharmacol. **20**, 519–534 (2010)
38. Goldenberg, D., Galván, A.: The use of functional and effective connectivity techniques to understand the developing brain. Dev. Cogn. Neurosci. **12**, 155–164 (2015)
39. Sporns, O., Chialvo, D.R., Kaiser, M., Hilgetag, C.C.: Organization, development and function of complex brain networks. Trends Cogn. Sci. **8**, 418–425 (2004)
40. van den Heuvel, M.P., Stam, C.J., Boersma, M., Hulshoff Pol, H.E.: Small-world and scale-free organization of voxel-based resting-state functional connectivity in the human brain. Neuroimage **43**, 528–539 (2008)
41. Bullmore, E., Sporns, O.: Complex brain networks: graph theoretical analysis of structural and functional systems. Nat. Rev. Neurosci. **10**, 186–198 (2009)
42. Jain, M.: A next-generation approach to the characterization of a non-model plant transcriptome. Curr. Sci. **101**, 1435–1439 (2011)
43. Eguíluz, V.M., Chialvo, D.R., Cecchi, G.A., Baliki, M., Apkarian, A.V.: Scale-free brain functional networks. Phys. Rev. Lett. **94**, 1–4 (2005)
44. Fair, D.A., Cohen, A.L., Power, J.D., Dosenbach, N.U.F., Church, J.A., Miezin, F.M., Schlaggar, B.L., Petersen, S.E.: Functional brain networks develop from a "local to distributed" organization. PLoS Comput. Biol. **5**, 14–23 (2009)
45. Power, J.D., Fair, D.A., Schlaggar, B.L., Petersen, S.E.: The development of human functional brain networks. Neuron **67**, 735–748 (2010)
46. Wu, X., Wen, X., Li, J., Yao, L.: A new dynamic Bayesian network approach for determining effective connectivity from fMRI data. Neural Comput. Appl. **24**, 91–97 (2014)
47. Granger, C.W.: Investigating causal relations by econometric models and cross-spectral methods. Econom. J. Econom. Soc. **37**, 424–438 (1969)
48. Wen, X., Rangarajan, G., Ding, M.: Is granger causality a viable technique for analyzing fMRI data? PLoS ONE **8**, e67428 (2013)
49. Dang, S., Chaudhury, S., Lall, B., Roy, P.K.: Learning effective connectivity from fMRI using autoregressive hidden Markov model with missing data. J. Neurosci. Methods **278**, 87–100 (2017)
50. Friston, K.J., Harrison, L., Penny, W.: Dynamic causal modelling. Neuroimage **19**, 1273–1302 (2003)
51. McIntosh, A.R., Gonzalez-Lima, F.: Structural equation modeling and its application to network analysis in functional brain imaging. HBM **2**, 2–22 (1994)
52. Fox, M.D., Raichle, M.E.: Spontaneous fluctuations in brain activity observed with functional magnetic resonance imaging. Nat. Rev. Neurosci. **8**, 700–711 (2007)

53. Friston, K.J., Li, B., Daunizeau, J., Stephan, K.E.: Network discovery with DCM. Neuroimage **56**, 1202–1221 (2011)
54. Ramsey, J.D., Hanson, S.J., Hanson, C., Halchenko, Y.O., Poldrack, R.A., Glymour, C.: Six problems for causal inference from fMRI. Neuroimage **49**, 1545–1558 (2010)
55. Seth, A.K., Barrett, A.B., Barnett, L.: Granger causality analysis in neuroscience and neuroimaging. J. Neurosci. **35**, 3293–3297 (2015)
56. Smith, S.M., Miller, K.L., Salimi-Khorshidi, G., Webster, M., Beckmann, C.F., Nichols, T. E., Ramsey, J.D., Woolrich, M.W.: Network modelling methods for FMRI. Neuroimage **54**, 875–891 (2011)
57. Smith, S.M., Bandettini, P.A., Miller, K.L., Behrens, T.E.J., Friston, K.J., David, O., Liu, T., Woolrich, M.W., Nichols, T.E.: The danger of systematic bias in group-level FMRI-lag-based causality estimation. Neuroimage **59**, 1228–1229 (2012)
58. Seth, A.K.: A MATLAB toolbox for granger causal connectivity analysis. J. Neurosci. Methods **186**, 262–273 (2010)
59. Friston, K., Moran, R., Seth, A.K.: Analysing connectivity with Granger causality and dynamic causal modelling. Curr. Opin. Neurobiol. **23**, 72–178 (2013)
60. Dhamala, M., Rangarajan, G., Ding, M.: Analyzing information flow in brain networks with nonparametric Granger causality. Neuroimage **41**, 354–362 (2008)
61. Roebroeck, A., Formisano, E., Goebel, R.: The identification of interacting networks in the brain using fMRI: model selection, causality and deconvolution. Neuroimage **58**, 296–302 (2011)
62. Lin, F.-H., Ahveninen, J., Raij, T., Witzel, T., Chu, Y.-H., Jääskeläinen, I.P., Tsai, K.W.-K., Kuo, W.-J., Belliveau, J.W.: Increasing fMRI sampling rate improves Granger causality estimates. PloS ONE **9**(9), e100319 (2014)
63. Friston, K.: Causal modelling and brain connectivity in functional magnetic resonance imaging. PLoS Biol. **7**, 0220–0225 (2009)
64. Stephan, K.E., Penny, W.D., Moran, R.J., den Ouden, H.E.M., Daunizeau, J., Friston, K.J.: Ten simple rules for dynamic causal modeling. Neuroimage **49**, 3099–3109 (2010)
65. Daunizeau, J., David, O., Stephan, K.E.: Dynamic causal modelling: a critical review of the biophysical and statistical foundations. Neuroimage **58**, 312–322 (2011)
66. Penny, W.D.: Comparing dynamic causal models using AIC, BIC and free energy. Neuroimage **59**, 319–330 (2012)
67. Bressler, S.L., Seth, A.K.: Wiener-Granger causality: a well established methodology. Neuroimage **58**, 323–329 (2011)
68. Institute of Mathematical Statistics is collaborating with JSTOR to digitize, preserve, and extend access to The Annals of Mathematical Statistics. ® www.jstor.org
69. Karahoca, D., Karahoca, A., Yavuz, Ö.: An early warning system approach for the identification of currency crises with data mining techniques. Neural Comput. Appl. **23**, 2471–2479 (2013)
70. Sohrabi, B., Mahmoudian, P., Raeesi, I.: A framework for improving e-commerce websites usability using a hybrid genetic algorithm and neural network system. Neural Comput. Appl. **21**, 1017–1029 (2012)
71. Li, R., Chen, K., Zhang, N., Fleisher, A.S., Li, Y., Wu, X.: Effective connectivity analysis of default mode network based on the Bayesian network learning approach. In: Proc. SPIE, vol. 7262, pp. 72621W–72621W–10 (2009)
72. Rajapakse, J.C., Zhou, J.: Learning effective brain connectivity with dynamic Bayesian networks. Neuroimage **37**, 749–760 (2007)
73. Zeng, Z., Ji, Q.: Knowledge based activity recognition with dynamic Bayesian network, pp. 532–546 (2010)

The Effect of Alcohol-Use and Sleep Deprivation on Quantitative Changes in EEG for Normal Young Adults During Multitasking Evaluation

Young-A Suh, Jung Hwan Kim, and Man-Sung Yim(✉)

Nuclear Energy Environment and Nuclear Security Laboratory,
Department of Nuclear and Quantum Engineering,
Korea Advanced Institute of Science and Technology (KAIST),
Daejeon, Republic of Korea
{dreameryounga,poxc,msyim}@kaist.ac.kr

Abstract. The Fitness-For-Duty (FFD) of a worker in high-reliability systems such as Nuclear Power Plants (NPPs) and the civilian aircraft industry has been largely recognized as a key indicator when identifying the causes for human-error related accidents/incidents. Specifically, an alcohol drinker and a sleep-deprived worker has been identified as fatal to nuclear safety. In NPPs, a reactor operator must be capable of multitasking to detect a system failure. Thus, the objective of this study is to understand the EEG variations underpinning the reactions of drinkers and drowsy persons during multitasking cognitive performance. An experimental task was performed which included multitasking, working memory and real time EEG signals recorded. Quantitative EEG (absolute and relative powers of the seven bands) was analyzed for 10 college students (five subjects with 0.05% Blood Alcohol Concentration (BAC) and five with less than one hour of sleep). The same experiment was performed again when each of the subjects had a normal health status (0% BAC and their typical sleep duration). The results indicated a statistical difference in cognitive performance depending on their physical status. These results can be used to investigate the feasibility of determining alcohol-use and sleep quality measurements using EEG indicators, as well as determining a worker's FFD related to cognitive performance.

Keywords: Electroencephalogram (EEG) · Alcohol-use · Sleep deprived Multitasking

1 Introduction

In 1997, a Korean Airline flew into a hillside in Guam and this accident caused the death of all 227 people aboard. According to the investigation, the accident occurred because of reduced situation awareness due to sleep-deprivation. In addition, the Three Miles Island (TMI) accidents also occurred resulting from a sleep-deprived worker. These historical cases lead that the importance of Fitness-For-Duty (FFD) programs highlighted.

H. Ayaz and L. Mazur (Eds.): AHFE 2018, AISC 775, pp. 113–123, 2019.
https://doi.org/10.1007/978-3-319-94866-9_11

The FFD refers to a worker's physical, physiological and psychological ability to perform their tasks competently and safely [1]. Implementing an effective FFD program provides reasonable assurance that a worker's performance will not pose a safety or security risk. The Nuclear Regulatory Commission's (NRC's) regulation 10 CFR part 26 highlights the importance of FFD programs in nuclear facilities [2]. To comply with the NRC regulation, nuclear industries implemented a drug and alcohol test and fatigue management tracking (through reporting working hours by licensees) [3]. However, current FFD programs are limited in self-evaluating psychological distress, time delay (sample collection to complete analysis), and no reliable tool to measure actual fatigue.

Unfit status associated with alcohol consumption and sleep-deprived lead to cognitive decrements [4]. In case of alcoholics, they showed significant impairments in concept formation, shifting, and working memory [5, 6]. However, there was little research on alcohol-related cognitive impairments in the general population [7, 8]. The effects of mental fatigue on cognitive performance [9, 10] have investigated. Otherwise, the impact of sleep-deprivation on multitasking performance remains mostly unknown.

Previous studies examined the effect of alcohol-use on EEG measurements [11, 12]. In the alcohol-use group, when compared to healthy controls, drunken subjects consistently reported higher absolute beta power. Some researchers found that alcohol consumption is associated with an increase in the production of EEG alpha activity [13–16]. Recent studies [17, 18] have shown the feasibility of assessing driver drowsiness using EEG signals. Corsi-Cabrera et al. [19, 20] reported sleep-deprived subjects have increased EEG absolute power in all frequency bands compared with normal status individuals.

Most studies examined the resting EEG characteristics in alcohol-use and sleep-deprived group. There are only a few studies assessing EEG signals to determine a worker's fitness status (sleep-deprived and alcohol-use) during cognitive performance. Therefore, this paper aims at understanding the difference on cognitive performance using EEG variation depending on an employee's specific unfitness status. It is helpful to understand why a worker having alcohol-use or sleep-deprived results in human error.

2 Materials and Method

2.1 Participants

EEG data from ten college students (five subjects with 0.05% Blood Alcohol Concentration (BAC) and five with less than one hour of sleep) was collected in the experiment. The same experiment was performed again when each of the subjects had a normal health status (0% BAC and their typical sleep duration). All subjects were in generally good health, took no medications, and had normal sleep habits. This study categorized the subjects into Alcohol-use group (five subjects) and Sleep-deprived group (five subjects). The criteria used to classify these groups were:

1. Above 0.05% Blood Alcohol Concentration (BAC),
2. Less than 1 or 2 h of sleep over a 48 h period,

If a subject participated in alcohol-use group, this subject also performed the same experimental tasks when the subject is normal status. The normal means the status of no caffeine, smoking, alcohol-use, or medications, and more than 8 h sleep. In case of sleep-deprived group, the same protocol was applied for five participants.

2.2 Experimental Tasks

Experimental tasks included computer-games from Lumosity (http://www.lumosity.com) which tested a subject's cognition capacity. These cognition tasks can be useful in evaluating similar demands on operators of nuclear power plants. The types of demands include, processing speed, working memory, attention, flexibility/cognitive-control, and problem solving.

The Cambridge Neuropsychological Test Automated Battery (CANTAB) (http://www.cambridgecognition.com/resources/cantab-test-selector) covers a wide range of cognitive domains. The validity of these tests has been repeatedly demonstrated in several research areas [21, 22]. As such, these tasks are suitable for evaluating the cognition capacity of NPP operators. This was demonstrated when the CANTAB was used and is now a required cognitive tests for determining healthy volunteers reaction time and spatial working memory ability. Lumosity tasks are very similar to the CANTAB required test. For instance, the tile matrix test from Lumosity is similar to the paired associates learning and spatial working memory tests from CANTAB (as depicted in Fig. 1). In addition, the train of thought game (Fig. 1) is related to reaction time, multitasking and working memory. Moreover, Oh and Lee [23] research in evaluating Nuclear Power Plants (NPPs) Operators' cognitive tasks, their defined tasks are similar to the tasks used in this study, as identified above.

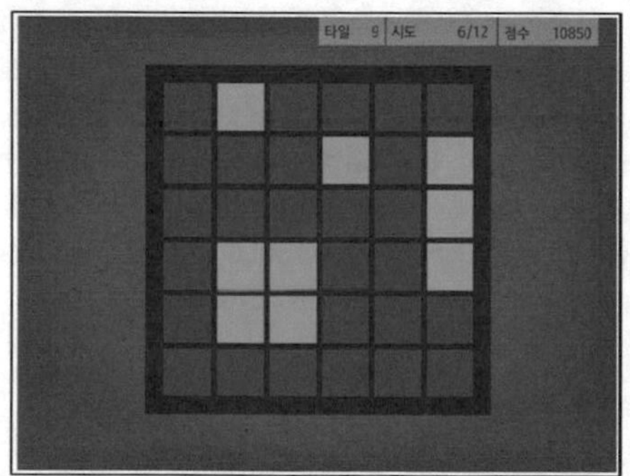

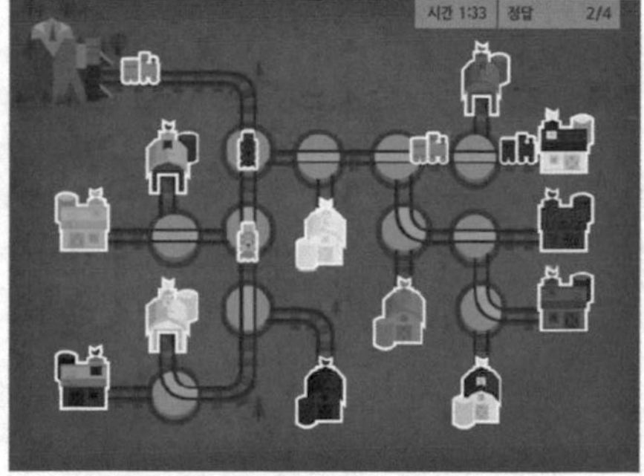

Fig. 1. The tile matrix (left) and train of thought game from lumosity

2.3 The EEG Recording and Analysis

An EEG system with 19 channels (BrainMaster Discovery 24ETM (Brain Master Technologies Inc.)) recorded EEG data with Linked ears reference (LE). The 19 channels are labeled Fp1, Fp2, F3, F4, F7, F8, Fz, C3, C4, Cz, T3, T4, T5, T6, P3, P4, Pz, O1, and O2. The electrodes are arranged in the international 10-20 system.

Participants wore an electro cap with matching channel positions. The EEG data was recorded in the resting status (eye closed, eye open). During the EEG recording, the impedance was kept below 5 kΩ. All channels of EEG are acquired with the sampling rate of 256 Hz, 24 bits resolution mark.

A Quantitative EEG (QEEG) analysis [24] is a computer analysis of the EEG signal using 19 or more channels of a simultaneous EEG recording. First, raw digital EEG raw data were subjected to a Fast Fourier Transform (FFT) algorithm to calculate the absolute (μV^2) power and relative (%) power and the FFT Power Ratio (Arb). Absolute Power is the actual power (voltage) in a subject's EEG database (Power is microvolts squared). Relative Power is the relative power of each given band/sum of power from 1 to 50 Hz. FFT Power Ratio is calculated by one given band/other given band.

EEG frequency bands were categorized into the following seven bands:

(1) Delta: 1–4 Hz, (2) Theta: 4–8 Hz, (3) Alpha: 8–12 Hz,
(4) Beta: 12–25 Hz, (5) High Beta: 25–30 Hz,
(6) Gamma: 30–40 Hz, and (7) High Gamma: 40–50 Hz.

2.4 Statistical Analysis

All data are presented as Means with Standard Deviation (M ± SD). The effects of condition (Sleep-deprived vs. Control and Alcohol-use vs. Control) on EEG variation were tested using Multivariate Analysis of Variance (MANOVA) analysis [25]. Significance (p-value) was set at 0.05 or 0.01 (2-tailed) for all analyses, which were conducted using the Statistical Package for the Social Sciences (SPSS), version 24.

3 Results

To categorize the differences between the each two groups, this study examined the multivariate test of significance. Figures 2, 3, 4 and 5(a) show the differences between the mean values of possible EEG indicators, based on the different cognitive performance. Figures 2, 3, 4 and 5(b–c) also represent the differences of seven EEG indicators, depending on a subject's FFD status. The significant differences ($P < 0.01$ and $P < 0.05$) between the two groups were marked as** and * in the figures.

3.1 The Effect of Sleep-Deprived on Cognitive Performance

Depending upon the type of tasks involved (Figs. 2(a) and 3(a)), the theta indicator of the subjects showed significant differences. "Theta" is associated with high mental stress (cognitive workload). Sleep-deprived subjects felt more stressful when performing the multitasking game rather than working memory. "Beta" is related to conscious focus and problem solving: Results show that the subjects' mental focus levels (cognitive workload) increased during the tasks. However, sleep-deprived subjects seems not easy to concentrate (relative power of beta decreased). Gamma brainwaves are associated with increased mental abilities. This implies performing the multitasking tasks is the highest mental abilities demanded. In addition, high gamma

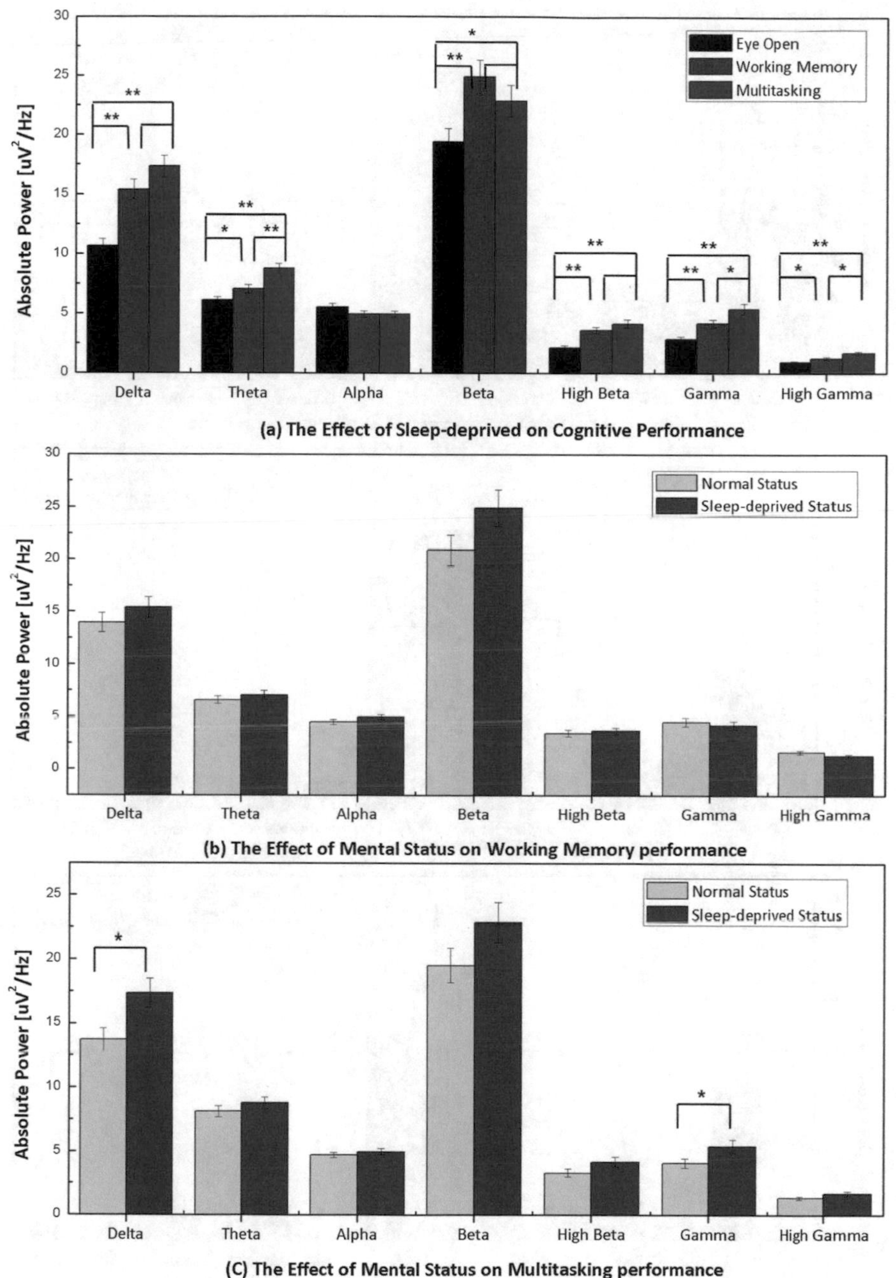

Fig. 2. *The Absolute Power differences of sleep-deprived group during cognitive performance* Group mean ($\pm S.E.$) Absolute Power difference (y-axis) between two groups for seven indicators (x-axis). The significant indicator ($P < 0.01$ and $P < 0.05$) differences between the two groups were marked as** and * in the figure

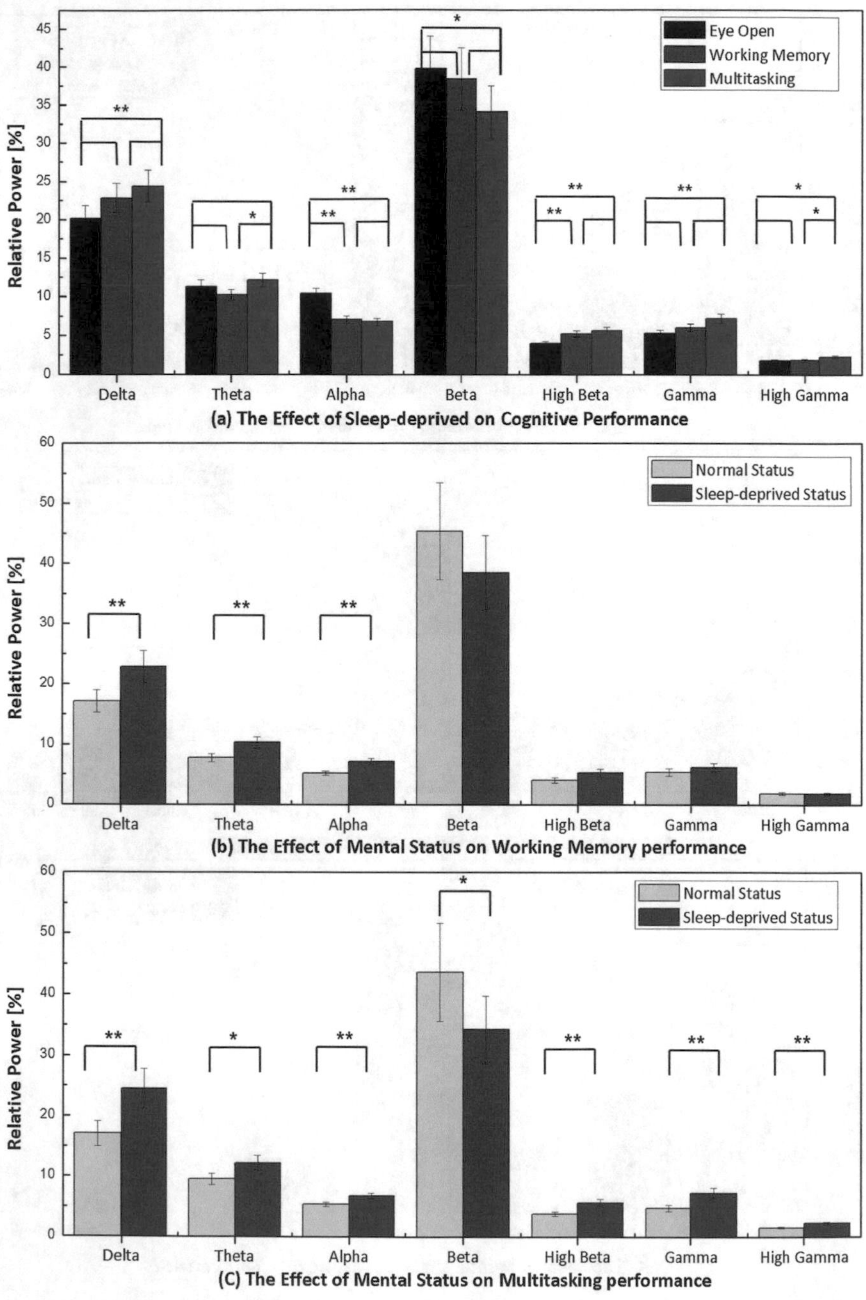

Fig. 3. *The Relative Power differences of sleep-deprived group during cognitive performance* Group mean (±*S.E.*) Absolute Power difference (y-axis) between two groups for seven indicators (x-axis). The significant indicator (P < 0.01 and P < 0.05) differences between the two groups were marked as** and * in the figure

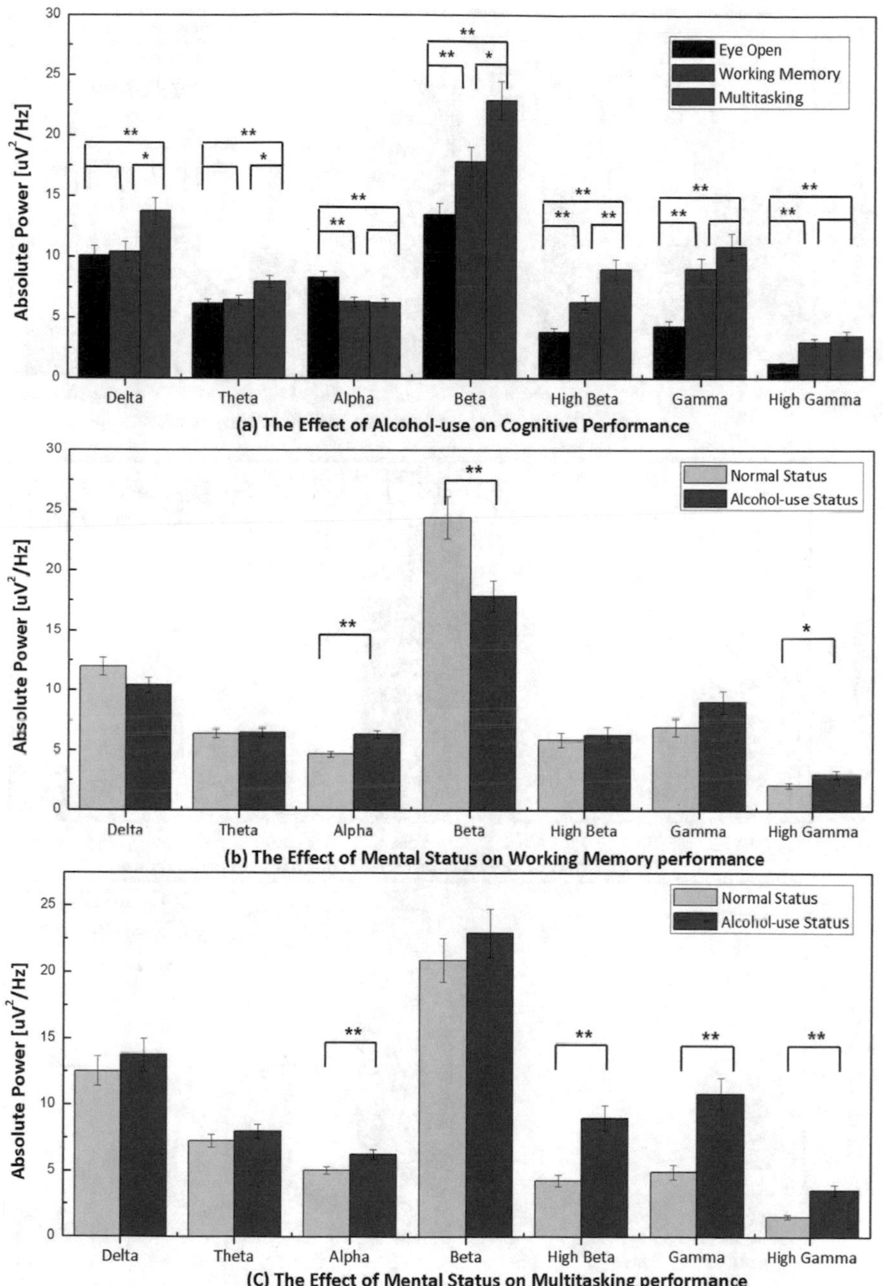

Fig. 4. *The Absolute Power differences of Alcohol-use group during cognitive performance* Group mean (±*S.E.*) Absolute Power difference (y-axis) between two groups for seven indicators (x-axis). The significant indicator (P < 0.01 and P < 0.05) differences between the two groups were marked as** and * in the figure

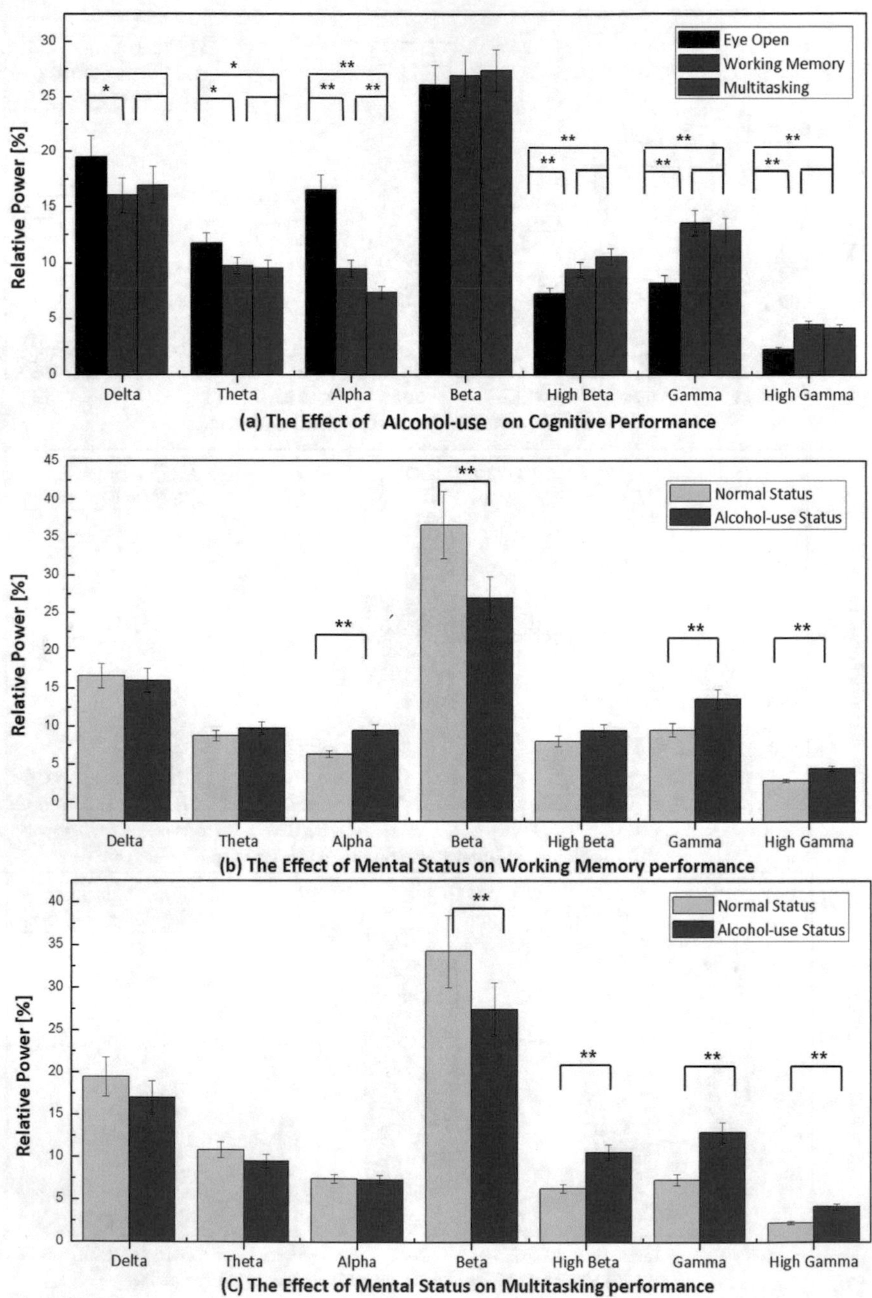

Fig. 5. *The Relative Power differences of Alcohol-use group during cognitive performance* Group mean (±*S.E.*) Absolute Power difference (y-axis) between two groups for seven indicators (x-axis). The significant indicator (P < 0.01 and P < 0.05) differences between the two groups were marked as** and * in the figure

activity helps to improve memory and perception. Despite of sleep-deprived status, when a subject performed the tasks, they increased efforts.

The delta brainwave range is associated with a decreased sense of awareness. In Fig. 3(b–c), delta power increased during the tasks when a subject did not have enough sleep. Alpha and theta band power increased and this result is similar to previous studies [26–31]. According these studies, this implies a decrease in arousal and high cognitive workload.

3.2 The Effect of Alcohol-Use on Cognitive Performance

In Figs. 4 and 5(b–c), the result of alcohol-use are similar to sleep-deprived results. Interestingly, the alcohol use group's beta power decreased and this is associated with decreased stress. This result can be evidence for an argument that drinking alcohol was a good way to release their stress. In Fig. 5(a), opposite to sleepy subjects, delta, theta and alpha band power decreased during multitasking or working memory tasks. This means alcohol-use subjects did not feel high mental workload during the tasks. It implies the alcohol-use employee could work in low concentration. The effects of alcohol-use on performance might be a major safety concern to air transportation, medicine, and industrial. This effect will be more serious than a sleep-deprived worker.

4 Conclusion

This study examined the effect of alcohol-use and sleep deprivation on quantitative EEG variation during working memory and multitasking tasks. Ten subjects' EEG data during cognitive performance was recorded. The statistical analysis was performed using the independent variables (the seven EEG absolute power indicators and the seven EEG relative power indicators). Depending on the cognitive status (Eye Open, Working Memory and Multitasking), this paper investigated the variation of EEG indicators in sleep-deprived and alcohol-use group. In addition, the results from the effect on working memory and multitasking showed that alpha power increased and beta power decreased in both groups. These indicators have a statistically significant difference for unfitness subjects compared to healthy subjects. This result warned the risk of alcohol-use and sleep-deprived workers' job performance. It can be used to investigate the feasibility of determining alcohol-use and sleep quality measurements using EEG indicators, as well as determining a worker's FFD related to cognitive performance.

Acknowledgments. This work was supported by the Nuclear Safety Research Program through the Korea Foundation of Nuclear Safety (KoFONS) using the financial resource granted by the Nuclear Safety and Security Commission(NSSC) of the Republic of Korea. (No. 1703009) This research was also supported by the KUSTAR-KAIST Institute, KAIST. This work was in part supported by the BK21 plus program through the National Research Foundation (NRF) funded by the Ministry of Education of Korea. This research was also supported by Basic Science Research Program through the National Research Foundation of Korea(NRF) funded by the Ministry of Science, ICT & Future Planning (NRF-2016R1A5A1013919).

References

1. Americans with Disabilities Act of 1990, as amended by the ADA Amendments Act of 2008, Pub. L. No. 110–325. http://www.ada.gov/pubs/adastatute08.htm. Accessed 9 Feb 2018
2. United States Nuclear Regulatory Commission: NRC regulations 10 CFR Part 26 Fitness 25 for Duty Programs (2008). https://www.nrc.gov/reading-rm/doc-collections/cfr/part026. Accessed 28 Feb 2018
3. Barnes, V., et al.: Fitness for duty in the nuclear power industry: a review of technical issues. No. NUREG/CR-5227; PNL-6652; BHARC-700/88/018. Nuclear Regulatory Commission, Washington, DC (USA). Div. of Reactor Inspection and Safeguards; Battelle Human Affairs Research Center, Seattle, WA (USA); Pacific Northwest Lab., Richland, WA (USA) (1988)
4. Lee, K., Miller, L., Hardt, F., et al.: Alcohol induced brain damage in young males. Lancet **2**, 759–761 (1979)
5. Parsons, O.A.: Neuropsychological deficits in alcoholics: facts and fancies. Alcoholism **1**, 51–56 (1977)
6. Ryan, C., Butters, N.: Learning and memory impairments in young and old alcoholics: evidence for the premature-aging hypothesis. Alcoholism **4**, 288–293 (1980)
7. Parker, E.S., Noble, E.P.: Alcohol consumption and cognitive functioning in social drinkers. J. Stud. Alcohol **38**, 1224–1232 (1977)
8. Parker, E.S., Bimbaum, I.M., Boyd, R., et al.: Neuropsychological decrements as a function of alcohol intake in male students. Alcoholism **4**, 330–334 (1980)
9. Caldwell Jr., J.A., et al.: The effects of 37 hours of continuous wakefulness on the physiological arousal, cognitive performance, self-reported mood, and simulator flight performance of F-117A pilots. Mil. Psychol. **16**(3), 163 (2004)
10. Lopez, N., et al.: Effects of sleep deprivation on cognitive performance by United States Air Force pilots. J. Appl. Res. Memory Cogn. **1**(1), 27–33 (2012)
11. Rangaswamy, M., Porjesz, B., Chorlian, D.B., Wang, K., Jones, K.A., Bauer, L.O., et al.: Beta power in the EEG of alcoholics. Biol. Psychiatr. **52**, 831–842 (2002)
12. Coutin-Churchman, P., Moreno, R., Anez, Y., Vergara, F.: Clinical correlates of quantitative EEG alterations in alcoholic patients. Clin. Neurophysiol. **117**, 740–751 (2006)
13. Doctcr, R., Niatoh, R., Smith, J.: Electroencephalographic changes and vigilance behavior during experimentally induced intoxication with alcoholic subjects. Psychosom. Med. **28**, 311–315 (1966)
14. Lukas, S.E., Mendelson, J.H., Benedikt, R.A., Joncs, B.: EEG alpha activity increases during transient episodes of ethanol-induced euphoria. Pharmacol. Biochem. Behav. **25**, 889–895 (1986)
15. Cohcn, H.L., Porjesz, B., Begleiter, H.: The effects of ethanol on EEG activity in males at risk for alcoholism. Electroenceph. Clin. Neurophysiol. **86**, 368–376 (1993)
16. Pollock, V.E., Volavka, J., Goodwin, D.W., Mednick, S.A., Gabrielli, W.F., Knop, J., Schulsinger, F.: The EEG after alcohol administration in men at risk for alcoholism. Arch. Gen. Psychiatr. **40**, 857–861 (1983)
17. Papadelis, C., Kourtidou-Papadeli, C., Bamidis, P.D., Chouvarda, I., Koufogiannis, D., Bekiaris, E., Maglaberas, N.: Indicators of sleepiness in an ambulatory EEG study of night driving. In: EMBS Proceedings, New York City, USA, 30 August–3 September 2006, pp. 6201–6204 (2006)
18. Lal, S.K., Craig, P., Boord, L., Kirkup, H., Nguyen, H.: Development of an algorithm for an EEG-based driver fatigue countermeasure device. J. Saf. Res. **34**(3), 321–328 (2003)

19. Corsi-Cabrera, M., Arce, C., Ramos, J., Lorenzo, I., Guevara, M.A.: Time course of reaction time and EEG while performing a vigilance task during total sleep deprivation. Sleep **19**, 563–569 (1996)
20. Corsi-Cabrera, M., Ramos, J., Arce, C., Guevara, M.A., Ponce-de Leon, M., Lorenzo, I.: Changes in the waking EEG as a consequence of sleep and sleep deprivation. Sleep **15**, 550–555 (1992)
21. Luciana, M., Nelson, C.A.: Neurodevelopmental assessment of cognitive function using the Cambridge Neuropsychological Testing Automated Battery (CANTAB): validation and future goals. In: Functional Neuroimaging in Child Psychiatry, pp. 379–397. Cambridge University Press, Cambridge (2000)
22. Fray, P.J., Robbins, T.W., Sahakian, B.J.: Neuorpsychiatyric applications of CANTAB. Int. J. Geriatr. Psychiatr. **11**(4), 329–336 (1996)
23. Oh, Y.J., Lee, Y.H.: Human error identification based on EEG analysis for the introduction of digital devices in nuclear power plants. J. Ergon. Soc. Korea **32**(1), 27–36 (2013)
24. Nuwer, M.: Assessment of digital EEG, quantitative EEG, and EEG brain mapping: report of the American Academy of Neurology and the American Clinical Neurophysiology Society. Neurology **49**(1), 277–292 (1997)
25. Morrison, D.F.: Multivariate Statistical Methods. McGraw-Hill, New York (1967)
26. Boksem, M.A., Meijman, T.F., Lorist, M.M.: Effects of mental fatigue on attention: an ERP study. Cogn. Brain. Res. **25**(1), 107–116 (2005)
27. Klimesh, W.: EEG alpha and theta oscillations reflect cognitive and memory performance: a review and analysis. Brain Res. Brain Res. Rev. **29**, 169–195 (1999)
28. Laufs, H., Kleinschmidt, A., Beyerle, A., Eger, E., Salek-Haddadi, A., Preibisch, C., Krakow, K.: EEG-correlated fMRI of human alpha activity. NeuroImage **19**, 1463–1476 (2003)
29. Oken, B.S., Salinsky, M.: Alertness and attention: basic science and electrophysiologic correlates. J. Clin. Neurophysiol. **9**(4), 480–494 (1992)
30. Paus, T., Zatorre, R.J., Hofle, N., Caramanos, Z., Gotman, J., Petrides, M., Evans, A.C.: Time-related changes in neural systems underlying attention and arousal during the performance of an auditory vigilance task. J. Cogn. Neurosci. **9**(3), 392–408 (1997)
31. Tanaka, H., Hayashi, M., Hori, T.: Topographical characteristics and principal component structure of the hypnagogic EEG. Sleep **20**, 523–534 (1997)

Comparison of Machine Learning Approaches for Motor Imagery Based Optical Brain Computer Interface

Lei Wang[1,2], Adrian Curtin[1,2,3], and Hasan Ayaz[1,2,4,5](✉)

[1] School of Biomedical Engineering, Science & Health Systems, Drexel University, Philadelphia, PA, USA
{lw474, abc48, ayaz}@drexel.edu

[2] Cognitive Neuroengineering and Quantitative Experimental Research (CONQUER) Collaborative, Drexel University, Philadelphia, PA, USA

[3] School of Biomedical Engineering, Shanghai Jiao Tong University, Shanghai, China

[4] Department of Family Medicine and Community Health, University of Pennsylvania, Philadelphia, PA, USA

[5] The Division of General Pediatrics, Children's Hospital of Philadelphia, Philadelphia, PA, USA

Abstract. A Brain-computer Interface (BCI) is a system that interprets specific patterns in human brain activity, such as the intention to perform motor functions, in order to generate a signal which can be used for communication or control. Functional near infrared spectroscopy (fNIRS) is an emerging optical neuroimaging technique which is a relatively new modality for BCI systems. As such, the optimal paradigms and classification techniques for the interpretation of fNIRS-BCI systems is an area of active investigation. Presently, most fNIRS BCIs have adopted Linear Discriminant Analysis (LDA) algorithm as the primary classification approach, however other alternative methods may offer increased performance. In order to compare different algorithms, a dataset from a four-class motor imagery-based fNIRS-BCI study was re-analyzed, and we systematically compared the performance of different machine learning algorithms: Naïve Bayes (NB), LDA, Logistic Regression (LR), Support Vector Machines (SVM) and Multi-layer Perception (MLP). Our findings suggest that the LR classifier slightly outperformed other classifiers, unlike most fNIRS-BCI studies which reported LDA or SVM as the best classifier. The results presented here suggest that an LR classifier could be a potential replacement for LDA classifiers in motor imagery tasks.

Keywords: Brain-Computer Interface (BCI)
Functional near-infrared spectroscopy (fNIRS) · Machine learning
Motor imagery

1 Introduction

A Brain-Computer Interface (BCI) is a method of communication based on neural activity generated by the brain, independent of the normal output pathways by way of peripheral nerves and muscles [1]. The goal of BCI is to provide a new channel of

H. Ayaz and L. Mazur (Eds.): AHFE 2018, AISC 775, pp. 124–134, 2019.
https://doi.org/10.1007/978-3-319-94866-9_12

output for the brain that allows for voluntary adaptive control by the user. Currently, the development of BCIs mainly targets usage as a neurorehabilitation tool to improve motor or cognitive performance for people with motor disorders, such as spinal cord injury, amyotrophic lateral sclerosis (ALS), or people in the persistent locked-in state (LIS) [1–3].

The neural activities used in BCI can be recorded by using either invasive or noninvasive techniques [1, 4]. While, invasive BCIs acquire signals from electrodes surgically implanted in or on the cortex or other brain areas, such as electrocorticographic (ECoG) based BCI [5–7], noninvasive BCIs typically measure brain activity with electroencephalography (EEG) sensors placed on the surface of the scalp [8, 9] or via induction using magnetoencephalography (MEG) [10, 11]. Nonelectrical neurophysiological signals also have been explored as a possible basis for BCI measure, such as the measurement of cortical hemodynamics using techniques like functional magnetic resonance imaging (fMRI) [12–14] and functional near-infrared spectroscopy (fNIRS) [15–19]. Of these two techniques, fNIRS has several distinct features which make it a good candidate for next generation BCIs. fNIRS uses near-infrared light to measure the concentration changes of oxygenated hemoglobin (HbO) and deoxygenated hemoglobin (HbR) [20, 21]. As an optical imaging modality, fNIRS is not susceptible to electrical noise unlike EEG and MEG. Additionally, when compared to fMRI, fNIRS has merits in terms of cost, portability, safety, system noise. It also provides a balanced trade-off between temporal and spatial resolution that sets it apart and presents unique opportunities for investigating new approaches for the development of new BCIs including mental tasks, information content and signal processing [22].

Motor imagery refers to the imagined movement of the body while keeping the muscles motionless, and is sometimes considered to be a conscious elicitation of otherwise unconscious preparation for actual movement. Due to its similarity to naturally produced motor signals and relatively intuitive task performance, motor imagery has been a popular choice for use in BCI studies [15–17, 23–26]. In a recent study, we have compared motor execution and motor imagery for upper and lower limb and demonstrated that fNIRS can be used to capture related brain activity [27]. EEG BCIs have shown success in classifying control signals with up to four classes [28–30]. Other studies have shown potential for EEG to detect difference between right and left foot/leg motor imagery tasks [31, 32] and even individual fingers [33]. Studies have also used fNIRS to detect motor imagery tasks, with many focusing on a left hand versus right hand/leg [19, 34, 35], or the performance of different motor imagery tasks versus rest state [36, 37]. Some studies have implied that a four-class motor imagery based fNIRS-BCI could be feasible with sufficient subject training [15, 17].

With the intention to further investigate the four-class motor imagery based fNIRS-BCI, we systematically compared the performance of several available machine learning algorithms. Previously, features to be used for machine learning has been investigated [38, 39]. Here we compared the performance of machine learning approaches using the same dataset. In the following sections, we first explain the general structure of the comparison framework, the introduced dataset, and finally different algorithms are compared through various performance metrics.

2 Methods

fNIRS data collected by Hitachi ETG-4000 systems for a previously reported motor imagery based BCI study was re-analyzed [15, 16]. During the experiment, subjects were instructed to imagine performing one of the four tasks while refraining from any muscle movement: right hand, left hand, right foot or left foot tapping. Data collected from 11 subjects were used. For each subject, 120 motor imagery tasks were used in the training session and 60 motor imagery tasks in the testing session.

2.1 Preprocessing

All data from training sessions and testing sessions for HbO were filtered using a FIR filter with a 0.1 Hz cutoff frequency. An automatic data-quality analysis was used to determine which optodes and trials should be removed due to poor data quality [40]. Areas with a standard deviation of 0 in a 2-s window of the data were considered to have been saturated, and artifacts were determined to be areas with a change of 0.15 μM during a 2-s period on HbO data after application of the low-pass filter [15]. The average HbO level for the first two seconds of the task was subtracted from 15 s task duration immediately after preprocessing.

2.2 Feature Extraction

A 5 s sliding window (instead of using the entire 15 s task period) was adopted for use in feature extraction. Henceforth, for each task, eleven 5 s segments of HbO data (i.e., 1–5, 2–6, 3–7, …, 11–15 s) were generated. Then the signal slope (SS) and the signal mean (SM) for each 5 s segment were computed. Therefore, feature matrix with 528 columns/features were obtained (11 segments × 2 values × 24 optodes).

Previously removed low quality data in the feature matrix were imputed with the mean of that column and each feature was standardized by removing the mean and scaling to unit variance. Next, principal component analysis (PCA) was performed on the feature space to project it to a lower dimensional space and enable classification algorithms to operate faster and more effectively.

2.3 Classification

In supervised learning, given a set of N training samples $\{(x_1, y_1), \ldots, (x_N, y_N)\}$, such that x_i is the feature vector and y_i is the label of the corresponding feature vector, the learning algorithm seeks a function: $g: X \rightarrow Y$, where X is the input matrix and Y is the output vector corresponding to the label of each row of X.

Naïve Bayes (NB). In probabilistic classification, instead of approximating the function g, we find a posterior probability $\boldsymbol{P}(\boldsymbol{c}|\boldsymbol{x})$, where c is the predicted class label and x is the feature vector, and eventually assign the label to the class with the highest probability. This probability can be calculated using Bayes rule. $\boldsymbol{P}(\boldsymbol{c}|\boldsymbol{x})$ is called the posterior and $\boldsymbol{P}(\boldsymbol{x}|\boldsymbol{c})$ is the class conditional density.

$$P(c|x) = \frac{P(x|c)P(c)}{P(x)}, \tag{1}$$

NB is a simple technique for constructing classifiers: models that assign class labels to problem instances, represented as vectors of feature values, where the class labels are drawn from some finite set. The NB classifier is especially appropriate when the dimension of the feature matrix is high, making density estimation unattractive [41]. NB classifier assumes that the value of a particular feature is independent of the value of any other feature, given the class variable. NB classifiers are generally easy to understand, fast and require only a single pass through the data if all attributes are discrete [42].

Linear Discriminant Analysis (LDA). LDA, the most commonly used classification algorithm in fNIRS-BCI studies, utilizes discriminant hyperplanes to separate data representing two or more classes. It is designed to maximize the difference between the means of different classes while minimizing the variance within classes in this linear projection of the input data. LDA assumes a normal data distribution along with an equal covariance matrix for both classes [43]. In an N-class problem (N > 2), several hyperplanes are used. The strategy generally used for multiclass BCI is the "one versus rest" (OVR) strategy which consists in separating each class from all the others. LDA has very low computational requirements, which makes the technique very popular in BCI applications [2, 43]. This classifier is also simple to use and generally provides good results.

Logistic Regression (LR). LR is a discriminant learning classifier that directly estimates the parameters of the posterior distribution function $P(c|x)$. LR algorithm assumes the distribution $P(c|x)$ is given by (2)

$$P(c = k|x) = \frac{\exp\left(w_k^T x\right)}{\sum_{j=1}^{K} \exp\left(w_j^T x\right)}, \tag{2}$$

where w_j is the parameter to estimate and K is the number of classes. LR models are usually fit by maximum likelihood, using conditional likelihood of $c = k$ given x to directly approximate w_j. As the hessian matrix for the logistic regression model is positive definite, the error function has a unique minimum [44]. Overfitting can occur in logistic regression when the feature matrix is sparse and of high dimensions (which is often the circumstance in BCIs) [44]. Therefore, Lasso Regression (L_1) and Ridge Regression (L_2) regularization were used to cope with the overfitting problem.

Support Vector Machines (SVM). SVM is a discriminate classification algorithm which also uses a discriminant hyperplane to identify classes [45, 46]. The selected hyperplane maximizes the distance (margins) to the nearest data points (support vectors) of each class helping to enhance the generalization capabilities of the classifier [45, 46]. The decision function of SVM is fully specified by a subset of the training data, which leads to a sparse solution for SVM. The cost function of SVM is a convex function that leads to an optimal solution for the optimization task [44].

The mathematical formulation of SVM renders the usage of kernel to map the original finite dimensional space into a destination space with much higher dimensions [44]. The kernel functions make it possible to have SVM with non-linear decision boundaries, such as the Gaussian or radial basis functions (RBF). Non-linear SVM provides a more flexible decision boundary that can result in an increased classification accuracy. The regularization parameter C in SVM allows for accommodating the outliers and therefore reduces errors on the training sets [45], however, the advantages are gained at the expense of a low speed of execution.

Multi-Layer Perception (MLP). The last classifier used is a feedforward neural network [47] with one hidden layer. It has been shown that an MLP with enough number of neurons in the hidden layer can approximate any function. Despite the flexibility and capability of this algorithm to approximate any nonlinear function, MLP algorithm can easily overfit and the cost function to optimize is a non-convex function. In the present study, L_2 regularization was used to avoid overfitting.

2.4 Performance Measure

Classification results are reported as accuracy (average number of correct classifications), macro-precision (positive prediction value: calculate metrics for each class/label, and find their unweighted mean) [48], macro-recall (sensitivity or true positive rate), macro-F-score (the balance between precision and recall) and area under the Receiver Operating Characteristic (ROC) curve (AUC) (Table 1).

Table 1. Measures for multi-class classification

Measure	Evaluation focus
Macro-precision	An average per-class agreement of the data class labels with those of a classifier
Macro-recall	An average per-class effectiveness of a classifier to identify class labels
Macro-F-score	Relations between data's positive labels and those given by a classifier based on a per-class average

3 Results

Table 2 shows the classification accuracy and AUC of each algorithm on the motor imagery based fNIRS BCI data. LR achieved higher average accuracy and AUC value than other classifiers with 31.02% overall accuracy and an average 0.55 AUC value. NB, SVM and MLP reached average accuracy around 30%, while LDA achieved 28.45% average accuracy, and all 4 classifiers reached an AUC of over 0.51. Four subjects (Subject 1, 6, 7 and 12) achieved an accuracy of 30% or higher using all five classifiers. Another two subjects (Subjects 5 and 8) also achieved accuracy over 30% in four classifiers except for LDA. Subject 12 achieved the highest accuracy with LR, 3 subjects (Subjects 7, 8 and 12) had the highest AUC under LR and 2 subjects (Subject 1 and 2) had two ties for the highest with LR and LDA.

Table 2. The accuracy and AUC of classifiers for all subjects

	Accuracy					AUC				
	NB	LDA	LR	SVM	MLP	NB	LDA	LR	SVM	MLP
S 1	30.00	33.33	35.00	36.67	33.33	0.52	0.56	0.56	0.55	0.51
S 2	38.33	25.00	30.00	25.00	25.00	0.54	0.54	0.54	0.50	0.50
S 4	20.00	28.33	26.67	33.33	30.00	0.46	0.50	0.50	0.54	0.51
S 5	40.00	26.67	31.67	33.33	33.33	0.54	0.57	0.56	0.52	0.53
S 6	33.33	30.00	35.00	38.33	33.33	0.55	0.59	0.59	0.60	0.57
S 7	35.00	31.67	33.33	31.67	30.00	0.50	0.53	0.59	0.54	0.52
S 8	30.00	23.33	33.33	33.33	40.00	0.53	0.57	0.66	0.60	0.56
S 9	21.67	31.67	28.33	26.67	30.00	0.47	0.60	0.59	0.55	0.53
S 10	20.00	18.33	21.67	20.00	23.33	0.48	0.45	0.40	0.43	0.47
S 11	31.15	26.23	26.23	21.31	24.59	0.49	0.47	0.48	0.50	0.47
S 12	31.67	38.33	40.00	36.67	30.00	0.52	0.54	0.60	0.58	0.56
Avg.	30.10	28.45	31.02	30.57	30.27	0.51	0.54	0.55	0.54	0.52

The precision, recall and F-score for each subject are detailed in Table 3, five classifier showed similar results in each measure. There is no statistically significant difference between classifiers on accuracy means, as determined by one-way ANOVA ($F(4, 50) = 0.317$, $p = 0.866$). However, paired sample t-test of different classifiers' accuracy revealed: LDA vs. LR: $t(10) = -2.322$, $p = 0.043$. Paired t-test on AUC values revealed: LDA vs. NB: $t(10) = -2.307$, $p = 0.044$; LR vs. NB: $t(10) = -2.317$, $p = 0.043$; LR vs. MLP: $t(10) = 2.287$, $p = 0.045$; SVM vs. MLP, $t(10) = 2.243$, $p = 0.049$.

For subject 1 and subject 6, both showed different patterns when using LR and NB classifiers and have confusion matrices that indicate a different focus on correct classification. The confusion matrix of LR classification results for subject 1 shows a relative strong diagonal pattern, as expected for a well-performing classifier. Interestedly, left hand and right hand are almost never misclassified as the opposite hand, the left hand and left foot task was also seldom misclassified as the other one. In this, LDA and SVM classification also showed patterns similar to LR. On the other hand, the NB classifier showed relative focus on correct classification of left side tasks compared to right side tasks. For subject 6, LR classifier showed stronger right-side tasks accuracy than left side tasks, while NB classifier displayed stronger left side tasks accuracy than right side tasks. LR, LDA and SVM classifier all showed a focus on correct classification on hand tasks compared to foot tasks. The confusion matrices are shown in Fig. 1.

Table 3. The precision, recall and F-score of classifiers for all subjects

	Precision					Recall					F-score				
	NB	LDA	LR	SVM	MLP	NB	LDA	LR	SVM	MLP	NB	LDA	LR	SVM	MLP
S 1	0.36	0.37	0.39	0.40	0.42	0.31	0.35	0.36	0.38	0.34	0.30	0.34	0.36	0.38	0.34
S 2	0.29	0.15	0.16	0.11	0.36	0.34	0.27	0.29	0.23	0.29	0.30	0.17	0.20	0.15	0.19
S 4	0.16	0.25	0.24	0.25	0.30	0.17	0.25	0.25	0.27	0.30	0.16	0.25	0.24	0.25	0.28
S 5	0.35	0.25	0.30	0.33	0.32	0.36	0.26	0.31	0.32	0.32	0.34	0.24	0.29	0.29	0.30
S 6	0.30	0.33	0.36	0.45	0.35	0.31	0.31	0.38	0.41	0.34	0.29	0.30	0.34	0.37	0.33
S 7	0.09	0.15	0.28	0.27	0.36	0.24	0.26	0.31	0.27	0.29	0.13	0.19	0.26	0.25	0.26
S 8	0.30	0.25	0.36	0.37	0.44	0.29	0.26	0.33	0.33	0.36	0.26	0.22	0.31	0.32	0.36
S 9	0.36	0.31	0.20	0.25	0.16	0.18	0.30	0.30	0.28	0.29	0.16	0.28	0.23	0.25	0.20
S 10	0.21	0.16	0.15	0.15	0.21	0.20	0.17	0.19	0.18	0.22	0.19	0.16	0.16	0.15	0.21
S 11	0.32	0.26	0.24	0.20	0.23	0.31	0.24	0.23	0.19	0.22	0.30	0.24	0.23	0.19	0.22
S 12	0.32	0.44	0.42	0.40	0.34	0.31	0.37	0.40	0.37	0.29	0.30	0.36	0.37	0.34	0.27
Avg.	0.28	0.27	0.28	0.29	0.32	0.28	0.28	0.30	0.29	0.30	0.25	0.25	0.27	0.27	0.27

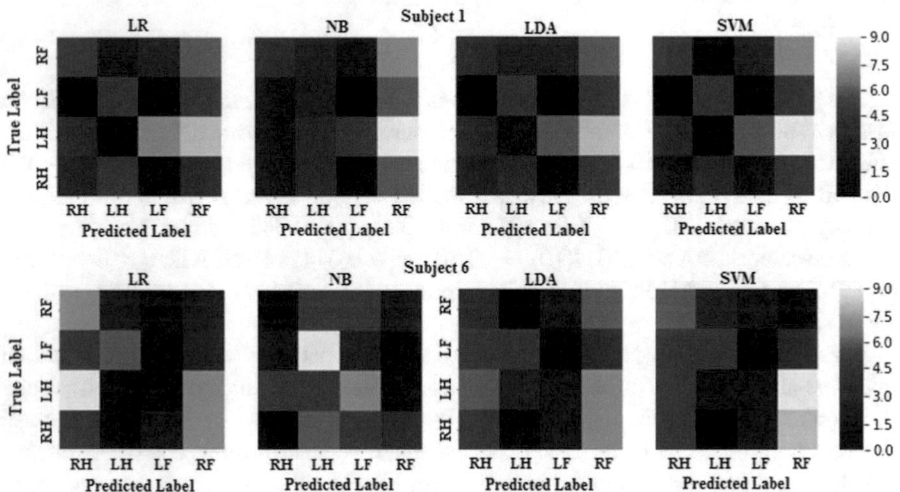

Fig. 1. Confusion matrices for the two subjects showing different pattern between LR, LDA, SVM vs NB classification results. The confusion matrices indicate different behaviors between classifiers: subject 1 and subject 6 show a focus on correct classification on hand task using LR, LDA and SVM classifier, display a focus on correct classification of left side tasks through NB classifier.

4 Discussion

In this work, we present the comparison of different machine learning algorithms for a four-class motor imagery based fNIRS-BCI. There was no significant difference in performance metrics (accuracy, precision, recall and F-score) among compared algorithms, with accuracy ranged 28.45–31.02%, AUC fell within 0.51–0.55. Of these algorithms, the LR classifier reached the highest accuracy and AUC showing that this method could be among the top performing classifiers. In the present study, we noted that the performance of LR was better than LDA. One potential reason for this may be due to the fact that the data in BCI is noisy and sometimes may have outliers. LDA has been reported to be more sensitive to outliers and may therefore not robust as LR [44]. The AUC value of LR and LDA classifiers showed significant difference in pairwise comparison with the NB classifier, but differences between them did not reach significance, which indicates that while LR might be a potential alternative to LDA, NB may not be qualified.

LR, LDA and SVM showed similar patterns in some subjects' confusion matrices, suggesting that LR could be a potential substitute classifier for either LDA or SVM. Even though these methods differ in their basic idea, where LR makes no assumptions on the distribution of the data, LDA has been developed for normally distributed variables [49]. Thus in the case of normality assumptions are fulfilled, LDA is expected to give better results than LR, but in all other situations, LR could be more appropriate than LDA [49]. For sample sizes above 50, the difference between LR and LDA could be negligible [49].

One limitation of the study is the sample size is relative small, only 11 subjects, 180 motor imagery tasks for each subject (training + testing, with 4 classes distributed in training session equally) included in the analysis. Increasing the sample sizes may reveal significant difference among these classifiers. The second limitation is only SM and SS were used for classification, the combination of several other features such as different time window segment, median of the segment and HbR data also could be utilized in feature space, and further enhance the classification performance.

5 Conclusion

We have made comparison on performance of different machine learning approaches for use in four-class motor imagery fNIRS-BCI. With this framework, we compared 5 classifiers: NB, LR, LDA, SVM and MLP on different performance measures: accuracy, precision, recall, F-score and AUC over 11 subjects.

Our result show that the LR is among the best performing classifiers and is recommended for inclusion in similar designs. LR, LDA and SVM (linear kernel) are linear classifiers, while MLP is a powerful nonlinear classifier. Both LR and MLP classifiers are prone to overfitting compared with other classifiers, regularization terms were included to avoid overfitting. The observation that LR was among the best classifiers suggested that the feature space of this four-class motor imagery fNIRS-BCI was somewhat linearly separable.

Unlike most fNIRS-BCI studies, which specifically selected LDA or SVM classifiers, out findings show that LR should also be considered as a candidate classification method in general BCI applications. In general, it is not common for an individual classifier to dramatically outperform all others in every situation, therefore several recommend classifiers should be tested in order to determine the best classifier for a particular application based-on cross validation results.

References

1. Vallabhaneni, A., Wang, T., He, B.: Brain—computer interface. In: Neural Engineering, pp. 85–121. Springer, Boston (2005)
2. Naseer, N., Hong, K.-S.: fNIRS-based brain-computer interfaces: a review. Front. Hum. Neurosci. **9**, 3 (2015)
3. Coyle, S., Ward, T., Markham, C., McDarby, G.: On the suitability of near-infrared (NIR) systems for next-generation brain–computer interfaces. Physiol. Meas. **25**, 815 (2004)
4. Allison, B.Z., Wolpaw, E.W., Wolpaw, J.R.: Brain–computer interface systems: progress and prospects. Expert Rev. Med. Devices **4**, 463–474 (2007)
5. Friehs, G., Penn, R.D., Park, M.C., Goldman, M., Zerris, V.A., Hochberg, L.R., Chen, D., Mukand, J., Donoghue, J.D.: Initial surgical experience with an intracortical microelectrode array for brain-computer interface applications 881. Neurosurgery **59**, 481 (2006)
6. Leuthardt, E.C., Miller, K.J., Schalk, G., Rao, R.P., Ojemann, J.G.: Electrocorticography-based brain computer interface-the Seattle experience. IEEE Trans. Neural Syst. Rehabil. Eng. **14**, 194–198 (2006)
7. Levine, S.P., Huggins, J.E., BeMent, S.L., Kushwaha, R.K., Schuh, L.A., Rohde, M.M., Passaro, E.A., Ross, D.A., Elisevich, K.V., Smith, B.J.: A direct brain interface based on event-related potentials. IEEE Trans. Rehabil. Eng. **8**, 180–185 (2000)
8. Wolpaw, J.R., Birbaumer, N., McFarland, D.J., Pfurtscheller, G., Vaughan, T.M.: Brain–computer interfaces for communication and control. Clin. Neurophysiol. **113**, 767–791 (2002)
9. Wang, D., Miao, D., Blohm, G.: Multi-class motor imagery EEG decoding for brain-computer interfaces. Front. Neurosci. **6**, 151 (2012)
10. Mellinger, J., Schalk, G., Braun, C., Preissl, H., Rosenstiel, W., Birbaumer, N., Kübler, A.: An MEG-based brain–computer interface (BCI). Neuroimage **36**, 581–593 (2007)
11. Birbaumer, N., Murguialday, A.R., Weber, C., Montoya, P.: Neurofeedback and brain–computer interface: clinical applications. Int. Rev. Neurobiol. **86**, 107–117 (2009)
12. Yoo, S.-S., Fairneny, T., Chen, N.-K., Choo, S.-E., Panych, L.P., Park, H., Lee, S.-Y., Jolesz, F.A.: Brain–computer interface using fMRI: spatial navigation by thoughts. Neuroreport **15**, 1591–1595 (2004)
13. Hinterberger, T., Weiskopf, N., Veit, R., Wilhelm, B., Betta, E., Birbaumer, N.: An EEG-driven brain-computer interface combined with functional magnetic resonance imaging (fMRI). IEEE Trans. Biomed. Eng. **51**, 971–974 (2004)
14. Weiskopf, N., Mathiak, K., Bock, S.W., Scharnowski, F., Veit, R., Grodd, W., Goebel, R., Birbaumer, N.: Principles of a brain-computer interface (BCI) based on real-time functional magnetic resonance imaging (fMRI). IEEE Trans. Biomed. Eng. **51**, 966–970 (2004)
15. Batula, A.M., Kim, Y.E., Ayaz, H.: Virtual and actual humanoid robot control with four-class motor-imagery-based optical brain-computer interface. In: BioMed Research International 2017 (2017)
16. Batula, A.M., Mark, J., Kim, Y.E., Ayaz, H.: Developing an optical brain-computer interface for humanoid robot control. In: International Conference on Augmented Cognition, pp. 3–13. Springer, Cham (2016)
17. Batula, A.M., Ayaz, H., Kim, Y.E.: Evaluating a four-class motor-imagery-based optical brain-computer interface. In: 2014 36th Annual International Conference on Engineering in Medicine and Biology Society (EMBC), pp. 2000–2003. IEEE (2014)
18. Matthews, F., Pearlmutter, B.A., Wards, T.E., Soraghan, C., Markham, C.: Hemodynamics for brain-computer interfaces. IEEE Sig. Process. Mag. **25**, 87–94 (2008)

19. Sitaram, R., Zhang, H., Guan, C., Thulasidas, M., Hoshi, Y., Ishikawa, A., Shimizu, K., Birbaumer, N.: Temporal classification of multichannel near-infrared spectroscopy signals of motor imagery for developing a brain–computer interface. NeuroImage **34**, 1416–1427 (2007)
20. Villringer, A., Chance, B.: Non-invasive optical spectroscopy and imaging of human brain function. Trends Neurosci. **20**, 435–442 (1997)
21. Strangman, G., Boas, D.A., Sutton, J.P.: Non-invasive neuroimaging using near-infrared light. Biol. Psychiatr. **52**, 679–693 (2002)
22. Gramann, K., Fairclough, S.H., Zander, T.O., Ayaz, H.: Trends in neuroergonomics. Front. Hum. Neurosci. **11**, 165 (2017)
23. Schlögl, A., Lee, F., Bischof, H., Pfurtscheller, G.: Characterization of four-class motor imagery EEG data for the BCI-competition 2005. J. Neural Eng. **2**, L14 (2005)
24. Townsend, G., Graimann, B., Pfurtscheller, G.: Continuous EEG classification during motor imagery-simulation of an asynchronous BCI. IEEE Trans. Neural Syst. Rehabil. Eng. **12**, 258–265 (2004)
25. Park, C., Looney, D., ur Rehman, N., Ahrabian, A., Mandic, D.P.: Classification of motor imagery BCI using multivariate empirical mode decomposition. IEEE Trans. Neural Syst. Rehabil. Eng. **21**, 10–22 (2013)
26. LaFleur, K., Cassady, K., Doud, A., Shades, K., Rogin, E., He, B.: Quadcopter control in three-dimensional space using a noninvasive motor imagery-based brain–computer interface. J. Neural Eng. **10**, 046003 (2013)
27. Batula, A.M., Mark, J.A., Kim, Y.E., Ayaz, H.: Comparison of brain activation during motor imagery and motor movement using fNIRS. Comput. Intell. Neurosci. **2017**, 12 (2017)
28. Barbosa, A.O., Achanccaray, D.R., Meggiolaro, M.A.: Activation of a mobile robot through a brain computer interface. In: 2010 IEEE International Conference on Robotics and Automation (ICRA), pp. 4815–4821. IEEE (2010)
29. Doud, A.J., Lucas, J.P., Pisansky, M.T., He, B.: Continuous three-dimensional control of a virtual helicopter using a motor imagery based brain-computer interface. PLoS ONE **6**, e26322 (2011)
30. Ge, S., Wang, R., Yu, D.: Classification of four-class motor imagery employing single-channel electroencephalography. PLoS ONE **9**, e98019 (2014)
31. Hashimoto, Y., Ushiba, J.: EEG-based classification of imaginary left and right foot movements using beta rebound. Clin. Neurophysiol. **124**, 2153–2160 (2013)
32. Hsu, W.-C., Lin, L.-F., Chou, C.-W., Hsiao, Y.-T., Liu, Y.-H.: EEG classification of imaginary lower limb stepping movements based on fuzzy support vector machine with kernel-induced membership function. Int. J. Fuzzy Syst. **19**, 566–579 (2017)
33. Stankevich, L., Sonkin, K.: Human-robot interaction using brain-computer interface based on eeg signal decoding. In: International Conference on Interactive Collaborative Robotics, pp. 99–106. Springer, Cham (2016)
34. Naseer, N., Hong, K.-S.: Classification of functional near-infrared spectroscopy signals corresponding to the right-and left-wrist motor imagery for development of a brain–computer interface. Neurosci. Lett. **553**, 84–89 (2013)
35. Shin, J., Jeong, J.: Multiclass classification of hemodynamic responses for performance improvement of functional near-infrared spectroscopy-based brain–computer interface. J. Biomed. Opt. 19, 067009 (2014)
36. Coyle, S.M., Ward, T.E., Markham, C.M.: Brain–computer interface using a simplified functional near-infrared spectroscopy system. J. Neural Eng. **4**, 219 (2007)

37. Ito, T., Akiyama, H., Hirano, T.: Brain machine interface using portable Near-InfraRed spectroscopy—improvement of classification performance based on ICA analysis and self-proliferating LVQ. In: 2013 IEEE/RSJ International Conference on Intelligent Robots and Systems (IROS), pp. 851–858. IEEE (2013)
38. Noori, F.M., Naseer, N., Qureshi, N.K., Nazeer, H., Khan, R.A.: Optimal feature selection from fNIRS signals using genetic algorithms for BCI. Neurosci. Lett. **647**, 61–66 (2017)
39. Naseer, N., Noori, F.M., Qureshi, N.K., Hong, K.S.: Determining optimal feature-combination for LDA classification of functional near-infrared spectroscopy signals in brain-computer interface application. Front. Hum. Neurosci. **10**, 237 (2016)
40. Takizawa, R., Kasai, K., Kawakubo, Y., Marumo, K., Kawasaki, S., Yamasue, H., Fukuda, M.: Reduced frontopolar activation during verbal fluency task in schizophrenia: a multi-channel near-infrared spectroscopy study. Schizophr. Res. **99**, 250–262 (2008)
41. Friedman, J., Hastie, T., Tibshirani, R.: The Elements of Statistical Learning. Springer Series in Statistics. Springer, New York (2001)
42. Kohavi, R.: Scaling up the accuracy of Naive-Bayes classifiers: a decision-tree hybrid accuracy scale-up: the learning. Data Min. Vis. no. Utgo 1988 **7**, 1–6 (1996)
43. Lotte, F., Congedo, M., Lécuyer, A., Lamarche, F., Arnaldi, B.: A review of classification algorithms for EEG-based brain–computer interfaces. J. Neural Eng. **4**, R1 (2007)
44. Bashashati, H., Ward, R.K., Birch, G.E., Bashashati, A.: Comparing different classifiers in sensory motor brain computer interfaces. PLoS ONE **10**, e0129435 (2015)
45. Burges, C.J.: A tutorial on support vector machines for pattern recognition. Data Min. Knowl. Disc. **2**, 121–167 (1998)
46. Bennett, K.P., Campbell, C.: Support vector machines: hype or hallelujah? ACM SIGKDD Explor. Newsl. **2**, 1–13 (2000)
47. Christopher, M.B.: Pattern Recognition and Machine Learning. Springer, New York (2016)
48. Sokolova, M., Lapalme, G.: A systematic analysis of performance measures for classification tasks. Inf. Process. Manag. **45**, 427–437 (2009)
49. Pohar, M., Blas, M., Turk, S.: Comparison of logistic regression and linear discriminant analysis: a simulation study. Metodoloski zvezki **1**, 143 (2004)

Autonomic Nervous System Approach to Measure Physiological Arousal and Scenario Difficulty in Simulation-Based Training Environment

Sinh Bui[1(✉)], Brian Veitch[1], and Sarah Power[1,2]

[1] Faculty of Engineering and Applied Science, Memorial University of Newfoundland, St. John's, NL A1B 3X5, Canada
{sdbui, bveitch, b09sdp}@mun.ca

[2] Faculty of Medicine, Memorial University of Newfoundland, St. John's, NL A1B 3V6, Canada

Abstract. Given the impossibility of exposing trainees to hazardous scenarios for ethical, financial, and logistical reasons, virtual-environment (VE) based simulation training has been adopted in various safety-critical industries. Through simulation, participants can be exposed to a variety of training scenarios to assess their performance under different conditions. Along with performance measures, physiological signals may provide useful information about the trainee's experience of the task. In this study, signals of the autonomic nervous system (ANS), specifically electrocardiogram (ECG), galvanic skin response (GSR), and respiration (RSP), were used to assess physiological arousal levels in 38 participants during 8 different conditions of an emergency evacuation task. On average, neutral and training conditions could be distinguished with a 79.4% average accuracy. In addition, arousal levels in different training scenarios were significantly different, and arousal level was negatively correlated with participant performance. This suggests ANS signals could be a useful measure of the scenario difficulty.

Keywords: Human factors · Physiological signal · Machine learning
Virtual reality · Simulation training

1 Introduction

On-site workers in a variety of disciplines are exposed to hazardous environments, where quick and incisive decisions must be made in the event an emergency situation arises. Therefore, it is critical for such personnel to receive comprehensive training in proper performance of emergency response procedures. However, exposing workers to realistic emergency conditions for the purposes of training is impossible for obvious ethical, financial, and logistical reasons. Therefore, virtual environments (VE) are often used to simulate these conditions, allowing participants to experience emergency

H. Ayaz and L. Mazur (Eds.): AHFE 2018, AISC 775, pp. 135–144, 2019.
https://doi.org/10.1007/978-3-319-94866-9_13

situations without actually being exposed to hazards. It is important that training scenarios be designed to have different difficulty levels in order to help trainees acquire skills effectively, as well as confirm their ability to respond to a variety of situations. Scoring trainees' performance through a rubric is the most common method to assess the difficulty of a particular training scenario. However, this approach is not comprehensive because it only assesses the number and type of errors the individual performs during the scenario, and ignores their cognitive and affective experience of the scenario. For example, a participant could achieve similar performance scores in two different scenarios, but could have experienced significantly different levels of stress and/or mental workload while performing them. This would indicate a difference in difficulty level between the scenarios that performance levels alone would not reveal.

Physiological signal changes - particularly those of the autonomic nervous system (ANS), including electrocardiogram [1–5], galvanic skin response [1–8], and respiration [4, 5] - have been shown to be good indicators of mental workload and stress in a variety of domains. In terms of virtual environment-based simulation training, these measures have been applied to investigate how immersive a virtual environment is [9–11], or the effectiveness of a training simulation in helping trainees better cope with the real situations in terms of mental state. In this study, we investigate the effectiveness of autonomic nervous system signals in providing information on the difficulty of various emergency response training scenarios completed by naive trainees in a simulation-based training environment. The study was conducted using a VE-based program called AVERT (All-hands Virtual Emergency Response Trainer) [11], which was developed for training emergency evacuation procedures for the offshore petroleum industry. AVERT is a software-based training program, where participants use a standard video game controller to direct an avatar through emergency evacuation procedures on a realistic offshore oil platform under various conditions (Fig. 1).

2 Methods

2.1 Participants

Data from 38 participants were collected during the experiment (28 males and 10 females, mean age 28 $\pm$ 9 years). Participants were excluded if they had any prior experience with the experiment procedure, or with the real-life offshore petroleum platform on which it is based. The subjects were asked not to consume alcohol 24 h before the experiment, and refrain from exercise, caffeine, smoking and fasting 2 h before the experimental session to ensure the physiological signals of interest would not be affected. Approval of the experimental protocol was obtained from the appropriate research ethics board at Memorial University of Newfoundland prior to study commencement.

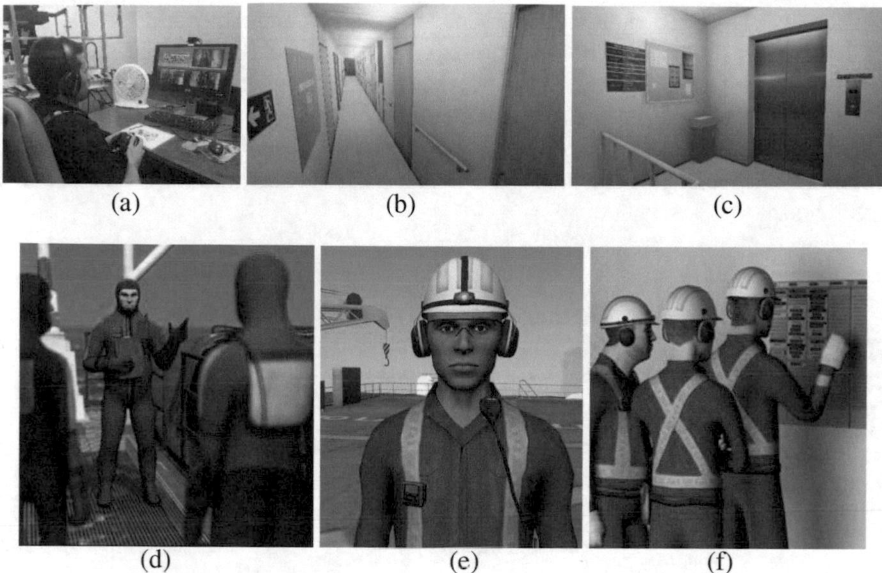

Fig. 1. AVERT training application: (a) participant is using the software (b)(c) simulated spaces on a vessel (d)(e)(f) simulated subjects on a vessel

2.2 Experimental Protocol

The experiment consisted of one session, divided into two phases. In phase 1, the participants were first trained to competence in basic emergency evacuation skills using AVERT. Specifically, they went through a series of modules consisting of instructional material and practice trials to become familiar with tasks such as recognizing various alarms and responding appropriately (e.g., prepare to abandon platform), travelling to an appropriate muster station from their cabin, and selecting appropriate personal protective equipment. Training involved the completion of four scenarios addressing different learning objectives. Participants were required to re-attempt each training scenario until they could complete it error-free before moving on to the next module and training scenario (i.e., a mastery learning approach was taken).

After the participants were trained in the basic evacuation procedures, and demonstrated a minimum level of competence, they were given a short break and then began the second phase of the experiment. In this phase, participants were asked to perform the evacuation procedures they had learned in phase 1, but this time under various new conditions. A 2^3 factorial design [12] was employed: participants completed eight (8) scenarios based on three (3) performance shaping factors (PSFs), each varied at two levels (low and high). A description of the PSFs is given in Table 1. The aim was to create scenarios with different levels of difficulty.

Before each of the eight training scenarios, participants completed a five-minute rest interval in order to give them a break, and to get a baseline measure of their physiological signals. Though there was considerable variation in the duration of scenarios across conditions and participants, the average duration of one training scenario was 237 ± 138 s. The order in which the scenarios were performed was randomized for each participant. Figure 2 shows the trial sequence for phase 2 of the experiment.

Table 1. Factors and levels in the 2^3 factorial design

PSF	Low level	High level
(1) Quality of information received over public announcement (PA) system during the scenario	The PA announcement is clear, concise, and includes all relevant information	The PA announcement is not clear and does not provide sufficient information
(2) Proximity to hazard	There is no hazard (e.g., fire, explosion, smoke)	There is close proximity to hazard (e.g., fire, explosion, smoke)
(3) Familiarity of environment	Scenario starts in familiar location (i.e., from Phase 1), participants take known route, and there is potential for known re-route	Scenario starts in unfamiliar location, there is potential for re-routing based on acquired information

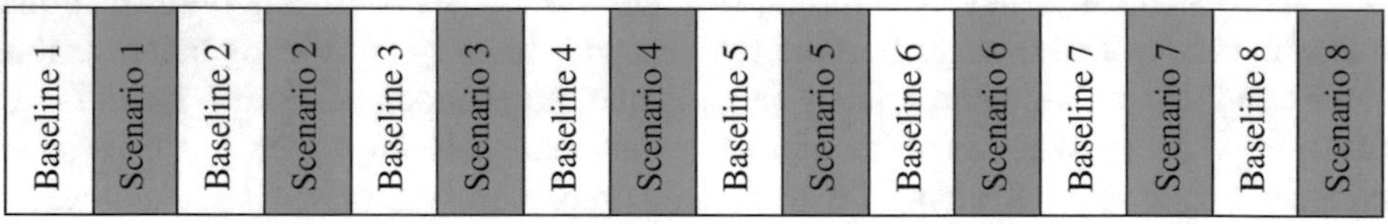

Fig. 2. Experiment baseline-scenario sequence

2.3 Physiological Signals

Three ANS signals - electrocardiogram (ECG), galvanic skin response (GSR), and respiration (RSP) - were collected during both baselines and training scenarios. The signals were collected using the Nexus-10 MarkII data acquisition system with Biotrace+ software. Sampling rates were 256 Hz for ECG and 32 Hz for GSR and RSP. The three ECG electrodes were placed on the left and right chest (just below the clavicles), and just below the last rib on the left side. The two GSR electrodes were placed on the middle phalanxes of the middle and ring fingers. The RSP sensor band was worn around the participant's rib cage. See Fig. 3 for sensor placement.

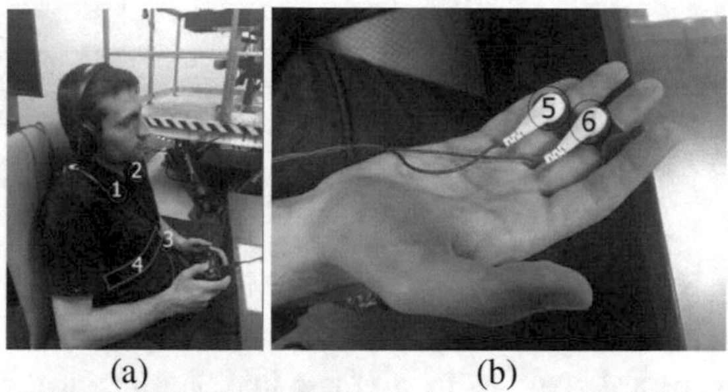

(a) (b)

Fig. 3. Participant performing AVERT scenarios with physiological sensors placed on hands and chest. (a) electrodes 1, 2, 3 – ECG, sensor 4 – RSP belt (b) electrodes 5, 6 – GSR.

2.4 Performance Score

To evaluate each participant's performance, the following information was collected for each scenario:

1. Alarm recognition: did the participant recognize the meanings of different alarm types and react accordingly
2. Identification of mustering announcement: did the participant muster at the correct location and perform the correct task after reaching the muster station (e.g., put on immersion suit);
3. Route selected: which route did the participant take in a given situation, and did they re-route appropriately when a hazard (e.g., fire, smoke) was encountered;
4. Observation of general safety rules: did the participant close all safety doors, and walk, not run, on the platform.

Based on this performance data, a performance score was calculated for each participant in each scenario.

2.5 Subjective Rating of Stress/Anxiety

The participant's subjective rating of stress/anxiety experienced during the performance of each scenario was recorded. In particular, following each trial, participants were asked the question: "How did you feel during the scenario you just completed?" and were asked to provide a rating from 1 (very relaxed) to 7 (very stressed).

3 Data Analysis

As an indicator of the level of stress or workload experienced by each participant in each of the eight scenarios, we used the classification accuracy of each scenario versus the baseline interval that preceded it. The baseline was assumed to represent a low arousal state (e.g., low stress). The higher the classification accuracy between a scenario and the baseline condition, then the more the physiological signals changed during the scenario, indicating a higher arousal state (e.g., high stress/workload). We

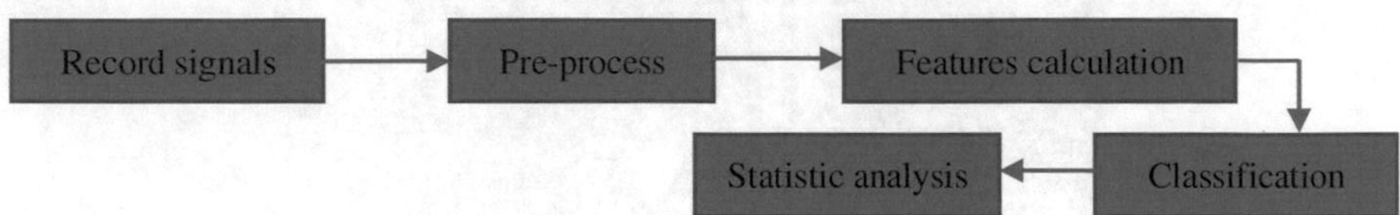

Fig. 4. Data analysis process

then performed statistical analyses to compare this physiological measure of arousal (e.g., stress/workload) with the participants' subjective stress ratings, as well as their performance scores (Fig. 4).

3.1 Signal Pre-processing

The ECG, GSR, and RSP signals were first pre-processed to remove unwanted noise and to prepare them for feature extraction and classification. The ECG signal was first filtered by a 5–15 Hz [13] 3rd-order Butterworth bandpass filter. The GSR signal was put through a 2nd-order Chebyshev lowpass filter with cutoff frequency of 1 Hz. The RSP signal was detrended to eliminate any linear trend, then smoothed to facilitate the respiration rate calculation.

3.2 Feature Calculation

After pre-processing, all signals were segmented into 3-s intervals for feature extraction. All features were calculated from these 3-s segments. From the pre-processed

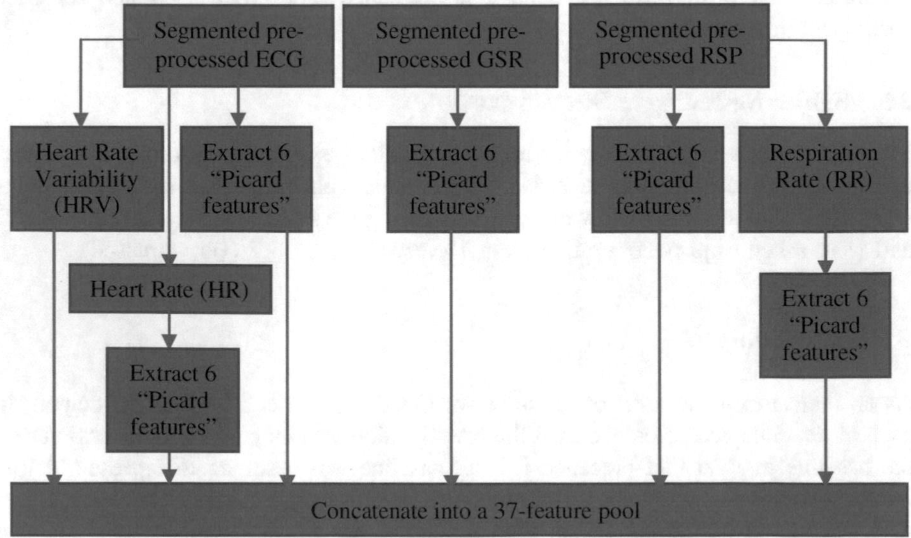

Fig. 5. Features extraction process

ECG, a heart rate (HR) signal was calculated using the modified Pan-Tompkins algorithm [13]. From this HR signal, seven different features of heart rate variability (HRV) were calculated [14]: (1) VLF (power in very low frequency range), (2) LF (power in low frequency range), (3) LF norm (LF power normalized), (4) HF (power in high frequency range), (5) HF norm (HF power normalized), (6) LF/HF, and 7) RMSSD (the square root of the mean of the sum of the squares of differences between adjacent normal-to-normal intervals).

From the pre-processed RSP signal, a respiration rate (RR) signal was calculated via a peak detection method developed by the authors. The overall feature pool considered for classification consisted of the seven HRV measures, plus the following six characteristics calculated based on Picard et al.'s method [15] for the pre-processed ECG, GSR, and RSP signals, as well as for the calculated HR and RR signals: (1) mean of the signal, (2) standard deviation of the signal, (3) mean of the absolute value of the first difference of the signal, (4) mean of the absolute value of the first difference of the normalized signal, (5) mean of the absolute value of the second difference of the signal, and (6) the mean of the absolute value of the second difference of the normalized signal. The resulting feature pool comprised 37 features. Figure 5 depicts the feature extraction process.

3.3 Classification

A support vector machine (SVM) algorithm with Gaussian kernel was employed to classify the data between each baseline and scenario for every participant individually. 30 runs of 5-fold cross validation were performed for each baseline versus scenario condition and the average adjusted accuracy across runs was calculated.

3.4 Statistical Analysis

After the three measures of interest were calculated (performance score, subjective stress rating, and classification accuracy), factor analysis was conducted to explore which performance shaping factors or interactions had significant effects on the responses, which would indicate the difficulty of the scenarios. Additionally, repeated measures ANOVA was implemented to investigate whether there was a significant difference in responses among scenarios, which would indicate that different scenarios led to different difficulty levels. Finally, correlation analysis was employed to find any relationships among the subjects' performance scores, physiological arousal levels (as indicated by classification accuracy between baseline and scenarios), and subjective measures of stress (Table 2).

4 Results and Discussion

In all scenarios, the classification accuracy of the ANS signals were significantly different from chance (mean 79.4%), indicating that all scenarios were effective in eliciting a physiological response. Furthermore, the classification accuracies were significantly correlated with the subjective ratings of stress ($\rho = 0.353$, $p = 0.0296$),

Table 2. Results summary

Data analysis process	Results
Classification between baseline and scenario	Average adjusted accuracy of 79.4%
Factors showing significant effects of responses	Only one factor – familiarity of the environment
Repeated ANOVA analysis	Significant difference found among scenarios in terms of performance score and classification accuracy (indicating physiological arousal), but not in subjective ratings
Correlation findings	Significant correlations found between: • Performance score and classification accuracy • Subjective ratings and classification accuracy

supporting their use as a measure of emotional arousal. Results from the factor analysis indicate that familiarity with the environment was the only factor that had significant effect on the difficulty of the task; this factor showed a significant effect on both the trainees' performance scores ($p < 0.0001$) and their level of physiological arousal ($p < 0.0001$), but not on the subjective ratings of stress. There were no significant interaction effects seen in any of the measures. Also, repeated measures ANOVA showed significant differences in both performance score and physiological arousal among scenarios ($p = 0.0001$ and $p = 10^{-7}$, respectively), further indicating that there were differences in the difficulty of the scenarios. Interestingly, again no significant difference was seen for the subjective ratings of stress, indicating that physiological signals may provide a more reliable measure of stress than do subjective ratings. In addition, arousal level was negatively correlated with participant performance ($\rho = -0.320$, $p = 0.0500$), suggesting that the more stressed the subjects felt, the worse they performed [9]. It is unclear if this indicates that trainees were performing badly because they were stressed, or that they were stressed because they were performing badly.

At the beginning of the experiment design, we predicted that all three performance shaping factors would have significant effect on the responses. However, the results showed that only familiarity with the environment significantly affected the participants' performance and their level of physiological arousal or stress. There might be several reasons for this. For example, the gap between high and low levels of PSFs 1 (quality of information received during scenario) and 2 (proximity to hazard) may not have been significant enough to make a difference; or those two factors might merely not be significant and should be ignored when designing training scenarios in the future. Further research should be conducted to confirm which one was the true reason.

Among the three measures of interest, performance score was objective and concise because it is based on a clear rubric applied to the performance data. Physiological arousal, as indicated by the classification accuracy of physiological signals between baseline and scenario intervals, was correlated with participant performance, supporting its use as a measure of scenario difficulty. On the other hand, subjective measure of stress was shown to be an ineffective indicator of difficulty, as it did not correlate with performance. This is not surprising, as such subjective ratings are known to be unreliable for a number of reasons, including subjective nature of self-evaluation, and inconsistency from person to person and also within a person [16]. Therefore, an objective measure of stress like physiological arousal may be helpful in estimating participants' feelings while doing virtual environment - or simulation-based training. For example, it could be used as an additional indicator of competency in the trained skills to complement the performance measure. In addition, such a measure could be a tool to evaluate a person's capability to work offshore. For example, different participants could derive the same results in emergency scenarios in terms of performance; their stress levels, however, could be much different. Because human failures are highly correlated to stress [17, 18], trainees who are more prone to this problem have potential to perform worse in real condition, where the emergency is completely real and they could feel the danger vividly. Therefore, applying stress detection in training might be a solution for organizations who need to choose people with solid performance during critical situations. Finally, the physiological arousal measurement could also be incorporated into VR applications to monitor users' feelings in real-time, allowing for modification of the scenarios accordingly to enhance each individual's experience while using VR applications.

5 Conclusion

Our results suggest that classification accuracy between physiological data collected during a training scenario and that collected during baseline can be a useful measure of the difficulty experienced by the trainee in a given training scenario to complement performance measures. Furthermore, physiological signals may be more reliable indicators of stress than subjective ratings. The findings from this research might be useful in a number of applications.

Acknowledgements. The authors acknowledge with gratitude the support of the NSERC/Husky Energy Industrial Research Chair in Safety at Sea.

References

1. Xu, Q., Nwe, T.L., Guan, C.: Cluster-based analysis for personalized stress evaluation using physiological signals. IEEE J. Biomed. Health Inform. **19**(1), 275–281 (2015)
2. Ollander, S., Godin, C., Campagne, A., Charbonnier, S.: A comparison of wearable and stationary sensors for stress detection. In: IEEE International Conference on Systems, Man, and Cybernetics, SMC 2016, 9–12 October, Budapest, Hungary (2016)
3. Sandulescu, V., Sally, A., Ellis, D., Bellotto, N., Mozos, O.M.: Stress detection wearable physiological sensors. Springer, Cham (2015). https://doi.org/10.1007/978-3-319-18914-7_55
4. Vries, J.J.G., Pauws, S.C., Biehl, M.: Insightful stress detection from physiology modalities using learning vector quantization. Neurocomputing **151**(Part 2), 873–882 (2015)
5. Sioni, R., Chittaro, L.: Stress detection using physiological sensors. IEEE Comput. **48**(10) (2015). https://doi.org/10.1109/MC.2015.316

6. Sano, A., Picard, R.W.: Stress recognition using wearable sensors and mobile phones. In: HUMAINE Association Conference on Affective Computing and Intelligent Interaction (2013). https://doi.org/10.1109/acii.2013.117
7. Zubair, M., Yoon, C., Kim, H., Kim, J., Kim, J.: Smart wearable band for stress detection. In: 5th International Conference on IT Convergence and Security (ICITCS), pp. 1–4 (2015)
8. Panju, S., Brian, J., Dupuis, A., Anagnostou, E., Kushki A.: Atypical sympathetic arousal in children with autism spectrum disorder and its association with anxiety symptomatology. Molecular Autism (2015). https://doi.org/10.1186/s13229-015-0057-5
9. Patton, D., Gamble, K.: Physiological measures of arousal during soldier-relevant tasks performed in a simulated environment. Springer, Cham (2016). https://doi.org/10.1007/978-3-319-3955-3_35
10. Lackey, S.J., Salcedo, J.N., Szalma, J.L., Hancock, P.A.: The stress and workload of virtual reality training: the effects of presence, immersion and flow. Ergonomics **59**(8), 1060–1072 (2016). https://doi.org/10.1080/00140139.2015.1122234
11. House, A.W.H., Smith, J., MacKinnon, S., Veitch, B.: Interactive simulation for training offshore workers. In: Proceedings of Oceans, St. John's (2014)
12. Montgomery, D.: Design and Analysis Of Experiments, 8th edn. Wiley, Hoboken (2013)
13. Pan, J., Tompkins, W.J.: A real-time QRS detection algorithm. IEEE Trans. Biomed. Eng. **BME-32**(3), 230–236 (1985)
14. Malik, M., Bigger, J.T., Camm, A.J., Kleiger, R.E., Malliani, A., Moss, A.J., Schwartz, P.J.: Eur. Heart J. **17**(3), 354–381 (1996). https://doi.org/10.1093/oxfordjournals.eurheartj.a014868Z
15. Picard, R.W., Vyzas, E., Healey, J.: Toward machine emotional intelligence: analysis of affective physiological state. IEEE Trans. Pattern Anal. Mach. Learn. **23**(10), 1175–1191 (2001)
16. Garcia-Ceja, E., Osmani, V., Mayora, O.: Automatic stress detection in working environments from smartphones' accelerometer data: a first step. IEEE J. Biomed. Health Inform. **20**(4), 1053–1060 (2016)
17. Cohen, S.: Aftereffects of stress on human performance and social behavior: a review of research and theory. Psychol. Bull. **88**(1), 82–108 (1980)
18. Robert, G., Hockey, J.: Compensatory control in the regulation of human performance under stress and high workload: a cognitive-energetical framework. Elsevier Biolog. Psychol. **45**(1–3), 73–93 (1997)

An Investigation of Human Error Identification Based on Bio-monitoring System (EEG and ECG Analysis)

Jung Hwan Kim, Young-A Suh, and Man-Sung Yim(✉)

Nuclear Energy Environment and Nuclear Security Laboratory, Department of Nuclear and Quantum Engineering, Korea Advanced Institute of Science and Technology (KAIST), 291, Daehak-ro, Yuseong-gu, Daejeon, Republic of Korea
{poxc, dreameryounga, msyim}@kaist.ac.kr

Abstract. Human error has been a critical issue at Nuclear Power Plants (NPP) as it accounts for a significant proportion of safety-related incidents. The objective of this research is to investigate the feasibility of using a bio-monitoring system (EEG and ECG) to predict and thereby minimize the risk of human error at NPPs. Ten subjects (8 male 2 female) with a mean age of 25 years participated in the experiment. Specifically, the Stroop test was used to measure each participant's accuracy in judgement and reaction time, in answering congruent and incongruent questions. Using these data, both heart rate and brain waves were recorded and analyzed via a power spectrum analysis, EEG and ECG indicators were investigated to determine their potential for identifying human error.

Keywords: Human error · Nuclear Power Plants · Bio-monitoring system EEG · ECG

1 Introduction

According to a study on human factors engineering, human error is the main cause for accidents in the workplace [1]. Human error causes not only severe economic loss [2] but also severe damage in various fields such as nuclear, petrochemistry industry, aviation, railroad, and marine [3, 4].

Human error has been a critical issue at nuclear power plants (NPP) as it accounts for a significant portion of safety-related incidents [5]. According to the literature, Three Mile Island, Chernobyl and Fukushima Daiichi nuclear power plant accidents can be classified as examples of human error [6]. There are four types of representative human errors that can occur in the NPP workplace; observation failure, task omission, information chaos, and hierarchical information memory failure [7]. Therefore, detecting and predicting human error can assist in preventing incidents at NPP facilities.

Recently, many researchers have studied bio-signals to understand the phenomenon of human information processing. In addition, electroencephalography (EEG), electrocardiogram (ECG), electromyography (EMG), galvanic skin response (GSR) are the

H. Ayaz and L. Mazur (Eds.): AHFE 2018, AISC 775, pp. 145–151, 2019.
https://doi.org/10.1007/978-3-319-94866-9_14

predominant bio-signals used [8]. The research focused on processing human error using EEG [3, 6, 9]. However, there appears to be a lack of research on detecting or predicting human error at NPP facilities, right before an incident occurs, using both EEG and ECG.

To overcome this problem, this research examines the possibility of using bio-signals to detect human error. EEG and ECG bio-signals were recorded and analyzed. EEG and ECG signals were selected since it is difficult for subjects to manipulate their signals and wearable devices are available at a reasonable price.

The objective of this research is to investigate the feasibility of using a bio-monitoring system (EEG and ECG) to predict and thereby minimize the risk of human error at NPPs.

2 Methods

This study records EEG and ECG signals during a simulated test protocol. Ten university student subjects (8 male 2 female) with a mean age of 25 years participated in the experiment. Prior to testing, we explained the experimental procedure to the subjects, as required by the KAIST Institutional Review Board (IRB) guideline. The subjects read an information sheet and signed an agreement regarding the data collection process. To set the parameters defining the Normal group, prior to testing, all participants were required to sleep more than 6 h and not to drink caffeine or alcohol for at least 24 h. Because EEG and ECG data collection is sensitive to light and sound, the experimental environment blocked both light and sound.

Using, the Stroop test each participant's accuracy in judgement and reaction time when answering congruent and incongruent questions was measured. The Stroop Color-Word Test is a reliable assessment tool used in psychology [9]. The Stroop test was designed to examine and provide a more comprehensive examination of selective attention [11]. When participants selected an answer to a test questions, their pre and post answer EEG and ECG signals were recorded.

Using the BrainMaster Discovery 24E wearable and portable device, each subjects EEG signals were recorded through 24 channels (A1, A2, C3, CZ, C4, F3, FZ, F4, F6, F7, O1, OZ, O2, P3, PZ, P4, T3, T4, T5, T6, FP1, FPZ, FP2, GND). Our frequency domain analysis (quantitative EEG analysis (qEEG)) resulted in categorizing the data into 14 categories. Using Multivariate analysis of variance (MANOVA), six categories were identified as key to determining the potential for human error. These six categories and their ranges were: High Beta (25.0–30.0 Hz), Gamma (30.0–40.0 Hz), Alpha 2 (10.0–12.0 Hz), Beta 1 (12.0–25.0 Hz), Gamma 1 (30.0–35.0 Hz), and Gamma 2 (35.0–40.0 Hz).

ECG data collection relied a lie detector polygraph to measure body reactions, such as heart rate, breathing, blood pressure and galvanic skin response. Again using MANOVA these data were analyzed to identify the five categories key to predicting human error. These categories are: heart rate, low frequency (LF, 0.04–0.15 Hz), high frequency (HF, 0.15–0.4 Hz), standard deviation of the normal-to-normal interval (SDNN), and the square root of the Mean Squared Differences of successive normal-to-normal intervals (RMSSD).

3 Results

This research evaluated: High Beta, Gamma, Alpha 2, Beta 1, Gamma 1, and Gamma 2 to analyze EEG. Alpha waves relate to the subject's relaxation state. In Fig. 1 we see the power of Alpha 2 decreased significantly for the Wrong answer (human error) versus the Correct answer. This means, the subjects were more relaxed and stable when there was no human error.

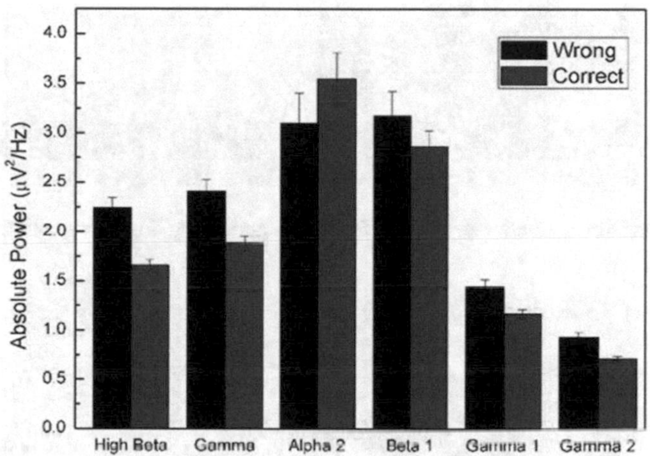

Fig. 1. Comparison of absolute power of EEG indicators during the Wrong and Correct answers

Beta waves relate to the subject's stress level. In Fig. 1 we see the power of Beta increased significantly for the Wrong answer (human error) versus the Correct answer. This means, the subjects were more tensed and anxious when there was human error.

Gamma waves relate to visual stimulation and movement [12]. In Fig. 1 we see the power of Gamma increased significantly for the Wrong answer (human error) versus the Correct answer. This means, the subject's feelings were heightened in response to their human error. Moreover, their bodies had what appeared to be a startled reaction at the same time.

Theta waves relate to the subject's level of stress or mental state [13]. As can be seen in Fig. 2, the relative power of Theta increased significantly for the Wrong answer (human error) versus the Correct answer. This means, the subjects were stressed when there was human error.

The Theta/Alpha ratio relates to the subject's stress or fatigue. As seen in Fig. 3, the Theta/Alpha ratio increased significantly for the Wrong answer (human error) versus the Correct answer. This means, the subject's level of stress increased and their relaxation decreased when there was human error.

The ECG data collection consisted of; heart rate, low frequency (LF), high frequency (HF), standard deviation of the normal-to-normal interval (SDNN), and the square root of the Mean Squared Differences of successive normal-to-normal intervals

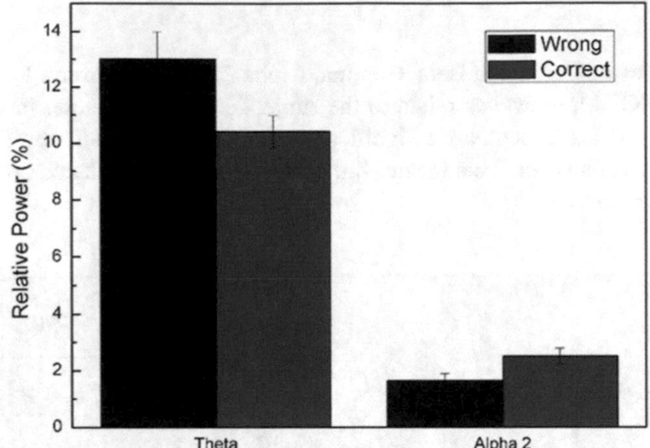

Fig. 2. Comparison of relative power of EEG indicators during the Wrong and Correct answers

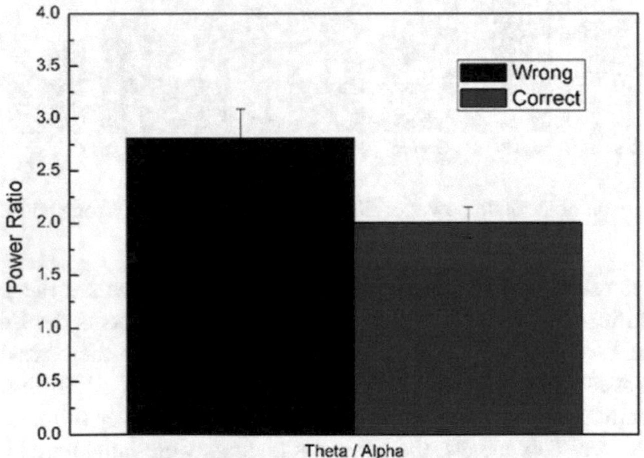

Fig. 3. Comparison of power ratio of EEG indicator during the Wrong and Correct answers

(RMSSD). ECG parameters can be analyzed by various methods, including time domain and frequency domain analyses.

Typically, SDNN and RMSSD are ECG parameters used to conduct a time domain analysis. SDNN includes all the elements that can contribute to heart rate variability. Since the SDNN value increases as heart rate variability becomes more irregular, this trend reflects a physiologically healthy state.

For more accurate short term analysis of HRV data, use of RMSSD is recommended. Typically, HRV parameters such as SDNN and RMSSD are transformed to log scale to display the data in a normal distribution.

LF and HF are representative parameters of the HRV Frequency domain analysis. In general, the HF represents a power value between 0.15 and 0.4 Hz, whereas the LF represents a power value between 0.04 and 0.15 Hz. HF is an indicator of the activity of the parasympathetic nervous system. In addition, there is a close relationship between HF and heart stability. Generally, heart rate variability increases as HF increases, and vice versa.

For short-term measurements, LF represents both sympathetic activity and parasympathetic activity simultaneously. LF is associated with mental stress. Commonly, HRV parameters such as HF and LF are transformed to log scale to display the data in a normal distribution.

As seen in the Table 1, both lnSDNN and lnRMSSD increased significantly. The lnSDNN and lnRMSSD values increase as heart rate variability becomes more irregular. Typically, the heart rate becomes more irregular when the subject selects a Wrong answer (human error). As a result, when human error occurs, we expect the subject's lnSDNN and lnRMSSD to increase.

Table 1. Comparison of HRV indicators during the Wrong and Correct answers

	Wrong (n = 10) (M ± SD)	Correct (n = 10) (M ± SD)	P-value
Mean heart rate (/min)	77.8 ± 2.6	80.0 ± 0.4	0.405
lnLF (log ms^2)	7.2 ± 0.7	6.3 ± 1.3	<0.0003
lnHF (log ms^2)	7.7 ± 0.8	6.7 ± 1.3	<0.00004
lnSDNN (log ms)	−0.9 ± 1.2	−1.7 ± 1.3	0.001
lnRMSSD (log ms)	−0.8 ± 1.1	−1.6 ± 1.2	0.001

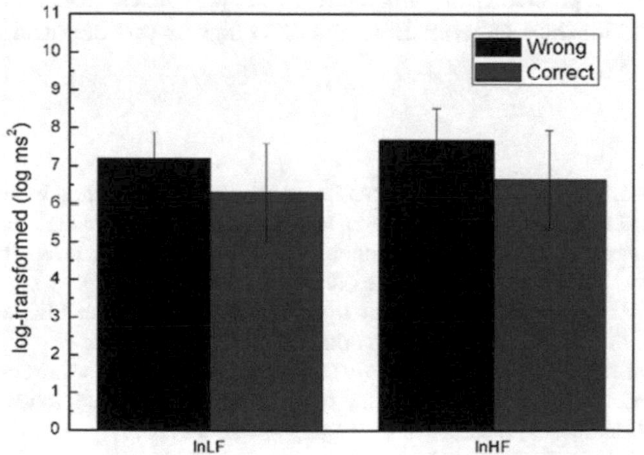

Fig. 4. Comparison of ECG indicators during the Wrong and Correct answers

Finally, when human error occurred, both lnLF and lnHF increased significantly. The lnLF represents the state of a subject's sympathetic nervous system. The subject's sympathetic nervous system became excited when human error occurred.

lnHF is related to subject's high emotional state and physical stress [14]. The subject was high emotional state when human error occurred.

4 Conclusion

This research investigated the use of EEG and ECG to identify human error. EEG and ECG data from 10 subjects were classified and analyzed. The subjects wore a 24 channel BrainMaster to measure their brain response, and were connected to a lie detector polygraph that measured their heart responses.

The Stroop test measured each participant's accuracy in judgement and reaction time when answering congruent and incongruent questions. When participants answered questions (correct or wrong), their EEG and ECG were recorded and analyzed. These data points were tested and validated, using a MANOVA statistical analysis.

Three different types of EEG indicators were used to identify human error: (1) the absolute power of Alpha, Beta, Gamma; (2) the relative power of Theta; (3) the power ratio of Theta/Alpha.

In addition, two different types of ECG indicators were analyzed: (1) the time domain analysis of lnSDNN and lnRMSSD; (2) the frequency domain analysis of lnLF and lnHF.

These indicators are useful in that they provide opportunities to identify human error.

Future work in predicting of human error, will apply machine learning. In addition, meaningful relationships between EEG and ECG signals will be identified.

References

1. Feyer, A.-M., Williamson, A.M.: Human factors in accident modelling. In: Encyclopaedia of Occupational Health and Safety, 4 th edn. International Labour Organisation, Geneva (1998)
2. Liu, H., Hwang, S.-L., Liu, T.-H.: Economic assessment of human errors in manufacturing environment. Saf. Sci. **47**(2), 170–182 (2009)
3. Chiang, Y.-C.: Predicting human error in industrial operation with EEG and data mining techniques. State University of New York. Buffalo (2011)
4. Alkhaldi, M., Pathirage, C., Kulatunga, U.: The role of human error in accidents within oil and gas industry in Bahrain. In: 13th International Postgraduate Research Conference (IPGRC): Conference Proceedings. University of Salford (2017)
5. Lee, D.-H., Byun, S.-N., Lee, Y.-H.: Short-term human factors engineering measures for minimizing human error in nuclear power facilities. J. Ergon. Soc. Korea **26**(4), 121–125 (2007)
6. Vucicevic, J.: Human Error–Crucial Factor in Nuclear Accidents. IAEA reports
7. Oh, Y.J., Lee, Y.H.: Human error identification based on EEG analysis for the introduction of digital devices in nuclear power plants. J. Ergon. Soc. Korea **32**(1), 27–36 (2013)

8. Stanford, V.: Biosignals offer potential for direct interfaces and health monitoring. IEEE Pervasive Comput. **3**(1), 99–103 (2004)
9. Holroyd, C.B., Coles, M.G.H.: The neural basis of human error processing: reinforcement learning, dopamine, and the error-related negativity. Psychol. Rev. **109**(4), 679 (2002)
10. Van der Elst, W., et al.: The Stroop color-word test: influence of age, sex, and education; and normative data for a large sample across the adult age range. Assessment **13**(1), 62–79 (2006)
11. Hiatt, K.D., Schmitt, W.A., Newman, J.P.: Stroop tasks reveal abnormal selective attention among psychopathic offenders. Neuropsychology **18**(1), 50 (2004)
12. Amo, C., et al.: Analysis of gamma-band activity from human eeg using empirical mode decomposition. Sensors **17**(5), 989 (2017)
13. Subhani, A.R., Xia, L., Malik, A.S.: EEG signals to measure mental stress. In: 2011 2nd International Conference on Behavioral, Cognitive and Psychological Sciences, Maldives (2012)
14. Kliszczewicz, B., et al.: Venipuncture procedure affects heart rate variability and chronotropic response. Pacing Clin. Electrophysiol. **40**(10), 1080–1086 (2017)

Neural Correlates of Math Anxiety of Consumer Choices on Price Promotions

Amanda Sargent[1(✉)], Atahan Agrali[1], Siddharth Bhatt[2], Hongjun Ye[2], Kurtulus Izzetoglu[1], Banu Onaral[1], Hasan Ayaz[1,3,4], and Rajneesh Suri[2]

[1] School of Biomedical Engineering, Science and Health Systems, Drexel University, 3141 Chestnut St, Philadelphia, PA 19104, USA
{as3625, saa76, ki25, bo26, ha45}@drexel.edu
[2] LeBow College of Business, Drexel University, 3141 Chestnut St, Philadelphia, PA 19104, USA
{shb56, hy368, surir}@drexel.edu
[3] Department of Family Medicine and Community Health, University of Pennsylvania, 3101 Walnut St, Philadelphia, PA 19104, USA
[4] Division of General Pediatrics, Children's Hospital of Philadelphia, 3401 Civic Center Blvd, Philadelphia, PA 19104, USA

Abstract. Math anxiety is a problem that faces most people in their everyday life. Past research suggests that math anxiety can lead consumers to make errors in numerical computations. In this study, we utilized brain-based measures for the assessment of math anxiety in relationship to price perceptions. Functional near-infrared spectroscopy (fNIRS) from the prefrontal cortex was measured to determine differences in responses between high and low math anxiety participants. Participants performed two tasks under load and no-load conditions performing math calculations in relation to a price promotion. Preliminary results indicate that there is a performance difference between low and high math anxiety participants in both oxygenated hemoglobin results and behavioral performance. This study outlines a new method for determining how math anxiety affects consumers' decisions regarding prices.

Keywords: Anxiety · Math anxiety · Functional near infrared spectroscopy Problem solving · Pricing · Promotions

1 Introduction

Anxiety is a construct that can be defined by fear or dread and causes feelings of uncertainty [1]. Math anxiety disrupts cognitive processing by compromising activity in working memory [2]. Math anxiety is a real phobia however math is something that is unavoidable in our everyday life [2]. Around 93% of Americans believe that they suffer from math anxiety to some degree [3]. People have a strong tendency to avoid math especially those who have math-anxiety [1–5]. People who suffer from math anxiety tend to have substantial performance differences because of it [6]. For people with math anxiety, having to do math problems or even attend a math class can trigger

H. Ayaz and L. Mazur (Eds.): AHFE 2018, AISC 775, pp. 152–160, 2019.
https://doi.org/10.1007/978-3-319-94866-9_15

a negative response, but that's not all. Activities such as reviewing a store receipt can send people into a panic [4]. When performing math computations, math anxious children showed reduced activity in the brain regions which support working memory and numerical processing such as the dorsolateral prefrontal cortex [7].

Most judgments and decisions rely on mathematical ability or skill especially when trying to decide to buy an item, however little research has been done to determine the role of numerical ability in decision making [8]. Retail promotions are used widely in advertising and marketing campaigns as a way to boost product sales [9–11]. People are more motivated to purchase an item when they believe they are getting a better deal or discount. Krishna et al. provide a meta-analysis of 20 publications on how price presentation affects consumers' perceived savings and thereby influences their probability to purchase a certain product [12]. Math anxiety can have an effect on the buyer's decision to purchase an item especially when a sale or promotion is involved [13]. Consumers can be sometimes misled by the way that the promotion or savings is presented to them [10, 11, 14]. For example, consumers may be informed of a "50% discount" or "1/2 price offer", and although these two promotions are the same, one may have a larger impact on the consumer and their behaviors [10].

Functional near-infrared spectroscopy (fNIRS) has emerged over the last decade as a new technique to measure brain activity non-invasively [15]. fNIRS uses near infrared light to monitor oxygenated and deoxygenated hemoglobin changes at the outer cortex [16]. It is a safe and portable device that can be used while wearers complete tasks in everyday environments [17]. Wearable sensors that include light sources and detectors are used over the scalp that shines photons to the tissue and detect a fraction of them that return back after going through the layers of tissue [18]. fNIRS systems use multiple wavelengths within the near infrared range 700–900 nm. The light sources project the light through the scalp and records the optical density fluctuations from metabolic changes that are due to neural activity through a mechanism called neurovascular coupling [19]. Overall, fNIRS can measure cortical regions such as prefrontal cortex or motor cortex relevant to the research interest such as cognitive or motor tasks. Signal processing algorithms specific to fNIRS have been developed to identify, eliminate and/or compensate for physiological or system noises [18]. fNIRS is relatively more resilient to motion artifacts compared to fMRI, allowing monitoring brain activity in ecologically valid natural environments [20].

Past research suggests that math anxiety not only increases consumers' tendencies to make computational errors but also influences their ability to make optimal numerical judgments [8, 21–23]. Math anxiety is a widespread problem that leads to poor math skills and we must do what we can to find ways to alleviate it [4, 6, 24, 25] However, little is known about how math anxiety affects consumers' evaluations of price promotions. This research examines the neural correlates and behavioral performance of consumers as they make judgments about price promotions. Using both behavioral and neural measures this research will provide a deeper understanding of the effect of math anxiety and numerical ability on day-to-day computation decisions.

In this study, we use fNIRS to assess the neural correlates of math anxiety under situations of low and high cognitive loads. Our aim is also to investigate if fNIRS can capture pre-frontal activation changes associated with math anxiety when consumers evaluate price promotions [6]. This is, to our knowledge, the first attempt to incorporate a neurophysiological approach to assess computation performance differences associated with anxiety and ability in an ecologically valid manner using price offers commonly used when promoting prices.

2 Method

2.1 Participants

Eighteen participants between the ages of 19 and 43 (8 females, mean age = 23 years) volunteered for the study. All confirmed that they met the eligibility requirements of being right-handed with vision correctable to 20/20, did not have a history of brain injury or psychological disorder, and were not on medication affecting brain activity. Prior to the study all participants signed consent forms approved by the Institutional Review Board of Drexel University.

2.2 Experimental Procedure

All participants completed three different surveys before the experiment day. These are the sMARS (short Mathematics Anxiety Rating Scale) [26], the need for cognition scale (NCS) [27] and the ETS mathematical proficiency test. The sMARS is a short version of MARS [28], which evaluates the level of apprehension and anxiety of people when they confront with a situation that is related to mathematics or mathematical calculations.

After arriving to the test site, participants got a full explanation from the experimenter, and then sat in a dimly lit room in front of the computer monitor and keyboard. Participants were asked to sit comfortably using the right-hand mouse and interact with the computer. The instructions were presented on the monitor and read aloud by the experimenter before the protocol started. The test was continued only after the participants agreed that they fully understand the instructions. After finishing the experimental protocol, the participants were compensated for their time with $15 for each hour they participated in the experiment.

The protocol consisted of two blocks (no load and with load) and each block consist of 18 trials. Participants completed two practice trials before starting each block. Before each trial, there was a fixation crosshair (+) shown on the screen for 15000 ms. Price (in dollar form) and rental days as a multiplier were given as variables. On the center of the display, below the variable information, two discount options were presented. First option was in dollar ($) form and the second one was in percentage (%) form (Fig. 1). Participants were asked to select the option that gives the highest discount value or the lowest total price.

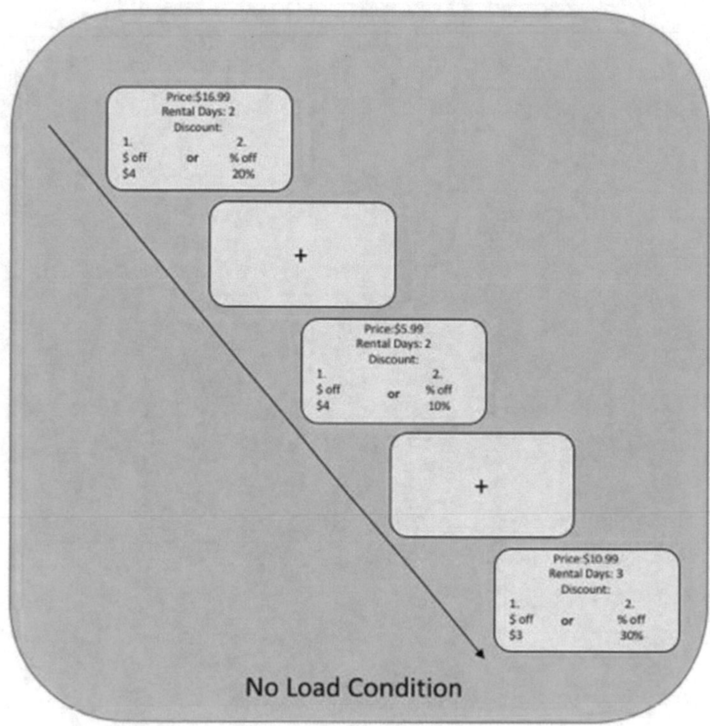

Fig. 1. No load condition shows the question with the dollar off and percentage off followed by the inter-stimulus time.

On block 2 (with load condition), each trial started with a display of five alphanumeric characters on the screen for 4000 ms. Participants were asked to memorize these alphanumeric characters during the discount task. After completion of each discount trial, three different sets of alphanumeric characters were shown on the computer screen. Participants were asked to recognize the set with the same characters that they were shown at the beginning of thc trial (Fig. 2).

2.3 fNIRS Acquisition and Analysis

Prefrontal cortex hemodynamics were measured using a continuous wave fNIRS system (fNIR Devices LLC, fnirdevices.com) that was described previously [29]. Light intensity at two near-infrared wavelengths of 730 and 850 nm was recorded at 2 Hz using COBI Studio software [30]. The headband system contains four LEDs and ten photodetectors for a total of sixteen optodes. Data was passed through a finite impulse response hamming filter of order 20 and cutoff frequency 0.1 Hz. Time synchronized blocks for each trial were processed with the Modified Beer-Lambert Law to calculate oxygenation for each optode. Statistics were calculated using repeated measures ANOVA.

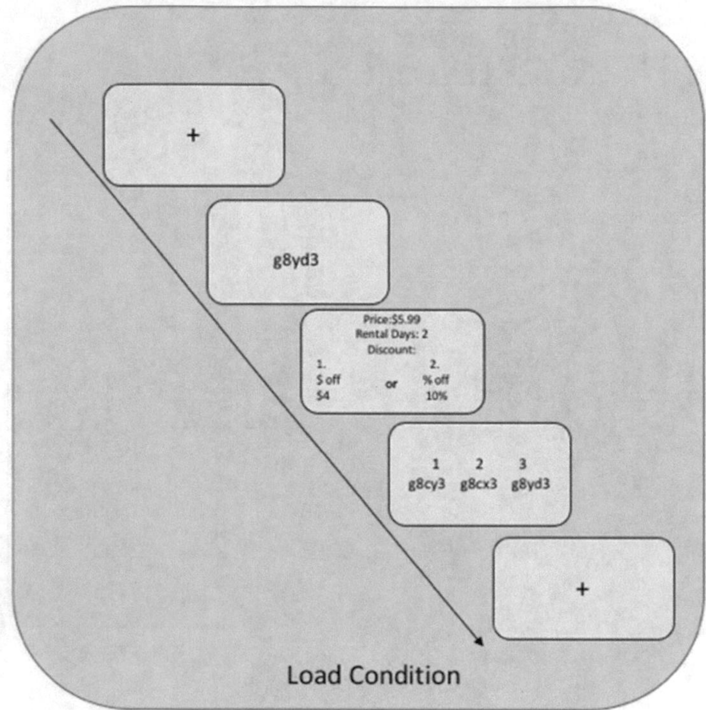

Fig. 2. Load condition shows alphanumeric characters followed by discount trial, followed by recall of the alphanumeric characters.

3 Results

Each participant performed the task under no load and load conditions. The results for the behavioral performance and prefrontal cortex hemodynamics is summarized below.

While analyzing the behavioral results we developed a composite score, 'Behavioral Efficiency'(BE), which is calculated by dividing the accuracy percentage of a participants to their average response times. BE increases as the number of correct trials increases and the average response times decrease. A participant is more efficient when they respond quickly and correctly. We compared three different variables based on the score of the sMARS math anxiety test, the need for cognition scale and the math competency test (math proficiency).

Behavioral efficiency for the load and no-load conditions can be seen in Table 1. We compared two variables: math anxiety levels ($F_{1,16}$ = .13, $p < 0.718$),) and need for cognition (NC) levels ($F_{1,16}$ = 1.62, $p < 0.22$). None of the results were significant.

Table 1. Behavioral efficiency. Comparison of math anxiety levels shows less math anxious participants had higher efficiency in both the no load and with load conditions. Comparison of NC shows overall that with load the performance is improving.

	Low anxiety	High anxiety	Low NC	High NC
No load	76.9 ± 14.2	70.9 ± 8.5	73.8 ± 14.5	73.9 ± 8.3
With load	108.6 ± 10.5	89.5 ± 7.6	109.4 ± 9.1	88.8 ± 9.1

Behavioral results for the load and no-load conditions average response times can be seen in Table 2. We compared two variables: math anxiety levels ($F1,16 = .58$, $p < 0.457$) and need for cognition levels ($F1,16 = 2.41$, $p < 0.14$). None of the results were significant.

Table 2. Response time. Comparison of math anxiety levels shows both groups had faster response times (s) in the load condition and slower response time in the no load condition. Comparison of NC shows high-NC participants had faster response rate in the no load condition compared to the low-NC group.

	Low anxiety	High anxiety	Low NC	High NC
No load	15.2 ± 2.9 s	15.2 ± 2.5 s	16.5 ± 2.9 s	13.9 ± 2.5 s
With load	10.3 ± 1.0 s	14.1 ± 2.6 s	9.9 ± 1.0 s	14.5 ± 2.6 s

Pre-frontal cortex hemodynamics results are summarized in Fig. 3. for the load and no-load conditions between low and high math anxiety levels. For the left hemisphere ($F1, 16 = 0.98$, $p < 0.337$), and right hemisphere ($F1, 16 = 0.01$, $p < 0.948$) there was no significant difference overall.

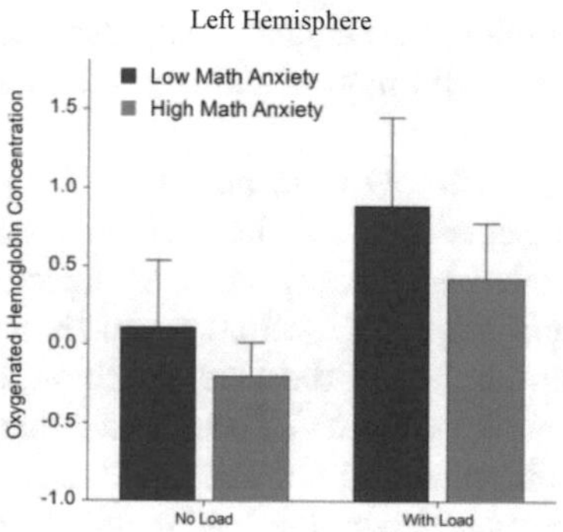

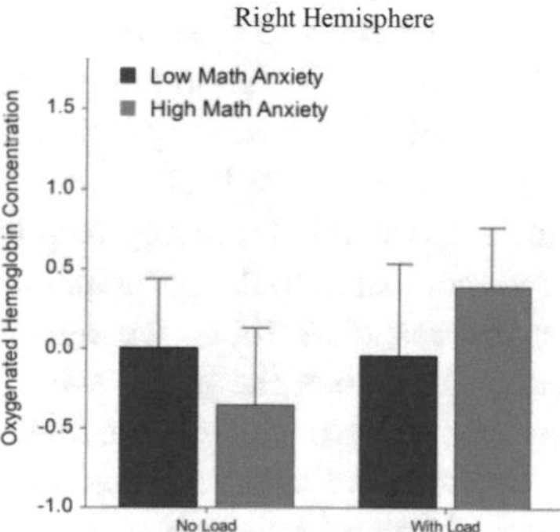

Fig. 3. Pre-frontal cortex oxygenation changes show under the no load condition and load condition that less math anxious participants had higher levels of oxygenated hemoglobin in the left hemisphere (left) where more math anxious participants had higher levels of oxygenated hemoglobin in the right hemisphere for the load condition (right). Whiskers are standard error of the mean (SEM).

Finally, pre-frontal cortex hemodynamics results are summarized in Fig. 4 for the load and no-load conditions between low and high NC. fNIRS based oxygenated hemoglobin changes did not have a significant effect in the left hemisphere between the low and high NFC groups ($F_{1,16} = 0.57$, $p < 0.462$). fNIRS based oxygenated hemoglobin changes also did not have a significant effect in the right hemisphere between the low and high NFC groups ($F_{1,16} = 0.61$, $p < 0.445$).

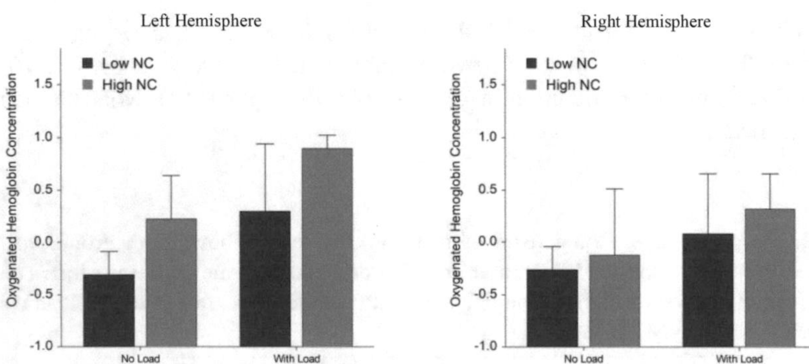

Fig. 4. Pre-frontal cortex oxygenation changes show in the left hemisphere (left) and the right hemisphere (right) the high need for cognition group had higher brain activation in both conditions. Whiskers are standard error of the mean (SEM).

4 Discussion

In this study, we have investigated the relationship between math anxiety and price promotions. Participants performed numerical computations under load and no-load conditions while their prefrontal cortex was monitored with fNIRS. For the no-load condition, participants were simply asked to determine which promotion was better, a dollar off or percentage deal. For the load condition, participants were also asked to remember a five-digit alphanumeric sequence before promotion and recall it after they responded to the promotion question.

Behavioral results indicated that participants with low math anxiety had a higher behavioral efficiency (accuracy/response rate) score in both the load and no-load conditions compared to the participants with a high math anxiety score, as expected [6].

For the response time, we again saw that the group with low math anxiety responded faster than the group with high math anxiety in the load condition, as we expected. In all groups, we see that both groups performed faster under the load condition compared to the no load condition, which we did not expect. This suggests that the participants were more engaged and motivated with the load condition of the task. They are not only faster but made the same number of mistakes under the load condition.

Prefrontal cortex hemodynamics results showed that participants with lower levels of math anxiety have higher levels of brain activation in the left hemisphere where the more math anxious participants had higher levels of activation in the right hemisphere.

In the need for cognition groups, we see that both groups have higher activation in the no load and load conditions, with the high need for cognition group having higher activation in both hemispheres, as we expected. This indicates that the high need for cognition group is requiring more effort to complete the questions.

In conclusion, this study demonstrated it is possible to detect math anxiety levels in brain activation using fNIRS. The study described here provides important albeit preliminary information about fNIRS measures of the PFC hemodynamic response and its relationship to math anxiety, pricing promotions, and performance in two different cognitive load scenarios. The level of math anxiety does appear to influence the hemodynamic response in the prefrontal cortices. Since fNIRS technology allows the development of miniaturized and wearable sensors, it has the potential to be utilized for monitoring while participants are in a real-world environment where consumers have to determine which promotions are best for making a purchase.

References

1. Hembree, R.: The nature, effects, and relief of mathematics anxiety. J. Res. Math. Educ. **21** (1), 33–46 (1990)
2. Ashcraft, M.: Math anxiety: personal, educational, and cognitive consequences. Curr. Dir. Psychol. Sci. **11**(5), 181–185 (2002)
3. Blazer, C.: Strategies for reducing math anxiety. Inf. Capsule **1102**, 1–8 (2011)
4. Maloney, E., Beilock, S.: Math anxiety: who has it, why it develops, and how to guard against it. Trends Cogn. Sci. **16**(8), 404–406 (2012)
5. Wang, Z., et al.: Who is afraid of math? Two sources of genetic variance for mathematical anxiety. J. Child Psychol. Psychiatr. **55**(9), 1056–1064 (2014)
6. Ashcraft, M., Kirk, E.: The relationships among working memory, math anxiety, and performance. J. Exp. Psychol. **130**(2), 224–237 (2001)
7. Young, C.B., Wu, S.S., Menon, V.: The neurodevelopmental basis of math anxiety. Psychol. Sci. **23**(5), 492–501 (2012)
8. Peters, E., Vastfjall, D., Slovic, P., Mertz, C.K., Mazzocco, K., Dickert, S.: Numeracy and decision making. Psychol. Sci. **17**(5), 407–413 (2006)
9. Bayer, R.-C., Ke, C.: Discounts and consumer search behavior: the role of framing. J. Econ. Psychol. **39**, 215–224 (2013)
10. McKechnie, J.D., Ennew, C., Smith, A.: Effects of discount framing in comparative price advertising. Eur. J. Mark. **46**(11/12), 1501–1522 (2012)
11. Darke, P.R., Chung, C.M.Y.: Effects of pricing and promotion on consumer perceptions: it depends on how you frame it. J. Retail. **81**(1), 35–47 (2005)
12. Krishna, A., Briesch, R., Lehmann, D., Yuan, H.: A meta-analysis of price presentation on perceived savings. J. Retail. **78**, 101–118 (2002)
13. Jones, W.J., Childers, T.L., Jiang, Y.: The shopping brain: math anxiety modulates brain responses to buying decisions. Biol. Psychol. **89**(1), 201–213 (2012)
14. Lowe, B., Maxwell, S.: Consumer perceptions of extra free product promotions and discounts: the moderating role of perceived performance risk. J. Prod. Brand Manag. **19**(7), 496–503 (2010)
15. Izzetoglu, K., Ayaz, H., Merzagora, A., Izzetoglu, M., Shewokis, P.A., Bunce, S.C., Pourrezaei, K., Rosen, A., Onaral, B.: The evolution of field deployable fNIR spectroscopy from bench to clinical settings. J. Innov. Opt. Health Sci. **4**, 1–12 (2011)

16. Ayaz, H., Shewokis, P.A., Bunce, S., Onaral, B.: An optical brain computer interface for environmental control. In: IEEE Engineering in Medicine and Biology Society, EMBC, pp. 6327–6330 (2011)
17. Ayaz, H., Onaral, B., Izzetoglu, K., Shewokis, P.A., McKendrick, R., Parasuraman, R.: Continuous monitoring of brain dynamics with functional near infrared spectroscopy as a tool for neuroergonomic research: empirical examples and a technological development. Front. Hum. Neurosci. **7**, 871 (2013)
18. Izzetoglu, M., Izzetoglu, K., Bunce, S., Ayaz, H., Devaraj, A., Onaral, B., Pourezzaei, K.: Functional near-infrared neuroimaging. IEEE Trans. Neural Syst. Rehabil. Eng. **12**(2), 153–159 (2005)
19. Fishburn, F.A., Norr, M.E., Medvedev, A.V., Vaidya, C.J.: Sensitivity of fNIRS to cognitive state and load. Front. Hum. Neurosci. **8**, 76 (2014)
20. McKendrick, R., et al.: Into the wild: neuroergonomic differentiation of hand-held and augmented reality wearable displays during outdoor navigation with functional near infrared spectroscopy. Front. Hum. Neurosci. **10**, 216 (2016)
21. Suri, R., Monroe, K.B., Koc, U.: Math anxiety and its effects on consumers' preference for price promotion formats. J. Acad. Mark. Sci. **41**(3), 271–282 (2012)
22. Gamliel, E., Herstein, R.: To save or to lose: does framing price promotion affect consumers' purchase intentions? J. Consum. Mark. **28**(2), 152–158 (2011)
23. Diamond, W.: Just what is a "dollar's worth"? consumer reactions to price discounts vs. extra product promotions. J. Retail. **68**(3), 254–270 (1992)
24. Beilock, S., Willingham, D.: Math anxiety: can teachers help students reduce it? Am. Educ. **38**, 28–32 (2014)
25. Ashcraft, M., Krause, J.: Working memory, math performance and math anxiety. Psychon. Bull. Rev. **14**(2), 243–248 (2007)
26. Alexander, L., Martray, C.R.: The development of an abbreviated version of the mathematics anxiety rating scale. Meas. Eval. Couns. Dev. **22**, 143–150 (1989)
27. Cacioppo, J.T., Petty, R.E.: The need for cognition. J. Pers. Soc. Psychol. **42**, 116–131 (1982)
28. Richardson, F.C., Suinn, R.M.: The mathematics anxiety rating scale. J. Couns. Psychol. **19**, 551–554 (1972)
29. Ayaz, H., Shewokis, P.A., Bunce, S., Izzetoglu, K., Willems, B., Onaral, B.: Optical brain monitoring for operator training and mental workload assessment. NeuroImage **59**(1), 36–47 (2012)
30. Ayaz, H., Shewokis, P.A., Curtin, A., Izzetoglu, M., Izzetoglu, K., Onaral, B.: Using MazeSuite and functional near infrared spectroscopy to study learning in spatial navigation. J. Vis. Exp. (56), 3443 (2011)

Research on Visual Sensing Capability Difference Between Designers and Non-designers Based on Difference Threshold Measurement

Sha Liu(✉), Runqi Li, and Jiaxin Dai

College of Engineering, China Agricultural University, 17 Qing Hua Dong Lu, Hai Dian District, Beijing 100083, China
shashaday@126.com, lius02@cau.edu.cn, 176536182@qq.com

Abstract. It is often found that designers are more sensitive to subtle differences in shape and forms. it seems easier for them to find these subtle differences than ordinary people. Can some of the designer's professional training enhance people's visual sensing capability? Based on this hypothesis, this study attempts to measure visual difference thresholds of two groups of design majors and non design majors. In the study, two groups of subjects were randomly selected among the 18–25 year old college students. A group of 20 students who had received design training for at least half a year (Design-group for short, DG). The other group of 25 students who did not have any design training (Non-design group for short, NDG). Then, based on the method of constant stimuli, a cell phone icon was chosen as standard stimulus whose size varied according to the percentage change. The subjects would compare and judge the change of the icon's size and the visual difference thresholds of the two groups were aquired. Finally, we conducted a questionnaire survey on all the subjects to investigate the types of visual training that they had received to figure out the possible factors which might enhance the designers' visual sensing capability.

Keywords: Difference threshold · Visual sensing capability · Designers The method of constant stimuli

1 Introduction

It is often found that designers are more sensitive to subtle differences in shape and forms. It seems easier for them to find these subtle differences than ordinary people. In other words, designers have a stronger visual sensing capability than ordinary people. Although designers generally show the trait, there is no empirical research that validates the phenomenon, let alone explores how the designers' stronger visual sensing capability is formed. Is the designers' stronger visual sensing capability is congenital or acquired by training, or the designer's professional habits which lead them more serious viewing the object and the visual perception is enhanced? These problems remain to be verified.

H. Ayaz and L. Mazur (Eds.): AHFE 2018, AISC 775, pp. 161–168, 2019.
https://doi.org/10.1007/978-3-319-94866-9_16

This research attempted to measure visual difference threshold of two groups of college students in design majors and non-design majors by the method of constant stimuli to explore whether design training may lead to the better visual sensing capability. In addition, a questionnaire survey was conducted to explore the possible causes of stronger visual sensing capability and the possible training methods to enhance it.

Sensory threshold is the core concept of the traditional psychophysics. Sensory threshold include two types: absolute threshold and difference threshold. Sensory Threshold and sensitivity are inversely proportional [1, 2]. The level of sensitivity represents the sensing capability. The lower the threshold value is, the higher the sensitivity and the stronger the sensing capability is. Vice versa. The formula is expressed as:

$$E = 1/R. \tag{1}$$

E represents the absolute sensitivity and R represents the absolute sensory threshold. The measurement of the absolute threshold is to measure the Minimum stimulation magnitude which just can be noticed. The measurement of difference threshold is to measure the difference between the magnitude of stimuli which just can cause a sense of difference.

The measurements of thresholds are based on the psychophysical methods (Fechner 1860), which include 3 basic methods:

1. The method of minimal-change, which is also known as the method of limits. That is, the stimulus are changed with the small and equal intervals in ascending or descending order and they are presented to the subject in turn to measure their thresholds.
2. The method of average error. That is, the subject adjust the stimulus himself and make it equal to the standard stimulus and the threshold value is achieved by the mean value of the difference between the adjusted stimulus and the standard stimulus.
3. The method of constant stimuli. That is: use a few (5–10) stimuli which remain unchanged during the experiment. Each stimulus was given the same number of times and the subjects were asked to respond to the stimulation. Then the threshold value was calculated by the proportion of stimuli that the subjects noticed [3, 4]. Psychologists proposed the definition of threshold measurement based on statistical principles: The absolute threshold is the stimuli can be noticed in 50% experiment times. The difference threshold is the difference between the stimuli which can be felt in 50% experiment times [5].

In the nineteenth Century, the psychophysical pioneers Ernst Weber and G. T. Fechner had begun measure thresholds by the psychophysical methods and put forward the “Weber-Fischer’s Law” according to these experiments. In recent years, measuring visual thresholds based the psychophysical methods has become one of the basic methods to research the relationship between visual sense and forms. Huang Xinfu used the method of limits to measure the difference threshold between the area of the four shapes and proposed the conclusion that the few edges the shape had, the bigger people felt the shape [6]. Yang Gangjun and Yu Suihuai et al. studied how to ensure the visual proportion of the furniture form base on the difference threshold measurement. In their experiments, they changed the size of furniture parts with different proportion several times at equal intervals as the stimuli and measured the difference threshold of the furniture parts' visual proportion. They proposed that the size change which was less than 1/5 can keep the visual proportion of furniture parts being felt unchanged [7]. Wu Jiaqian et al. used the limit method to measure the difference threshold of the proportion difference of geometries. Based on Weber's law, a method of calculating the Weber's coefficient of the visual difference threshold of the proportion difference of geometries was proposed [8, 9].

These previous studies provide useful cues to our research: using the simple form/color as stimulus and measure human's thresholds can quantify human's perception ability and effectively explore the visual law of some specific visual elements or some specific population.

Based on the above, our study will use the common simple graphics as stimuli and select designers and non-designer groups as subjects. Base on The method of constant stimuli, we will change the size of the standard stimulus in accordance with the proportion of less than 1/5. The difference threshold (ΔI) between the standard stimuli(I_0) and the comparative stimuli($I_{1\sim n}$) will be measured and then discuss whether there is a significant difference between the visual sensing capabilities of the two groups. If there is the significant difference between the visual sensing capabilities of the two groups, the questionnaire will be used to explore the possible causes of the differences.

2 The Experimental Design

2.1 The Experimental Preparation

Subjects: A total of 45 undergraduates were randomly selected, including 19 design majors and 26 non design/art majors. All subjects are aged 18 to 25 years old, with normal visual acuity or corrected visual acuity and they are all the first time to participate in such experiment.

Experiment Preparation: We chose a 4.7-in smart phone as the experimental device. The screen kept full-screen displaying. According to the laws in human factors engineering, we set up observation port to control the visual distance of 50 cm and the subjects could face the center of the screen. We chose a simple and clear phone icon as the stimulus. The phone screen background was treated as a uniform 50% medium gray and the background of the icon was single white for easier to identify.

2.2 The Experimental Process

As shown in Fig. 1, we placed the white icon in the prominent position of the screen. In the experiment, the white bottom icon, centered at the middle point, was resized to 92% (I_1), 94%(I_2), 96%(I_3), 98%(I_4), 100%(I_5), 102%(I_6), 104%(I_7), 106%(I_8), 108%(I_9) of the original size(I_0). And then we showed a pair of icons (one has the original size, the another has variable size) to the subjects one after another. The order of presenting the pairs was random.

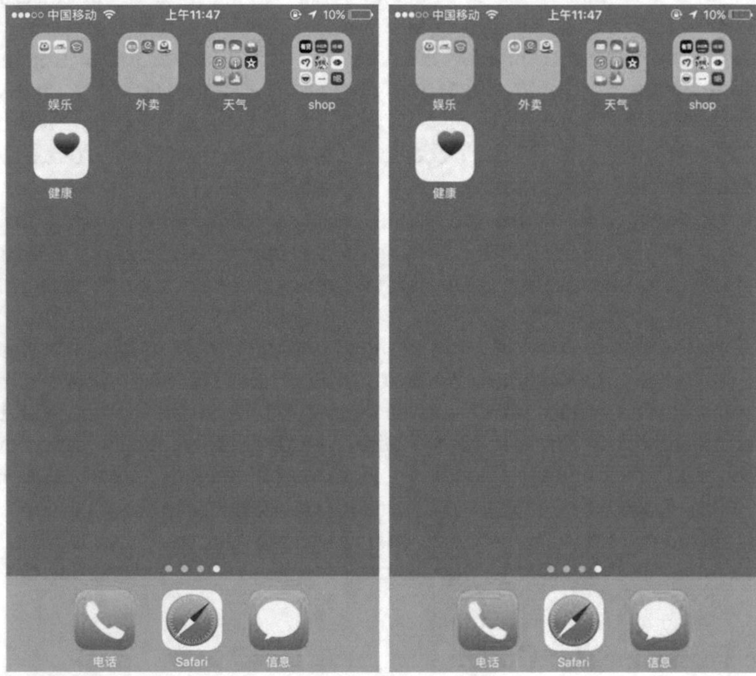

Fig. 1. The phone screen and the icons as stimuli. The right is the standard stimulus icon (I), the left is the comparison stimulus icon (I_1) which was resized to 92%.

The method of constant stimuli was used to acquire data, which is one of the three main methods proposed by Fechner to measure the threshold and it is also the most accurate and widely accepted method to measure absolute threshold, difference threshold and other psychological magnitude in psychophysics. The constant stimulation method presented a few constant stimuli at the same times and the threshold is determined by the number of times that each stimulus being noticed by the subject.

In our experiment, the standard stimuli (I) and the comparative stimuli ($I_{1\sim9}$) were appeared one after another, then the subjects reported whether the two icons were the same size or not. Compare 8 times for each pair of icons. In order to eliminate the error caused by the appearance orders, four of the eight times firstly presented the standard stimulus and the other four firstly showed the comparative stimulus. Each subject would compared the icons 72 times. Subjects observed each stimulus more than 5 s.

The subjects' reports were divided into three categories: (1) When the comparative stimulus was larger than the standard icon, "big" was reported and the data record was "+"; (2) When the comparative stimulus was smaller than the original stimulus, "small" was reported and the data recorded as "−"; (3) "equal" was reported and recorded as "−" when the comparative stimulus was equal to the standard stimulus or no difference was noticed. For in the two cases, it both meant that the subject did not notice the change of the icon's size. The subjects were asked to make judgments after each comparison.

3 Processing and Analysis of the Experimental Data

There were 19 validated data in the design group and 19 in the non designer group. Of these, 26% of the data in the non designer group exceeded the threshold ranges given by the experiment and were therefore considered invalid data and while there were no such cases observed in the design group.

The method of linear interpolation is used to calculate the experimental data. Firstly each subjects' reports were counted, then the correct rates of the 8 times of comparing each pair of icons were calculated. We put the comparative stimuli as the X-axis, and the correct rates of the reports as the Y-axis and made the fold line chart. Then the straight line with the correct rate of 50% was made. The intersection of the straight line and the fold line is the upper and lower limits of the threshold. Thus the threshold value was obtained.

For example, as shown in Table 1 and Fig. 2, the two points of (98,0.75) and (100,0.125) should be chose as the lower limit of the subject and the two points of (100,0.375) and (102,0.75) be chosen as the upper limit. According to the calculation, the lower limit was 98.8, the upper limit was 100.67, and the threshold is 0.933.

Table 1. The experimental data of a subject

Scale	+	=	–
92	0	0	1
94	0	0	1
96	0	0.125	0.875
98	0.125	0.125	0.75
100	0.375	0.5	0.125
102	0.75	0.25	0
104	1	0	0
106	1	0	0
108	1	0	0

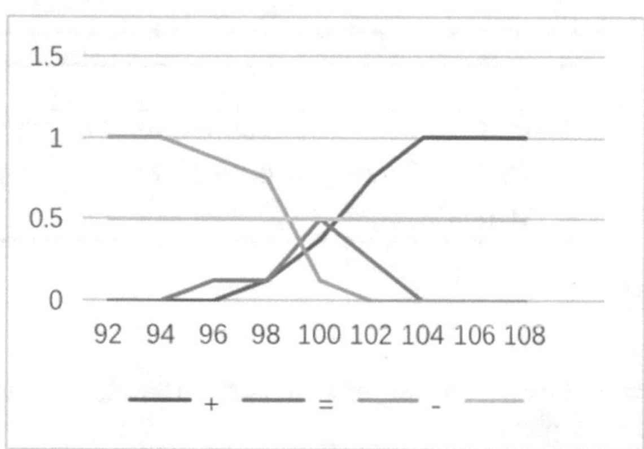

Fig. 2. The data of all the subjects were calculated and presented as Table 2.

The data of all the subjects were calculated and presented as Table 2.

The statistical result is: The average difference threshold of design majors is 1.80%. The maximum is 3.00% and the minimum is 0.93%. The average difference threshold of non-design major students is 3.28%. The maximum is 5.80% and the minimum is 1.50%. In the non designers group, 26% of the subjects have the threshold beyond the

Table 2. The difference threshold value of the DG and NDG groups

Design group				Non designers group			
Subjects	Threshold	Subjects	Threshold	Subjects	Threshold	Subjects	Threshold
1	1.268	11	1.571	1	4.667	11	2.228
2	2.178	12	1.000	2	3.500	12	2.072
3	1.500	13	1.900	3	3.332	13	3.808
4	1.333	14	1.950	4	3.250	14	3.750
5	2.000	15	1.133	5	3.349	15	3.000
6	2.833	16	1.500	6	2.750	16	2.500
7	3.000	17	2.932	7	5.800	17	4.000
8	1.200	18	3.000	8	3.272	18	1.500
9	0.933	19	1.800	9	4.400	19	3.000
10	1.150			10	2.200		
Mean		1.799		Mean		3.283	

Table 3. The results of variance analysis of the visual difference threshold of two groups

Group	**Observation number**	**Sum.**	**Mean**	**variance**
1	19	34.18375	1.799144737	0.488691173
2	19	62.376325	3.282964474	1.030623447

SS.	df.	MS	F	P-value	F crit
20.91634961	1	20.91634961	27.53392791	7.04364E-06	4.113165219
27.34766315	36	0.75965731			
48.26401276	37				

range given in this experiment (92%–108%), while there was no such situation in the design group.

By one-way analysis of variance (T-test), P-value = 0.000017, far less than 0.05, excluding the null hypothesis. There was a significant difference in thresholds of visual difference between the two groups (Table 3).

Then because the proportion of female in the designer group was too high than male, in order to exclude the possible impact of gender, 8 female subjects were randomly selected from the subjects. Their different thresholds were compared with the 8 male subjects and the variance analysis was carried out.

As Table 4 shows, the mean value of the different threshold of male group is 3.31 and female group is 3.30. The P-value is 0.985 > 0.05. It showed that there was no significant difference in the threshold between the two groups, which excluded the possible impact of gender to the visual sensing capability.

Finally, we conducted a questionnaire survey on all the subjects to investigate the types of visual training that they had received. The data from the questionnaire shows that the main training of DG includes sketching and computer art design (PS and 3D modeling), while 98% of the non designers group did not have any art and modeling training or related hobbies. To sum up, this study proves that designers are indeed more sensitive to visual perception than non designers, and to a large extent may be related to designers' professional training.

Table 4. The results of variance analysis of the visual difference threshold of male group and Female group

Subjects	Male	Subjects	Female	P-value
1	3.500	1	4.667	0.985011758
2	3.332	2	3.272	
3	3.250	3	4.400	
4	3.349	4	2.200	
5	2.750	5	2.228	
6	5.800	6	2.072	
7	3.000	7	3.808	
8	1.500	8	3.750	
Mean	3.310125	Mean	3.299625	

4 Conclusion

On the research, base on The method of constant stimuli, a simple cell phone icon was used as the standard stimulus and the standard stimulus was resized by 2% variables to obtain a series of comparative stimuli. The pair of the standard stimulus and a comparative stimulus were shown to the subjects of the design group and non-designers groups and their different visual thresholds were measured.

The experimental results show that:

1. the average difference threshold of DG is 1.8%.The maximum value is 3.00% and the minimum value is 0.93%. The average difference threshold of NDG is 3.28%. The maximum value is 5.8% and the minimum value is 1.5%. T-test was used to prove that there were significant differences between the visual difference threshold values of the two groups. Moreover, the subjects in the DG often can quickly judge the results than the subjects in the NDG. The data also shows that the difference threshold of 26% subjects is beyond the range given in this experiment, that is, in the range of changes provided by this experiment, they could not accurately identify the difference between these icons. These subjects are all in the NDG.

To sum up, the designers do have stronger visual sensing capability than non designers.

2. The data from the questionnaire shows that the main training of DG includes sketching and computer art design (PS and 3D modeling), while non-design students are lack the training of sketching and modeling. Therefore it can be inferred that sketching and 3D modeling training are likely to have the effect on enhancing the human's visual sensing capability.

Acknowledgements. This research is supported by the National Key R & D program "R & D of Multifunction Cane and Beet Harvest Technology and Equipment" (2016YFD0701200), Sub-project "Research on Key Technologies of Efficient Harvesting and Intelligent control for Cane and Beet Efficient Harvesting Equipments" (2016YFD0701201) and "The research and development of the intelligent fining production management technology and equipment for livestock and poultry breeding" (2017YFD0701605-2).

References

1. Guo, X.Y., Yang, Z.L.: Experimental Psychology. People's Education Press, Beijing (2004). (in Chinese)
2. Zhang, S.: Basic Psychology, pp. 248–257. Educational Science Press, Beijing (2001). (in Chinese)
3. Kantowitz, B.H., Roediger, H.L., Elmes, D.G.: Experimental Psychology, pp. 208–211. West Publishing Company, St. Paul (1997)
4. Hu, Z.-J.: The method of constant stimuli. J. Mu Dan Jiang Normal Coll. (Philos. Soc. Sci. Ed.) (3), 79–80 (1995). (in Chinese)
5. Richard, J.G., Philip, G.Z.: Psychology and Life. Allyn and Bacon, A Pearson Education Company, Boston (2002)
6. Huang, X.: A research on human perception of the area of the BasicGeometry. Institute of Industrial Design, Yunlin National University of Science and Technology, Taiwan (2004). (in Chinese)
7. Yang, G., Yu, S., Chu, J.: Practical significance and accomplishment method for innovative design of scaled-down Ming and Qing Dynasty furniture with Tenon and Mortise structure. J. Northwest For. Univ. **24**(1), 170–172 (2008). (in Chinese)
8. Wu, J., Yu, S., Yang, G., Zhang, X.: Difference threshold of proportion of a human being on configuration. Mech. Sci. Technol. Aerosp. Eng. **30**(5), 767–769 (2011). (in Chinese)
9. Wang, Z.-Q.: Traditional psychophysics method and signal detection theory. J. Liaoning Teach. Coll. (Soc. Sci. Ed.) **29**(5), 73–75 (2003)

Auditory Evoked Potentials and PET-Scan: Early and Late Mechanisms of Selective Attention

Sergey Lytaev[1,2(✉)], Mikhail Aleksandrov[2], Tatjana Popovich[3], and Mikhail Lytaev[4]

[1] Saint Petersburg State Pediatric Medical University, Saint Petersburg, Russia
salytaev@gmail.com
[2] Almazov National Medical Research Centre, St. Petersburg, Russia
[3] SPIIRAS Hi Tech Research and Development Office Ltd., St. Petersburg, Russia
[4] Bonch-Bruevich Saint Petersburg State University of Telecommunications, St. Petersburg, Russia

Abstract. Auditory evoked potentials (AEP) mapping in 19 unipolar points using the technique of extracting relevant signals from deviant ones were applied. To determine the brain functional status positron emission tomography (PET) scan with fluorodeoxyglucose was performed. Neuropsychological testing, the scale of mental status assessment, tests of "frontal dysfunction battery" and "drawing hours" were performed. 103 patients with neurologic attention deficit disorder and 26 healthy persons were examined. The paper discusses the dynamics of the AEP amplitude-time parameters over a time interval of 15–400 ms in the performance of the task for attention. The results of the PET study have established the 4-factor's structure of the attention system: dorsal (selective attention), ventral (involuntary attention), operative rest and visual attention system.

Keywords: Auditory evoked potentials · Perception · Selective attention
Positron emission tomography

1 Introduction

Neurosciences in the 21st century are characterized by simultaneous development of both new technologies of operative neurosurgical interventions and visualization systems (neuronavigation, neuromonitoring and mapping) of the functional brain state [1–3]. Preoperative morphological diagnosis using magnetic resonance imaging (MRI), computed tomography, angiography, and MRI angiography is performed. Physiological support is provided by functional MRI, positron emission tomography, magnetic-encephalography [4–6]. Ultrasonic dopplerography, functional MRI, PET, functional stereotaxis, EEG and evoked potentials (with the help of dipole localization methods, neuronavigation and the three-dimensional Lissajous trajectory), neurovideo endoscopy allow registering the brain state in real time [7–9].

H. Ayaz and L. Mazur (Eds.): AHFE 2018, AISC 775, pp. 169–178, 2019.
https://doi.org/10.1007/978-3-319-94866-9_17

The study of intercortical interaction is one of the most important tasks of modern neuroscience. Methods of functional neuroimaging, such as positron emission tomography, electroencephalography, functional magnetic resonance imaging, allow in vivo to study intercortical connections forming large anatomical and physiological networks involving various parts of the brain in their work [10–12].

To evaluate early and late mechanisms of selective attention, it is advisable to study the network pattern of intercortical interaction with formation of executive cognitive functions of cluster's based on auditory evoked potentials mapping and positron emission tomography.

2 Methods

103 patients (65 men and 38 women) with neurologic attention deficit disorder were examined. 103 people with neurologic attention deficit disorder syndrome (65 men and 38 women) were examined and treated at the N.P. Bekhtereva Human Brain Institute. 26 healthy persons made up a control group.

AEPs mapping in 19 unipolar points using the technique of extracting relevant signals from deviant ones were applied. At the AEPs recording technique to highlight the relevant background deviant signals having different frequency tone was used. Discriminant analysis ($F > 4.0$) and MDL algorithm were applied.

To determine the brain functional status PET-scan with fluorodeoxyglucose was performed. The rate of glucose metabolism in different parts of the brain was determined.

The 2-fluoro-2-deoxy-D-glucose radiopharmaceutical synthesized in the radiochemical laboratory of the Human Brain Institute was administered intravenously at a dose of 1–5 mCi in 8 ml of physiological saline [5]. Scanning lasting 20 min began 30–40 min after the administration of the drug. At the end of the scan, a visual and semi-quantitative evaluation of the PET image was carried out in three projections (axial, frontal and sagittal) and comparison of the obtained data with MRI pictures.

Neuropsychological testing, the scale of mental status assessment, tests of "frontal dysfunction battery" and "drawing hours" were performed.

The studies were carried out on clinical indications with written informed consent of patients, approved by a local ethical committee.

3 Results

According to the AEP mapping the acoustic signal is involved in brain structures processing after 16–18 ms from the moment of its presentation (Figs. 1 and 2). The greatest potential difference with stimulation with the target setting is more characteristic for the frontal parts of the left hemisphere, at this time in the occipital and parietal sites there is a slight symmetrical activation.

Among the middle latency AEPs, 4 waves – P16, N30, P40 and N60 corresponding to the components Po, Na, Pa and Nb were the most stable in both samples. The amplitude-time characteristics of the Pa/Nb complex did not differ significantly in

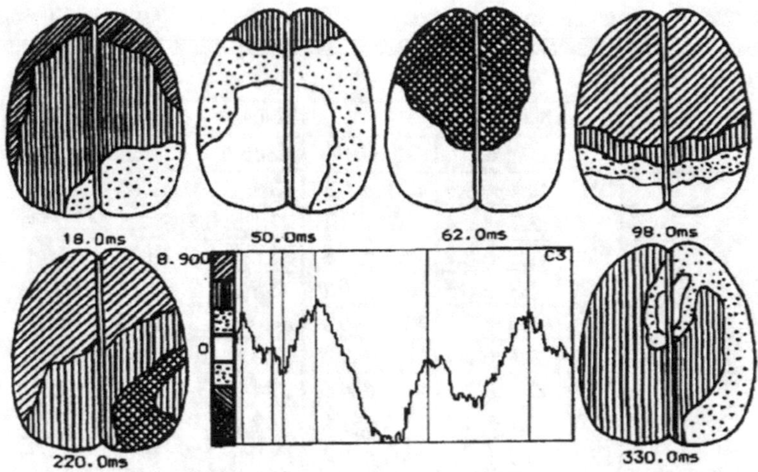

Fig. 1. AEP topomaps in healthy subject in a rhythmic auditory signals, ms.

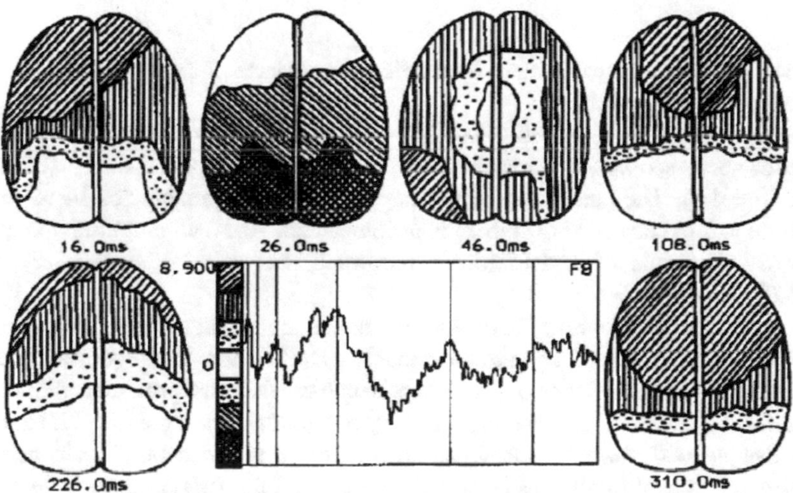

Fig. 2. AEP topomaps in healthy subject in the attention activation, ms.

different test variants (Table 1). Some peculiarities were observed in the development of the Po and Na waves under the conditions of presentation of the relevant stimulus. Thus, the Po amplitude in the parietal (P3 and P4) as well as the left hemispheric C3 and F3 leads (F > 2.0) increased significantly.

A similar trend persists even in the time interval for the formation of the Na wave in left-handed P3 and F3 (F > 2.0) registration points, as well as in the vertex. In addition, under the conditions of the relevant signal there is a tendency to shorten the peak latency (PL) which in the sites P4 and O2. Brain mapping (Fig. 2, 26 ms) shows a distinct symmetry with maximum activation of the occipital parietal lobes, in a lesser

Table 1. The amplitude-time characteristics (A, uV/T, ms) of the N40 and N90 components AEPs during auditory stimulation with the target signals in patients when compared with the control group

Parameters	Sites	N40		N90	
		Control	Pathology	Control	Pathology
T	P4	40.4 ± 9.6	42.3 ± 9.5	83.8 ± 10.0	98.0 ± 14.0*
A		2.7 ± 0.4	1.7 ± 1.1	2.1 ± 0.7	2.5 ± 1.7
T	P3	42.6 ± 8.2	44.0 ± 8.8	81.8 ± 8.1	95.0 ± 13.6*
A		2.5 ± 0.5	0.9 ± 0.2*	1.5 ± 1.2	2.7 ± 1.1
T	Cz	41.0 ± 7.7	43.6 ± 9.0	84.0 ± 6.3	93.6 ± 11.5*
A		2.1 ± 0.7	1.6 ± 1.0	1.3 ± 0.6	3.4 ± 1.6 *
T	F4	43.0 ± 7.1	41.3 ± 4.3	87.4 ± 5.5	95.3 ± 9.2 *
A		2.3 ± 1.1	1.3 ± 1.0	1.7 ± 1.0	4.6 ± 1.1*
T	F3	43.0 ± 7.0	41.0 ± 2.4	88.0 ± 3.7	94.3 ± 13.4
A		2.2 ± 0.4	1.9 ± 1.2	1.3 ± 0.3	3.9 ± 1.6**

Notes. F-statistic value by compared to the control: * - F > 4.0, 2 * - F > 10.0; in other, the differences are insignificant - F < 4.0.

degree of central departments and, in practice, the absence of excitability of the frontal sections of the neocortex.

Rhythmic auditory stimulation has little effect on the N90 component. N145 wave amplitude increases in left parietal site (F = 4.6), and the peak latency (PL) in the vertex (F = 4.4). The perception of the target signal by a generalized increase of an amplitude and latency N90 (Table 1) is accompanied. At N145 amplitude component in almost all points of registration is recovered, but remains elevated PL in C3 (F = 5.0) and C4 (F = 6.3).

Positive AEP wave with a PL of 90–110 ms has the maximum amplitude among all the oscillations of the AEP and the most stable manifestation. In the present study, the amplitude-time characteristics of this component have the most pronounced differences in practically all brain areas, accompanied by two tendencies. First, the PL shortening in the sample with the relevant signal in the posterior and central brain parts and, secondly, significant in all sites (F > 2.0) by reducing the P90 component in this task (Table 1). The marking of these waves gives in both cases symmetrical maps with more noticeable activation of the cortex in a situation without a target stimulus (Figs. 1 and 2).

PL values for the late waves both for the target and without the target stimulus are practically equalized. Only for the P250 wave, the PL is significantly lower with the action of the relevant signal (F > 2.0). In addition, for this component, attention is drawn to the tendency for all sites to lower the amplitude under the conditions of the target task. Significant differences are characteristic for the left-handed P3 and C3 sites and the vertex area. The noted facts in the brain mapping (Fig. 1, 220 ms, Fig. 2, 226 ms) are reflected. However, in this case, attention is drawn to the noticeable hemispheric symmetry in the development of excitation processes in the target task performance.

An even more pronounced symmetrical excitation in the sample with the relevant stimulus is observed when mapping the component with LP 300–350 ms. The analysis of the space-time characteristics in this task marks a significant alleviation of the amplitude of this wave in all sites except parietal points ($F > 2.0$, Figs. 1, 2).

In the group of patients with Parkinson's disease with cognitive impairment the level of fixation of the radiopharmaceutical in the middle third of the frontal lobes of the right and left hemisphere (Brodmann field 8) was significantly lower than in patients from the Parkinson's group without cognitive impairment (Fig. 3). In the parietal lobes there were also significant differences in the level of glucose metabolism between patients with Parkinson's disease without cognitive impairment and Parkinson's disease with cognitive impairment. In the group of patients with Parkinson's disease with cognitive impairment the metabolism of glucose in the posterior and lower halves of the convective cortex of parietal lobes (Brodmann fields 7, 39, 40) was significantly lower in both hemispheres of the large brain.

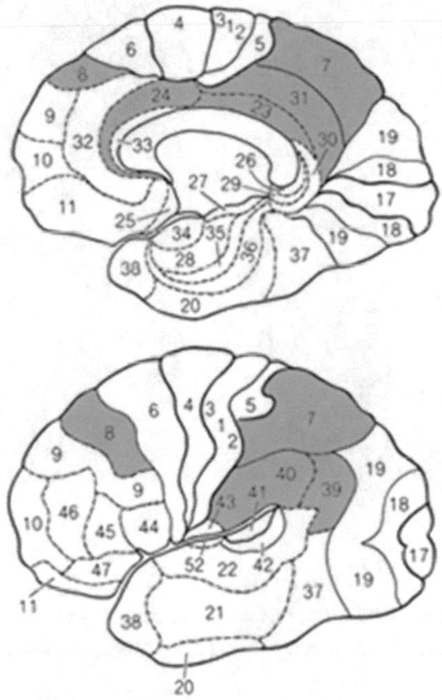

Fig. 3. Areas of interest in reducing the rate of glucose metabolism in Parkinson's disease with minimal cognitive impairment (up) compared to patients without cognitive impairment (below), $p \leq 0.05$.

Based on the results of the PET study with fluorodeoxyglucose a pattern characterized by a bilateral reduction in the rate of glucose metabolism in the posterior third of the frontal lobes and upper parietal areas was determined in patients with the syndrome of executive function impairment. To determine their relationship factorial analysis was applied.

The study uses a factor analysis of the method of principal components using varimax rotation. The factor was included in the factor solution if the Kaiser criterion was fulfilled (eigenvalue > 1). A variable with a load >0.45 was regarded as a factor.

As part of the continuation of the study of the intercortical interaction of the frontal and parietal lobes based on the analysis of PET data on glucose metabolism, the study of the self-organizing work of functional neuroanatomical systems was continued. The following systems were studied: the system of operational rest (the default system of the brain), the dorsal attention system, the ventral system of attention and the visual system in norm and in patients with cognitive impairments of varying severity (Fig. 4).

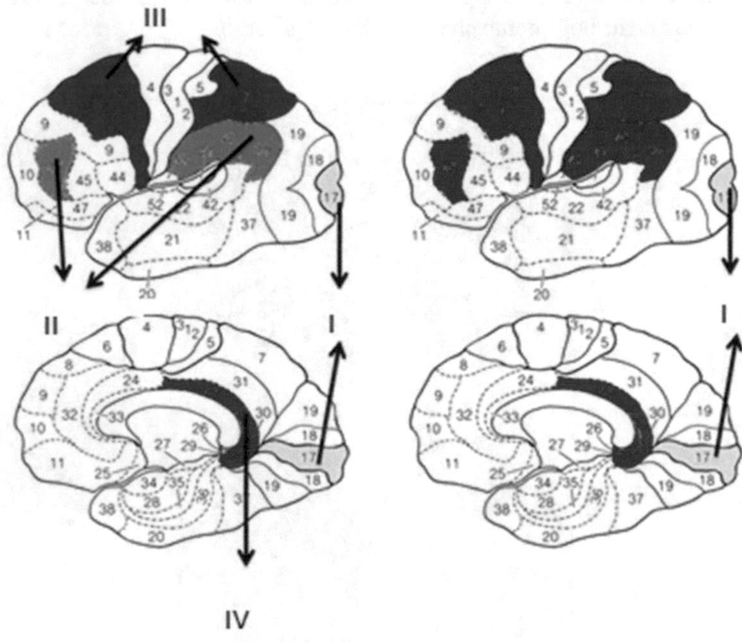

Fig. 4. PET-scan, factorial analysis. Systems of attention: I – visual system, II – ventral system, III – dorsal system, IV – operational rest system. Left – absence of executive functions violations, right – violation of executive functions.

It has been established that the factor structure of data on regional levels of glucose metabolism in the examined different nosologies, but without cognitive impairment, is stable from the point of view of the composition of the factors entering into the factor solution. These factors can be interpreted (in accordance with the belonging of the brain areas having the greatest burden on the factor) to a certain neuroanatomical system.

The first factor is 8, 6 and 7 cytoarchitectonic fields of Brodmann (the precentral cortex of the frontal lobes and the upper half of the parietal lobes of both large hemispheres). The second factor includes 46 and 40, 39 Brodmann fields (the anterior third of the convectional frontal lobes and the lower half of the parietal lobes). The third factor includes 23, 36, 29 and 30 cytoarchitectonic fields of Brodmann (the back part of the cingulate gyrus). The fourth factor is the 17th Brodmann field (primary visual cortex) (Fig. 4).

The obtained results proved that in the structure of the intercortical interaction of the frontal and parietal lobes four functional systems are stably recorded: the dorsal and ventral attention systems, the operative rest system and the visual system. This structure of intercortical interaction is consistently registered both in healthy volunteers and in patients without cognitive impairment. In patients with initial cognitive impairments, a reorganization of the structure of the intercortical interaction of the frontal and parietal lobes in the form of the decay of functional systems of operative rest, dorsal and ventral attention systems is revealed.

4 Discussion

There is quite contradictory information about middle latency AEPs (10–60 ms), their source and localization. These waves can reflect the activity of the first relay neurons of the auditory way or refer to the responses of the primary auditory cortex. There are data that AEPs with PL 12–37 ms from the Hirschlian gyrus of both temporal lobes were recorded. The same components may indicate the inclusion of subcortical delay mechanisms [3, 8].

Our data from registering middle latency AEPs (MLAEPs) in the time interval 15–40 ms under the action of the target signals in healthy subjects showed a significant increase in the amplitude of these waves ($F > 4.0$) in the parietal, left temporal and frontal areas of the neocortex. This fact testifies about intensification of consciousness in the form of a mechanism of reverse influence of these parts of the brain cortex to brainstem structures that generate MLAEPs in conditions of activation of attention. Brainstem pathology increases the involvement of parietal cortex in providing feedback afferent mechanisms. This is confirmed by hyperactivation N18 amplitude with predominance in the parietal cortex and the central fields.

The solution of back tasks in order to establish the functional significance of brainstem and subcortical structures in the literature data are controversial. IOM data recorded from the midbrain structures show the presence of negative waves in the time sequence following the brainstem auditory EPs. AEPs near-field sequentially recorded at the level of the brain stem and from scalp. These results confirm the importance of brainstem formations in generation MLAEPs. It can be assumed about the localization of the generator of these waves on the brainstem-thalamic level, where cortex is given function of regulator for "volley" of the deep-generated potentials [2, 13].

If the changes in the components of the MLAEPs in the time interval of 15–35 ms in the control group were reliable, the amplitude-time parameters of N40 and P60 did not practically differ in both tests. These facts were regarded either from the point of the identity of the mechanisms that ensure the transformation of auditory signals of varying complexity, or the absence of influence of overlying formations on this interval of time on the attention processes. N40 wave in brain pathology is generalized reduced with rhythmic stimulation. However, the amplitude of the next component (P60) under these conditions increases with brain damage mainly in the vertex. In the control group, a noticeable reduction of the N90 AEP can be observed under the conditions of presentation of the target task in all brain areas. In contrast, brainstem pathology, regardless of the nature and location of the lesion, is accompanied by a generalized increase in the N90 amplitude [14, 15].

The mechanism considered in many respects explains the results obtained by us, where the amplitude of the N90 AEP in normal much higher with simple rhythmic stimulation. Obviously, under such conditions the summation of the allocated signal is more adequate. On the contrary, when the target signal is allocated with brainstem pathology, the amplification of the N90 amplitude is recorded. Thus, for the auditory system there is evidence of a possible selective corticofugal modulation already at the level of the switching brainstem neurons [16–18].

Mechanisms for the formation of the syndrome of violation of executive functions mainly based on the assessment of each component executive functions are considered. Based on the results of factor analysis, one can confirm the assumption that the basis of the syndrome of violation of executive functions is a general process. And also that the reorganization of the ventral and dorsal attention systems underlies the formation of the syndrome of impaired executive functions in Parkinson's disease [19].

Our results proved that four functional systems are stably registered in the structure of the intercortical interaction of the frontal and parietal lobes: the dorsal and ventral systems of attention, the system of operative rest and the visual system. This structure of intercortical interaction is consistently registered both in healthy volunteers and in patients without cognitive impairment. In patients with initial cognitive impairments reorganization of the structure of intercortical interaction between the frontal and parietal lobes in the form of decay of anatomical and functional systems of operative rest, dorsal and ventral attention systems is revealed [5].

This study confirms that aggressiveness and hostility by different neuroanatomical structures are provided. Physical and verbal aggression which are external manifestations of the inner motivations of the subject, positively correlate with the level of glucose metabolism in the parietal cortex [20]. Irritability, negativism, suspiciousness and hostility negatively correlate with the level of glucose metabolism in the orbit-frontal cortex. The obtained results confirm the idea that when the glucose metabolism in the orbitofrontal cortex of the frontal lobes decreases the level of inhibitory control over the emotional system decreases, which leads to increased hostility.

5 Conclusion

The registration of auditory evoked potentials represents a model for studying both early and late mechanisms of selective attention. In this connection, the PET study allows us to speak only of late mechanisms tied to the fields of the cerebral cortex.

The simplification of amplitude of the AEP components with peak latency 15–40 ms with target signals indicates the inverse effect of brain cortex regions to brainstem in early mechanisms of selective attention. Parameters of the P40 and N60 AEP components are not related with stimulation conditions. The amplitude of the P90 AEP component is significantly higher in the absence of a relevant signal. The late positive wave with a latency of about 250 ms in the target task is reduced. For the P350 wave a similar trend is observed with a more pronounced symmetry of topomaps. However, this component in all but the parietal regions is recorded with increasing amplitude during task with the relevant signal.

The results of the PET study in neurological patients without the syndrome of executive cognitive impairment established the 4-factor's structure of the attention system: dorsal (selective attention), ventral (involuntary attention), operative rest and visual attention system. The same factor structure in the control group of healthy young persons was recorded.

References

1. Gnezditskiy, V.V., Korepina, O.S., Chatskaya, A.V., Klochkova, O.I.: Memory, cognition and the endogenous evoked potentials of the brain: the estimation of the disturbance of cognitive functions and capacity of working memory without the psychological testing. Usp. Fiziol. Nauk **48**(1), 3–23 (2017)
2. Khil'ko, V.A., Lytaev, S.A., Ostreiko, L.M.: Clinical physiological significance of intraoperative evoked potentials monitoring. Hum. Physiol. **28**(5), 617–624 (2002). https://doi.org/10.1023/A:1020295322474
3. Lytaev, S., Aleksandrov, M., Ulitin, A.: Psychophysiological and Intraoperative AEPs and SEPs monitoring for perception, attention and cognition. In: Communications in Computer and Information Science, vol. 713, pp. 229–236. Springer, Cham (2017). https://doi.org/10.1007/978-3-319-58750-9_33
4. Lytaev, S., Susin, D., Aleksandrov, M., Shevchenko, S.: PET-scan & neuropsychological diagnostics of cognitive disorders in Parkinson's disease. J. Int. Psychophysiol. **108**, 114 (2016). https://doi.org/10.1016/j.ijpsycho.2016.07.342
5. Lytaev, S.A., Susin, D.S.: PET-diagnostics of cognitive impairment in patients with Parkinson's disease. Pediatrician, **7**, 63–68 (2016). http://dx.doi.org/1017816/PED7263-68
6. Nojszewska, M., Pilczuk, B., Zakrzewska-Pniewska, B., et al.: The auditory system involvement in Parkinson disease: electrophysiological and neuropsychological correlations. J. Clin. Neurophysiol. **26**(6), 430–437 (2009). https://doi.org/10.1097/wnp.0b013e3181c2bcc8
7. Cohen, O.S., Vakil, E., Tanne, D., et al.: The frontal assessment battery as a tool for evaluation of frontal lobe dysfunction in patients with Parkinson disease. J. Geriatr. Psychiatry Neurol. **25**, 71–77 (2012). https://doi.org/10.1177/0891988712445087
8. Lytaev, S., Belskaya, K.: Integration and disintegration of auditory images perception. In: LNCS (LNAI), vol. 9183, pp. 470–480. Springer, Cham (2015). https://doi.org/10.1007/978-3-319-20816-9_45
9. Patterson, J.V., Sandman, C.A., Jin, Y., et al.: Gating of a novel brain potential is associated with perceptual anomalies in bipolar disorder. Bipolar Disord. **15**, 314–325 (2013)
10. Leh, S.E., Petrides, M., Strafella, A.P.: The neural circuitry of executive functions in healthy subjects and Parkinson's disease. Neuropsychopharmacology **35**, 70–85 (2010). https://doi.org/10.1038/npp.2009.88
11. Posner, M.I., Raichle, M.E.: Images of Mind. Scientific American Library/Scientific American Books, New York (1994)

12. Poston, K.L., Eidelberg, D.: FDG PET in the evaluation of Parkinson's disease. PET Clin. **1**, 55–64 (2010). https://doi.org/10.1016/j.cpet.2009.12.004
13. Razavi, B., O'Neill, W.E., Paige, G.D.: Auditory spatial perception dynamically realigns with changing eye position. J. Neurosci. **27**(38), 10249–10258 (2007)
14. Sanguebuche, T.R., Peixe, B.P., Bruno, R.S., et al.: Speech-evoked brainstem auditory responses and auditory processing skills: a correlation in adults with hearing loss. Int. Arch. Otorhinolaryngol. **22**(1), 38–44 (2018). https://doi.org/10.1055/s-0037-1603109
15. Touge, T., Gonzalez, D., Wu, J., et al.: The interaction between somatosensory and auditory cognitive processing assessed with event-related potentials. J. Clin. Neurophysiol. **25**(2), 90–97 (2008)
16. De Letter, M., Aerts, A., Van Borsel, J., et al.: Electrophysiological registration of phonological perception in the subthalamic nucleus of patients with Parkinson's disease. Brain Lang. **138**, 19–26 (2014). https://doi.org/10.1016/j.bandl.2014.08.008
17. Jones, S.J., Vaz Pato, M., Spraque, L., et al.: Auditory evoked potentials to spectro-temporal modulation of complex tones in normal subjects and patients with severe brain injury. Brain **123**(5), 1007–1016 (2000)
18. Melo, Â., Mezzomo, C.L., Garcia, M.V., Biaggio E.P.V.: Computerized auditory training in students: electrophysiological and subjective analysis of therapeutic effectiveness. Int. Arch. Otorhinolaryngol. **22**(1), 23–32 (2018). https://doi.org/10.1055/s-0037-1600121
19. Liu, C., Zhang, Y., Tang, W., et al.: Evoked potential changes in patients with Parkinson's disease. Brain Behav. **7**(5), e00703 (2017). https://doi.org/10.1002/brb3.703
20. Naskar, S., Sood, S.K., Goyal, V.: Effect of acute deep brain stimulation of the subthalamic nucleus on auditory event-related potentials in Parkinson's disease. Parkinsonism Relat. Disord. **16**(4), 256–260 (2010). https://doi.org/10.1016/j.parkreldis.2009.12.006

Brain-Computer Interfaces

A Proof of Concept that Stroke Patients Can Steer a Robotic System at Paretic Side with Myo-Electric Signals

Stijn Verwulgen[1(✉)], Wim Saeys[2,3], Lex Biemans[1], Annelies Goossens[1], Gido Grooten[1], Joris Ketting[1], Aurélie Van Iseghem[1], Brecht Vermeesch[1], Erik Haring[1], Kristof Vaes[1], and Steven Truijen[3]

[1] Department of Product Development, Faculty of Design Sciences, University of Antwerp, Antwerp, Belgium
stijn.verwulgen@uantwerpen.be
[2] Revarte Rehabilitation Hospital, Wilrijk, Belgium
[3] MOVANT, Faculty of Health and Health Sciences, University of Antwerp, Antwerp, Belgium

Abstract. An exoskeleton can be a possible aid for faster recovery for stroke survivors. Controlling such a device safe and accurately is mandatory for applying this rehabilitation aid outside a highly controlled lab environment. MYO electric signals have the potential to provide such steering, as they could command the exoskeleton with the same underlying muscle movement. As stroke survivors have severe motor impairments, the question arises whether the electric signals that steer their muscles at paretic side, can be picked up and classified with sensitive electrodes and machine learning techniques. Six paretic stroke survivors were evaluated in this first pilot study. An of-the shelf wearable myo electric device (MYO armband, Thalmic labs) and matching machine learning and control system was used. Patients were assessed for muscular control and manual motor functions via a modified Lovett test, in a five point Likert scale. Hands and forearms at unaffected and paretic side were used as baseline and inter-subject comparison. Calibration was successful in all subjects at unaffected forearm and in most subjects at paretic forearm. These subjects had moderate to good motor performance. The subjects that were unable to calibrate the device at paretic forearm exhibit inferior motor performance. One subject was able to perform full gesture control at paretic forearm. Most striking is the observation that in one patient, the functionality offered by the MYO armband, outperformed the subject's muscular control, notably without prior practice.

Keywords: Stroke survivors · Gesture control · Myo electric signals

1 Introduction

Full hand control has always been important, but maybe even more so in contemporary setting. Many features of everyday modern products, like mobile phones and laptops, expect the user to perform precise movements with their fingers and hands. Stroke survivors may perceive problems with such fine handlings in their day-to-day life.

H. Ayaz and L. Mazur (Eds.): AHFE 2018, AISC 775, pp. 181–188, 2019.
https://doi.org/10.1007/978-3-319-94866-9_18

Rehabilitation methods should be optimized to maximize fine motoric functions in stroke survivors. Each rehabilitation process is different and it is challenging to pinpoint individual needs and requirements. In each stage of the rehabilitation process, patients should use proper tools for optimal recovery. New technologies can potentially help patients in an earlier stadium of the rehabilitation process. By obtaining insights of muscle activity data in an early stage of the rehabilitation process, requirements could be mapped for patient to use these new technologies. For example, tools and techniques could be developed for stimulating myo electric signals and practicing muscular control.

The MYO armband created by Thalmic Labs [1], is a tool for gesture control. It uses Electromyography (EMG) sensors that can measure muscle activity. If this technology is capable of measuring sufficient signals in the arms of stroke survivors, it could potentially be used as an input signal to drive external products like exoskeletons, virtual reality training and other rehabilitation tools. This would be a cheaper and more accessible method than the current products on the market, for example the Amadeo [2]. The Amadeo helps patients with performing their rehabilitation exercises in all the phases of the process. Its dimensions are big and it is very expensive, so the only way to make use of this product is in the highly specialized rehabilitation centers. This is also the only moment the patients can train with professional devices. That is why the MYO armband, combined with an exoskeleton or other device, could give patients the opportunity to train at home.

Aim and Goal. This research aims to determine the weakest muscle activity (Read by EMG sensors), for using a MYO armband to remotely control an exoskeleton or other rehabilitation device by stroke survivors. This is done by using a MYO armband to measure the difference between paretic and non-paretic muscle activity in the forearm of stroke survivors.

First the level of arm-muscle control is measured by using a Lovett based scale [3]. This scale ranks the arm-muscle activity going from zero (no muscular activity) to five (normal). Difference between paretic and non-paretic muscle activity in the arm for stroke survivors is used as a benchmark. From this data the test subjects are labeled and then divided to get better insight in which patient can or cannot use the MYO armband.

2 Materials and Methods.

2.1 MYO Armband

The armband (Fig. 1, left) allows the wearer to control different pieces of (integrated) technology with arm movements and gestures. These physical gestures and movements cause muscle activity in the forearm. Underlying electrical pulses inducing this activity are detected by eight EMG sensors in the MYO armband, all located in an individual module of the armband. The acquired EMG signals are converted to digital data and send via Bluetooth to an electronic device that converts this data into pre-programmed actions. For example: if a user makes a fist, he will turn the lights on in his room. If he spreads his fingers he will turn them off. In Fig. 1, right, the 7 recognizable gestures are shown [4].

 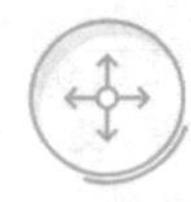  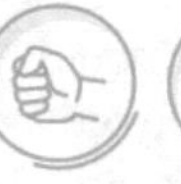    

Fig. 1. MYO armband (left) with possible gestures that can be recognized [4].

2.2 Data Capture

Thalmic Labs Calibration Software. The calibration software from Thalmic Labs is be used for calibration an training gesture recognition. At later stage, such activity can be used to control rehabilitation aids such as a wireless controlled exoskeleton.

The goal is detecting the weakest muscle activity (Read by EMG sensors) that is necessary for using a MYO armband to remotely control an exoskeleton. Difference between paretic and non-paretic muscle activity in the arm for stroke survivors is used as a benchmark.

With the software of Thalmic Labs, it is possible to calibrate the armband for each individual forearm, for muscular activity at paretic and non-paretic arm. When calibration is completed, it is known whether or not the MYO armband detects an EMG signal and if it can convert that signal into a specific preprogrammed gesture. If it converts the acquired signal in the right pre-programmed gesture, this will be displayed (within the Thalmic Labs software) by showing an equal image of the preformed gesture (Fig. 1, right). If there is no image or if the wrong image is shown, then the acquired signal is not detected properly or not converted rightly, because it does not correspond well enough with the calibration data.

Thalmic Labs Data Capture Software. Besides the calibration software, EMG data is sampled to compare amplitudes and average activity at both paretic and non-paretic arm.

2.3 Test Subjects

The test subjects are stroke survivors with lesser muscle activity in one of their arms due to a stroke, were the other arm functions normally. A detailed explanation will follow later. Six subjects will be tested, most of them have different levels on the Lovett based scale. During testing both arms will be tested to see the influence of lower muscle activity on using the MYO armband.

2.4 Modified Lovett Scale

The Lovett scale is a scaling system specifically used for muscle activity measurements. In this research a real Lovett scale with 100% accuracy is not applicable, therefore a modified Lovett scale is used, tailored to subject specific needs and limitations.

The proposed measure is a six grade scale ranging from 0 to 5 in which each number has its own explanation of strength of muscle activity at the upper arm.

Grade 0: There is no movement at all measured in the muscle.
Grade 1: Only a small trace or twitch of movement is monitored in the muscle.
Grade 2: The muscle can only be moved if the resistance of gravity is removed, e.g. the arm can be fully flexed only in a horizontal plane.
Grade 3: The joint can be moved against gravity without any help from the examiner.
Grade 4: Muscle strength is reduced, but muscle contraction can still move the joint against light resistance.
Grade 5: Muscle contracts normally against full resistance, as the non-paretic muscle.

2.5 Test Initiation

After informed consent, each participant was asked to perform a step-by-step test protocol. This protocol starts with determining what Lovett based scale grade a participant has. In the second step, the non-paretic arm is tested. First by calibrating it to recognize gestures (Fig. 1, right) and letting the participant get used to a cycle of gestures that will be repeated during the entire test. In the third step, monitoring and capturing a participants' muscle data throughout a cycle of gestures is started. Subjects were instructed to hold each gesture for 5 s. Lastly, step two and three are repeated at paretic arm.

2.6 Test Setup

Step 1: Lovett Based Scale. A patient's scale grade is determined by performing a cycle of the following gestures;

- Making a fist.
- Spreading all fingers, making a flexion movement between hand and wrist as far as possible, making an extension movement between their hand and wrist as far as possible.

The results of this quick test are evaluated and compared with the predefined scale, in this way patients can be labeled with a grade and individuals can be compared easier.

Step 2: Calibration at Non-paretic Arm. The calibration setup that Thalmic Labs uses for the MYO armband is carried out with software that can be downloaded from the Thalmic Labs website. The calibration asks the user to perform 4 gestures, the same gestures as in the modified Lovett scale test, and one period of rest. In this way the EMG sensors of the MYO armband can pick up the muscle activity for each gesture made or not made by a specific individual. After these five 5 steps, the user is asked to establish a synchronization between the user and the MYO armband by repeating the primed gesture.

The EMG data after the calibration ranged from 0 to 120 μV. Even though a stroke survivor is not able to perform an entire gesture in a correct way, indication for muscle stretch will always be 0 to 120 μV. Signals above 120 μV (rare) are made truncated to 120 μV.

Five gestures and instructions provided to the subjects are displayed in Table 1.

Table 1. Gestures used in the test protocol, at both non-paretic and paretic arm

Gesture 1, fist	The test subject is asked to leave his or her arm in a relaxed state flat on a horizontal surface, in this case a table. Next, the subject is asked to make a fist within the timeframe shown on display (MYO calibration software). This can be seen as one gesture captured data. Conditions stay the same for the next gestures
Gesture 2, open hand	The subject is asked to relax her or his arm and then flex all fingers within the timeframe shown on display
Gesture 3, bend hand towards inside	The test subject is then asked to relax her or his arm and then make a flexion movement between their hand and wrist (as far as possible). The fingers will stay in a relaxed state, within the timeframe shown on display
Gesture 4, bend hand towards outside	The test subject is then asked to relax her or his arm and then make an extension movement between their hand and wrist (as far as possible). The fingers will stay in a relaxed state, within the timeframe shown on display
Gesture 5, rest	The test subject is asked to relax the hand and muscles within the timeframe shown on display

Step 3: Data Capture. The participant arc instructed to repeat all gestures described in Table 1 for a duration of 5 s. The gesture rest is excluded from this cycle. Afterwards the data is controlled and converted into graphs. If there are any abnormalities or missing gestures in the data that were unexpected, the participant is asked to repeat the entire cycle.

Step 4: Paretic Arm Assessment. Steps 2 and 3 are repeated at the paretic arm. This allows comparison per test subject and between test subjects.

3 Results

All participants were able to successfully complete the calibration process with their non-paretic arm.

At paretic arm, all subjects with grade 3 or more on the Lovett based scale could calibrate the MYO band. Based on visual inspection of logged EMG signals, we could recommend grade 4 or up to provide sensitive and specific commands. The completed tests gave as enough data for comparison. A selection of data from experiments is graphically evaluated.

It is interesting to show the difference in data output between a paretic and a non-paretic arm while performing a gesture, for different grades (Fig. 2). It is clearly visible that there is a difference in EMG output at non-paretic: lower grade yield less expressed patterns and lower voltage. Also, EMG signals are more expressed at non-paretic arm compared to paretic arm.

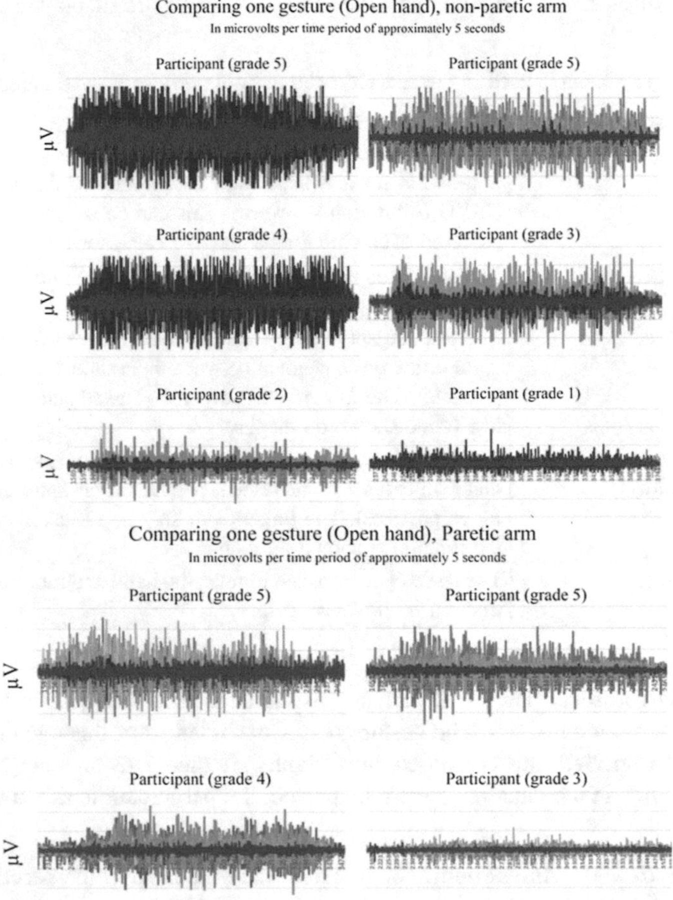

Fig. 2. EMG data from different subjects captured with the MYO armband. Different EEG channels are displayed in different colors

Two of the participants were not able to calibrate the MYO armband due to the lack of muscle strength that is necessary to use the MYO armband.

The most striking observation was that one participant was able to successfully complete the initial calibration stage with the armband with paretic arm, although she was not able to perform the gestures physically, due to the lack of muscle strength. Therefore, she could not confirm her gestures. In the current setup, this confirmation is required for successfully controlling external an exoskeleton or rehabilitation aid.

In another subject, gestures with paretic arm were actually performed, but some weak EMG amplitudes while making the gestures can be seen. Figure 3 is a selection of the acquired data that when performing those gestures.

They were not strong enough to be recognized by the MYO software. At the paretic side most amplitudes have an average value of 30 µV, with the non-paretic arm this number is 120 µV (maximum).

All EMG SENSORS
In microvolts per time period of approximately 20 seconds
Gestures made from left to right; fist, open hand, Bend hand inside and bent hand outside.

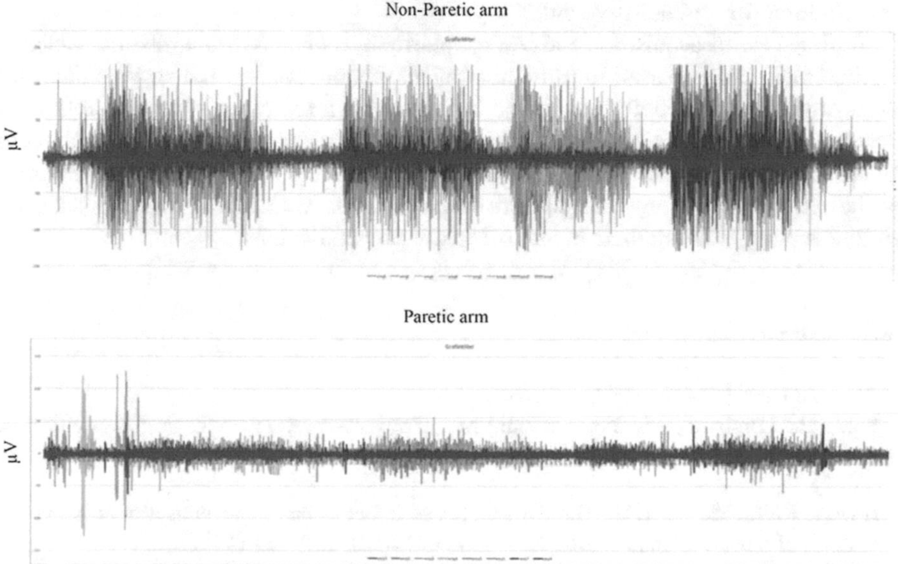

Fig. 3. Comparison of EMG data at non-paretic and paretic arm while performing the same sequence of gestures. Some patterns could be recognized, although not strong enough to be recognize gestures by the current classification algorithm.

4 Discussion and Conclusion

In this research a modified Lovett based is used to classify the test subjects into different levels based on essential actions with the MYO armband without necessary knowledge. This modified Lovett scale could be used by laymen or by stroke survivors for self-assessment, for because of its simplicity.

The MYO armband could detect muscle activity from all test subjects. Some signals or composed patterns are too weak to detect the actual gesture or intended gesture. Not every participant could complete the calibration stage, mostly because of the confirmation gesture. Without confirmation, usage of a MYO armband is excluded in the current setting. A dedicated setting, in which confirmation of intended gestures is organized alternatively, taking account of specific requirements and abilities of stroke survivors, is recommended. This first pilot and qualitative graphical assessment of output data indicates that such set up is feasible.

Because of the build-in calibration system it was not possible to increase the sensitivity of the EMG sensors. If the sensitivity could be adjusted to a higher setting, stroke survivors could benefit. Also, dedicated and more sensitive classifiers could be developed to extract instructions from weak patterns, e.g. as in Fig. 3. Further research is required construct data bases of EMG commands from stroke survivors that could serve as training sets for signal extraction, to identify pattern thresholds, and to map sensitivity and specificity of such commands. This could pinpoint safety, usability and feasibility to use myo electric signals in stroke survivors at paretic arm, to practice rehabilitation for and assistive aid.

With the aid of acquisition and control platform, such as MYO armband, devices as exoskeletons could be used for a more efficient rehabilitation at home. Wearables that assess myo electric activity could also be used as a monitoring system to follow the process of rehabilitation or current muscle activity. For example, patients could get the opportunity to monitor their muscle strength by comparing their EMG signals to former results. In this way, they can confirm improvements with actual (digital) data, e.g. without having to leave their home to consult physiotherapist.

References

1. Thalmic Labs. https://www.thalmic.com/
2. Sale, P., Lombardi, V., Franceschini, M.: Hand robotics rehabilitation: feasibility and preliminary results of a robotic treatment in patients with hemiparesis. Stroke Res. Treat. (2012)
3. Lovett, R.W., Martin, E.G.: Certain aspects of infantile paralysis: with a description of a method of muscle testing. J. Am. Med. Assoc. **66**(10), 729–733 (1916)
4. MYO band. https://www.myo.com/. Mar 2018

Implementing the Horizontal Vestibular Ocular Reflex Test While Using an Eye-Tracker as an Assessment Tool for Concussions Diagnosis

Atefeh Katrahmani(✉) and Matthew Romoser

Western New England University, Springfield, MA, USA
atefeh.katrahmani@gmail.com, matthew.romoser@wne.edu

Abstract. Since ocular impairment is present in up to 90% of concussed patients, investigation of visual processing can be very helpful for concussion diagnosis. The goal was to determine whether eye-tracking measures can be utilized as a means to test horizontal vestibular ocular reflex in patients with concussion symptoms. Twenty-two participants were recruited and assigned to one of two groups: teens with concussion symptoms group and teens without concussion symptoms group. Concussed participants put on an eye-tracker while doing a horizontal vestibular ocular reflex test. While the non-concussed group was more successful in looking and concentrating on the target point, the concussed group had more undeliberate eye shakings. The variance between the fixation points of the target points was significantly greater in the concussed group in comparison with the non-concussed group. Concussed patients demonstrated more longitudinal and latitudinal deviations. In conclusion, the eye-tracking method is a useful tool to measure the severity of eye movement impairment.

Keywords: Concussions · Eye-tracker · Human factors · Horizontal vestibular ocular reflex · Saccades · Concussion diagnosis

1 Introduction

In the United States, between 1.6 and 3.8 million sports-related concussive injuries occur yearly [1, 2]. A concussion is known to be a type of Mild Traumatic Brain Injury (mTBI). This type of brain injury occurs from a sudden biomechanical force on the brain, which causes the brain to hit the inside wall of the skull. Complicating matters is the fact that concussed patients resist a positive diagnosis because they fear the impact such a diagnosis will have on their day-to-day lives or it will keep them from preferred activities such as sports [3] or driving. Research shows 43% of athletes with a history of concussions hide symptoms in order to be able to return to playing [4, 5]. The cognitive tests that are commonly used for concussion assessment have some issues. One of the main issues is the lack of a baseline test for some test methods.

While there is a functional disturbance and trauma to the brain due to the concussive event, follow up neuro-imaging typically shows no structural impairment to the

H. Ayaz and L. Mazur (Eds.): AHFE 2018, AISC 775, pp. 189–195, 2019.
https://doi.org/10.1007/978-3-319-94866-9_19

brain. Therefore, it is difficult to diagnose concussions because neuro-images of the brain usually do not show signs of any physical brain damage. Concussion diagnosis has always been a challenge; it does not necessarily include a loss of consciousness [6]. However, while physical damage is usually not found, there are certain psychological and physiological symptoms and signs that manifest themselves. Among these are vision related symptoms, which can be measured easily [4, 7].

Almost 50% of brain circuits are dedicated to vision [8]. Many of these circuits, which are fronto-parietal and responsible for eye movement [9], are those most frequently involved in concussions [10]. Since ocular impairment is existing in up to 90% of concussed patients [11], investigation of visual processing can be very helpful for concussion diagnosis. Patients with a concussion show some visual impairments, such as issues in anti-saccades, prolonged saccadic latencies, higher directional errors, poorer spatial accuracy, and impaired memory guided saccades [12, 13], even after 10 days. Mucha et al. [14] implemented a Vestibular/Ocular Motor Screening Assessment test in which they stimulated their subjects with vestibular and ocular stimulation. They found that the concussed group reported symptoms after researchers stimulated the underlying vestibular–ocular reflex or visual motion sensitivity.

Recently, eye-tracking has been used in concussion-related research [15–17]. The eye movement quality can be measured as an assessment criterion of the healing progress. Eye-tracking measures how well the patient's eyes are capable of moving, rather than what one chooses to look at [11].

Using an eye-tracker in executing cognitive tests as an assessment tool for a concussion diagnosis and healing progress can be beneficial. There are not many research studies executed for concussed patients. The existing research has been performed while the patients were not engaged in doing a cognitive task; instead, they were sitting and watching a movie or a video game [11, 18]. Caplan et al. [15] and Suh et al. [19] implemented research on concussed military participants and TBI patients, respectively, with limited saccadic cognitive tests.

2 Method

The goal of the current research study is to determine whether eye-tracking measurements can be utilized as a means to test horizontal vestibular ocular reflex in patients with concussion symptoms. A horizontal vestibular ocular reflex test was designed to determine whether a concussed group behaves differently when compared to a non-concussed group.

2.1 Participants

Twenty-two participants were recruited for this study, the teen cohort with concussions symptoms (11 concussed) and the teens without concussion symptoms cohort (11 concussed). Participants with history of other neurological or eye disorders like Parkinson's or colorblindness were excluded from the study. The concussed teen's cohort participants were among the patients who have been referred to Connecticut Children's Medical Center (CCMC) in Farmington, CT.

2.2 Eye-Tracker

A Tobii Pro-II mobile eye-tracker recorded participants' scanning pattern. The eye-tracker consists of a lightweight pair of glasses with the scene camera and eye cameras integrated into the frame. The system was calibrated via a wireless connection to the analysis laptop and video is recorded on a 3″ × 5″ × 0.5″ video recording unit clipped to the participant's belt. The participant's point of gaze, represented as a red circle, was superimposed upon the scene camera video image. The eye-tracker used in this study is shown in Fig. 1.

Fig. 1. Tobii Pro-II mobile eye-tracker. Reprinted from Tobbipro.com, from http://www.tobiipro.com/fields-of-use/psychology-and-neuroscience/customer-cases/audi-attitudes/. Copyright © 2017 Tobii AB. Reprinted with permission.

2.3 Experimental Procedure

The study took place in a single session. During this session, participants met the study administrator in a lab at Connecticut Children's Medical Center (CCMC) in Farmington, CT. Informed consents were received at this time, and the participants were given the opportunity to ask any questions.

Participants were asked to sit at a desk in front of a monitor. A stationary cross appeared in the center of the screen. The participants were asked to fixate on the cross with their eyes and then slowly rotate their head side to side while keeping their eyes focused on the cross for ten seconds (Fig. 2).

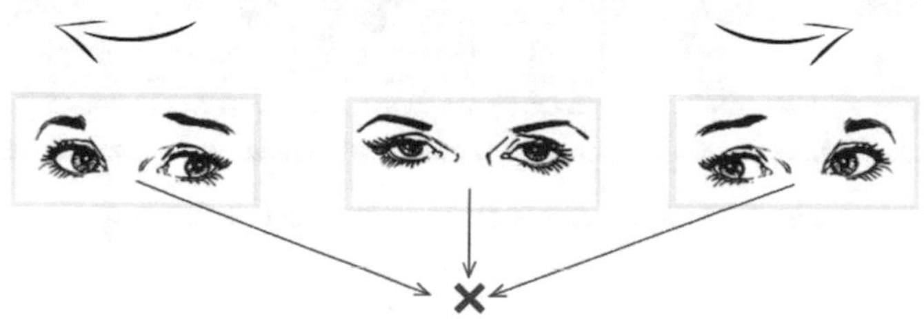

Fig. 2. Horizontal vestibular ocular reflex test

3 Results

For this experiment, the distances of each participant's fixation points from the target point were calculated. The distances include the longitudinal distance (X), the latitudinal distance (Y) and the total distance from the target point (D). The SPSS results show a significant difference between the fixation point's distances (X, Y and D) of the concussed group and the non-concussed group. This means that the participants with concussions symptoms were less able to focus on the target point and more likely to have longitudinal and latitudinal fixation points around the target points (Figs. 3, 4 and Table 1).

Table 1. ANOVA statistical analysis

		df	Mean square	F	Sig.
X	Between groups	1	192301987	1585.635	.000
	Within groups	67746	121277		
	Total	67747			
Y	Between groups	1	11196112	2005.345	.000
	Within groups	67746	5583		
	Total	67747			
D	Between groups	1	10584933	176.704	.000
	Within groups	67673	59902		
	Total	67674			

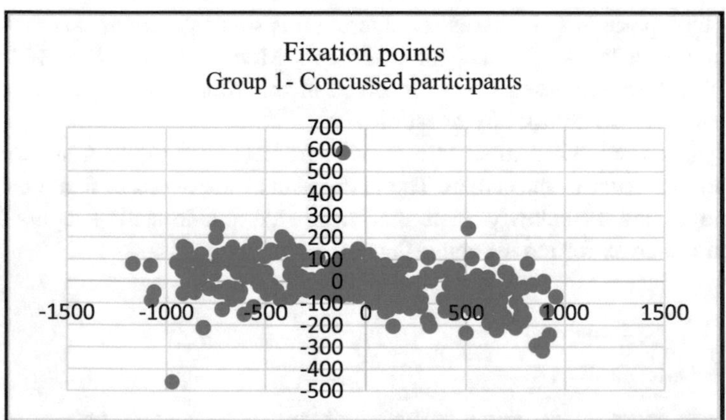

Fig. 3. Fixation points for the concussed participants

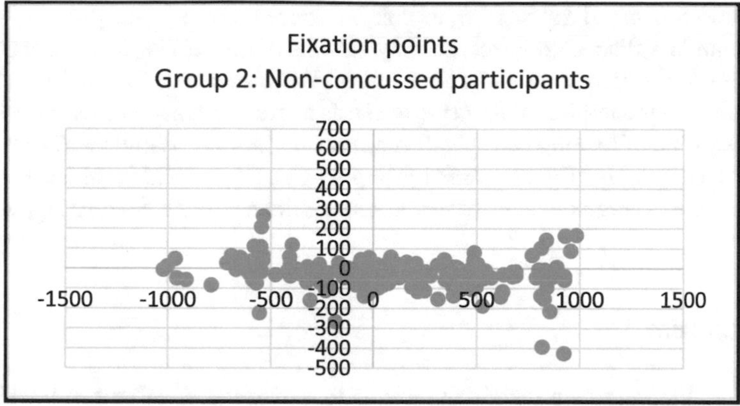

Fig. 4. Fixation points for the non-concussed participants

As visually demonstrated, the heat map and gaze plot of a concussed and a non-concussed participant is shown in Figs. 5 and 6. While the concussed patient has more random gaze points and a more wide-spread heat map, the non-concussed participant was able to focus at the target point and had less saccadic eye movements.

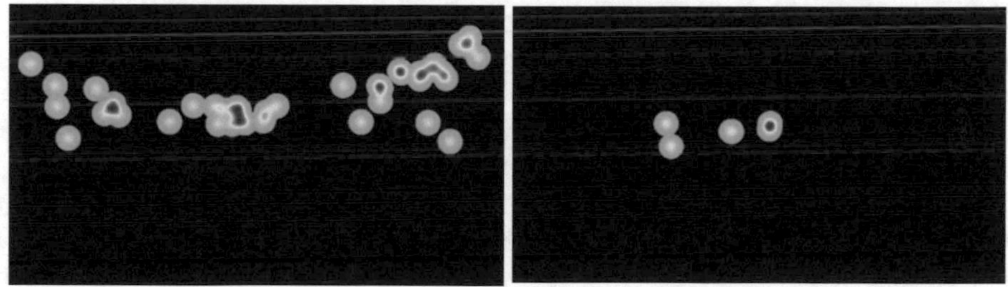

Fig. 5. Heat map of the concussed (left) and non-concussed (right) participants

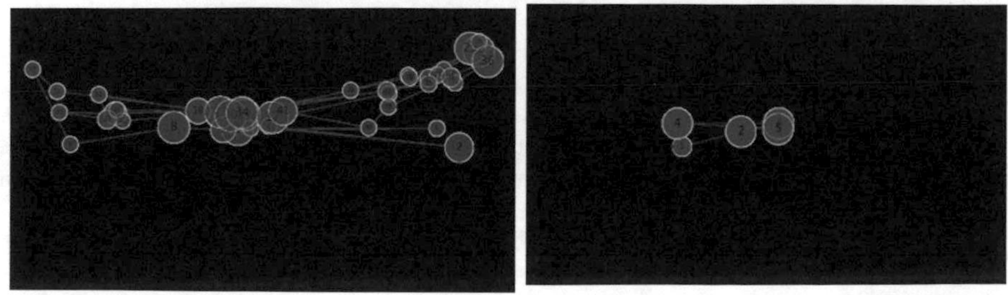

Fig. 6. Gaze points of the concussed (left) and non-concussed (right) participants

The non-concussed subjects were able to concentrate at each point more appropriately with less distracted gazes, so they showed a greater frequency, a mean center point closer to the zero point and a smaller standard distance deviation. This shows the level of their concentration ability due to the fact that the non- concussed participants are more capable of fixating at a specified point with less eye vibrations. The average of standard deviations for the concussed group was equal to 233.89, which is relatively higher than the average of standard deviations for the non-concussed group, which was 195.06.

4 Conclusion

The goal of the current research study was to determine whether eye-tracking measurements can be utilized as a means to test horizontal vestibular ocular reflex in patients with concussion symptoms. The expectation was to determine whether the concussed group behaves differently when compared to the non-concussed group and whether this method can be used in a concussion diagnosis.

In general, the eye-tracking results show that the patients with concussion symptoms showed a relatively higher level of undeliberate eye movements while doing a horizontal vestibular ocular reflex test. The concussed group showed significant differences in their fixation points' distances, either longitudinal, latitudinal or total distances.

While the non-concussed patients were more successful in keeping their eyes focused at the target, the concussed group had more saccadic eye movements and a lesser ability to keep their focus to one point. The concussed group had more unintended fixation and gaze points while the non-concussed group had fewer fixation and gaze points, which were closer to the target point.

The contribution of this study is that the doctors would be able to diagnose a concussion by using an eye-tracker while implementing cognitive tests such as a horizontal vestibular ocular test. In addition, they will be able to assess the healing trend of concussed patients by using this method in the follow-up test.

References

1. Langlois, J.A., Rutland-Brown, W., Wald, M.M.: The epidemiology and impact of traumatic brain injury: a brief overview. J. Head Trauma Rehabil. **21**, 375–378 (2006)
2. McCrory, P., Meeuwisse, W.H., Aubry, M., Cantu, B., Dvořák, J., Echemendia, R.J., Engebretsen, L., Johnston, K., Kutcher, J.S., Raftery, M.: Consensus statement on concussion in sport: : the 4th international conference on concussion in sport held in Zurich, November 2012. Br. J. Sports Med. **47**, 250–258 (2013)
3. Chrisman, S.P., Quitiquit, C., Rivara, F.P.: Qualitative study of barriers to concussive symptom reporting in high school athletics. J. Adolesc. Health **52**, 330–335 (2013)
4. Register-Mihalik, J.K., Guskiewicz, K.M., McLeod, T.C.V., Linnan, L.A., Mueller, F.O., Marshall, S.W.: Knowledge, attitude, and concussion-reporting behaviors among high school athletes: a preliminary study. J. Athl. Train. **48**, 645–653 (2013)

5. Torres, D.M., Galetta, K.M., Phillips, H.W., Dziemianowicz, E.M.S., Wilson, J.A., Dorman, E.S., Laudano, E., Galetta, S.L., Balcer, L.J.: Sports-related concussion anonymous survey of a collegiate cohort. Neurol. Clin. Pract. **3**, 279–287 (2013)
6. Carroll, L., Cassidy, J.D., Peloso, P., Borg, J., Von Holst, H., Holm, L., Paniak, C., Pépin, M.: Prognosis for mild traumatic brain injury: results of the WHO Collaborating Centre Task Force on Mild Traumatic Brain Injury. J. Rehabil. Med. **36**, 84–105 (2004)
7. McCrory, P., Meeuwisse, W., Johnston, K., Dvorak, J., Aubry, M., Molloy, M., Cantu, R.: Consensus statement on concussion in sport-the 3rd international conference on concussion in sport held in Zurich, November 2008: consensus. South Afr. J. Sports Med. **21**, 36–46 (2009)
8. Felleman, D.J., Van Essen, D.C.: Distributed hierarchical processing in the primate cerebral cortex. Cereb. Cortex **1**, 1–47 (1991)
9. White, O.B., Fielding, J.: Cognition and eye movements: assessment of cerebral dysfunction. J. Neuroophthalmol. **32**, 266–273 (2012)
10. Galetta, K.M., Morganroth, J., Moehringer, N., Mueller, B., Hasanaj, L., Webb, N., Civitano, C., Cardone, D.A., Silverio, A., Galetta, S.L., Balcer, L.J.: Adding vision to concussion testing: a prospective study of sideline testing in youth and collegiate athletes. J. Neuroophthalmol. **35**, 235–241 (2015)
11. Samadani, U., Ritlop, R., Reyes, M., Nehrbass, E., Li, M., Lamm, E., Schneider, J., Shimunov, D., Sava, M., Kolecki, R., Burris, P., Altomare, L., Mehmood, T., Smith, T., Huang, J.H., McStay, C., Todd, S.R., Qian, M., Kondziolka, D., Wall, S., Huang, P.: Eye tracking detects disconjugate eye movements associated with structural traumatic brain injury and concussion. J. Neurotrauma **32**, 548–556 (2015)
12. Heitger, M.H., Anderson, T.J., Jones, R.D., Dalrymple-Alford, J.C., Frampton, C.M., Ardagh, M.W.: Eye movement and visuomotor arm movement deficits following mild closed head injury. Brain **127**, 575–590 (2004)
13. Heitger, M.H., Anderson, T.J., Jones, R.D.: Saccade sequences as markers for cerebral dysfunction following mild closed head injury. Prog. Brain Res. **140**, 433–448 (2002)
14. Mucha, A., Collins, M.W., Elbin, R.J., Furman, J.M., Troutman-Enseki, C., DeWolf, R.M., Marchetti, G., Kontos, A.P.: A brief vestibular/ocular motor screening (VOMS) assessment to evaluate concussions: preliminary findings. Am. J. Sports Med. **42**, 2479–2486 (2014)
15. Caplan, B., Bogner, J., Brenner, L., Cifu, D.X., Wares, J.R., Hoke, K.W., Wetzel, P.A., Gitchel, G., Carne, W.: Differential eye movements in mild traumatic brain injury versus normal controls. J. Head Trauma Rehabil. **30**, 21–28 (2015)
16. Cifu, D.X., Gitchel, G.: Effects of hyperbaric oxygen on eye tracking abnormalities in males after mild traumatic brain injury. J. Rehabil. Res. Dev. **51**, 1047 (2014)
17. Heitger, M.H., Jones, R.D., Macleod, A.D., Snell, D.L., Frampton, C.M., Anderson, T.J.: Impaired eye movements in post-concussion syndrome indicate suboptimal brain function beyond the influence of depression, malingering or intellectual ability. Brain **132**, 2850–2870 (2009)
18. Lau, B.C., Collins, M.W., Lovell, M.R.: Sensitivity and specificity of subacute computerized neurocognitive testing and symptom evaluation in predicting outcomes after sports-related concussion. Am. J. Sports Med. **39**, 1209–1216 (2011)
19. Suh, M., Basu, S., Kolster, R., Sarkar, R., McCandliss, B., Ghajar, J.: Cognitive and Neurobiological Research Consortium, et al. Increased oculomotor deficits during target blanking as an indicator of mild traumatic brain injury. Neurosci. Lett. **410**, 203–207 (2006)

Research on the Brain Mechanism of Visual-Audio Interface Channel Modes Affecting User Cognition

Wenqing Xi[1], Lei Zhou[2(✉)], Huijuan Chen[1], Jian Ma[2], and Yueting Chen[2]

[1] Science and Technology on Avionics Integration Laboratory, Chinese Aeronautical Radio Electronics Research Institute (CARERI), 432 Gui Ping Road, Shanghai, China

[2] School of Mechanical Engineering, Southeast University, 2 Dong Nan Da Xue Road, Jiangning District, Nanjing, China

zhoulei@seu.edu.cn, jjjj0823@sina.com

Abstract. In multi-channel design, the combination of visual-audio dual channels is the most widely used way. This article elaborated on the characteristics of visual and audio channel, clarified the complementary relationship between visual and audio channels, and analysed the combination methods of visual and audio channels. EEG/ERP technology was used to carry out experiments and some alarm information designs of a real-time monitoring software system were used as an example to compare the cognitive differences caused by the combination of visual and audio channels' combination. Through the quantitative analysis of indicators such as reaction time, brain wave peak value, and latency, the interface information design and evaluation methods for visual-audio dual channels were developed.

Keywords: Visual · Audio · Cognition · Brain wave · Interface design

1 Introduction

The process of human perception of external stimuli is multi-channel parallel. Human cognitive information mainly relies on visual channels. Its usage rate reaches 70%, and the auditory channel usage rate reaches 20%. The remaining 10% is derived from touch, smell and taste. The stimulating effect produced by mobilizing multiple channels are much greater than that presented by a single channel [1–3]. Much sensory collaboration makes the human experience in the cognitive process full and three-dimensional. The purpose of the multi-channel human-computer interface design is to improve the naturalness and reliability of the information acquisition by rationally assigning and invoking perception channels [4–7]. However, the parallelism of the channels is not superimposed mechanically. How to coordinate the cooperation mechanism between channels and achieve the best perceptual effect is the difficulty in the current multi-channel design research [8–10].

H. Ayaz and L. Mazur (Eds.): AHFE 2018, AISC 775, pp. 196–204, 2019.
https://doi.org/10.1007/978-3-319-94866-9_20

2 Information Design Methodology of Visual-Audio Interface Channel Modes

2.1 Visual Channel

Vision has the following advantages in the information acquisition: direct information acquisition, high information perception efficiency, and fast acceptance; it can perceive the color, size, position, texture, sense of space, and movement tendency of the target. The visual channel has strong contrast ability to the picture perception; the visual has the selective attention feature; the vision has higher sensitivity and adaptability. Its disadvantages are: limitation of sight span, strong dependence on light, visual fatigue, misinterpretation; visual organs are easily damaged, and recovery after injury is difficult; Vision requires active attention, and intentional attention can be applied to perceptual processing; visual stimuli are limited by azimuth. For visually normal users, vision is the main channel in human-computer interaction, and there are also certain rules in practical applications.

2.2 Audio Channel

Hearing is the most important channel to experience outside information in addition to the visual sense. Hearing occurs because the ears are stimulated by a certain intensity of sound. From the perspective of human-computer interaction design, auditory stimuli are divided into two categories, speech auditory and non-voice auditory. Speech sound depends on language and is a unique human product with language function. Non-voice auditory research objects include Earcons and auditory icons, which are used to represent the sound stimuli of structured information and the sound stimuli associated with life memories. Some scholars conducted a comparative study of the speech auditory interface and the non-voice auditory interface, and considered that the independence of speech auditory is poor and restricted by language culture. Non-voice auditory has better independence and versatility while it is not affected by speech speed. At the same time Earcons as a non-verbal auditory in information interaction has a wide range of applications in the interface design [4–6]. The auditory perception capability has the following advantages: fast transmission, large amount of information, and strong coherence; reception of auditory stimulation is not limited by the space of the sound source and does not require physical contact; people can perceive the sense of orientation and spatial sense of stimulation. People have strong ability to analyze, distinguish and adapt to sound stimulation; auditory organs have strong fatigue resistance. The disadvantages are: hearing has a masking effect and is easily disturbed by environmental noise; it cannot be duplicated and copied; the hierarchical structure of sound stimulation is not obvious; people receive language information when they are restricted by language and culture.

2.3 Visual-Audio Dual Channel Design Method

Multi-channel human-computer interaction emphasizes the fit between channels. If there is no tacit agreement and support between channels, it can not only effectively increase the bandwidth of human-machine interaction, but may cause users' cognitive burden. The visual channel is often used to convey complex, abstract, information that does not require urgent transmission, and hearing can often be more easily noticed in information acquisition and have a higher timeliness. The combination of vision and hearing is the most common in the current multi-channel human-computer interaction interface, and the difficulty of achieving the technology is low [7–10]. Table 1 summarizes the combination of visual and auditory elements.

Table 1. Combination methods of visual and audio channels

No.	Visual elements	Auditory elements	Applicable conditions	Examples	Limitation
1	Color	Language hearing	1. Hysteresis or state changes in the span of the visual area 2. Alarm information 3. Qualitative judgment	"Pair" (green) and "Wrong" (red) indications for operational hints	Not suitable for characterizing a wide variety of properties
2	Text	Listening	1. Operational hints for cognition of text and audio	Right click on the computer to select "Delete" and confirm the sound of rubbing waste paper. The two form a cognitive relationship	Needs user pre-training for best results
3	Color/text	Earcons	1. State prompts (start, finish, etc.) 2. Out of sight information reminder	Intelligent rice cookers use indicator colors and panel text to display rice cooking status and use voice to remind users to pay attention	Limited amount of information
4	Text	Language hearing	1. Information guidance 2. Intuitive information emphasized 3. Quantification Information Tips 4. Alarm Information	The leading mode of combined text and speech in ATM machine operation	Limited by language, speed, and text information synchronization
5	Text/shape	Earcons (broken tone, frequency change)	1. Qualitative cues with tropism 2. Characterization of task status changes	Combining text and line graphs in the ventilator, using "drip" sound frequency changes to characterize vital signs and dangerous states	Needs user pre-training for best results

3 ERP Experiment Design of Visual and Audio Dual Channel Information Coding Modes Affecting User Cognition

3.1 Experiment Content

A real-time monitoring system software interface was taken as the experimental background. As shown in Fig. 1, the subjects were used as monitoring personnel to find out the fault points in the interface and respond through keystrokes. The presentation of failure points is divided into two types: a. Visual presentation, the normal site is a blue site, and in the event of a failure, the site turns red, as shown in Fig. 1; b. Visual + auditory presentation. When a fault occurs, the station turns red and the system sounds a "di" alarm, which ends with the completion of the keystroke. c. As the visual + auditory presentation, when the fault occurs, the station size changes and the system issues a "di" alarm sound, which ends with the completion of the keystroke. By comparing the response time, brain wave peak value, latency and other indicators to evaluate, so as to get the best information presentation mode of the alarm information in the digital interface.

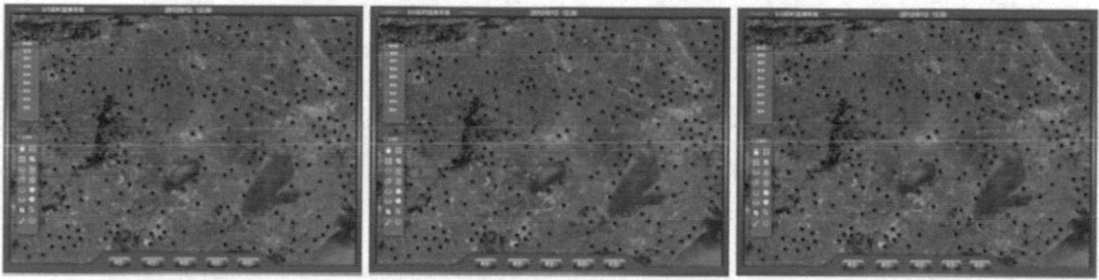

Fig. 1. Experiment samples. The first scene was the normal status, the second scene was the alarming status with the abnormal point turn red, the third scene was the alarming status with the abnormal point turn bigger, and a "di" sound, the second scene also used in another status with an additional "di" sound.

A total of 120 fault samples were presented in this experiment (The color change alarm was presented 40 times, the color change + prompt tone alarm was presented 40 times, the size change + prompt tone alarm was presented 40 times). Each time there was a point of failure, the point of failure was presented randomly. The subjects needed to find out the point of failure and hit a key to respond.

3.2 Experimental Instruments

The NeuroOne EEG/ERP 32 guidance system developed by Mega was used in this experiment, which was produced by a Finnish company, with 24 bit high resolution, 0–10000 Hz filter band-pass, 4000 Hz per point sampling frequency, electrode and scalp contact resistance less than 5000 Ω. The reference electrodes (electrode with a relatively zero potential on the body) were placed on the bilateral papillae (dual prominences) connection. The ground electrode was at the midpoint of the line between FPZ and FZ electrodes, and the vertical eye-wave was recorded.

3.3 Experimental Subjects

The test users were 25 students (13 males and 12 females) of Southeast University, including undergraduates, postgraduates and doctoral students. Among them were 7 undergraduates, 10 masters, and 8 doctoral students, between 20–35 years old, the average age was 24 years old, no color blindness, color weakness, corrected visual acuity above 1.0. Subjects need to practice training before the experiment, and should be familiar with the experimental process and operation requirements. The experiment was carried out in the ERP laboratory. After the subject took up the electrode cap and sat comfortably in front of the screen, the eyes were 550–600 mm away from the screen. During the experiment, both the horizontal and vertical viewing angles of the subjects were controlled within 2.3°.

3.4 Experimental Procedure

After 500 ms blank screen with a cross in the center, an ordinary system picture appeared and lasted for 2000 ms, then the fault point appears, the subject needed to find out the fault point, and make a judgment, pressing “A” meant the stimuli be in the left half of the interface, while pressing “L” meant be in the right half, a blank screen appeared and visual residue was removed. The next test sample was entered and 120 samples were presented randomly. The experimental paradigm was shown in Fig. 2.

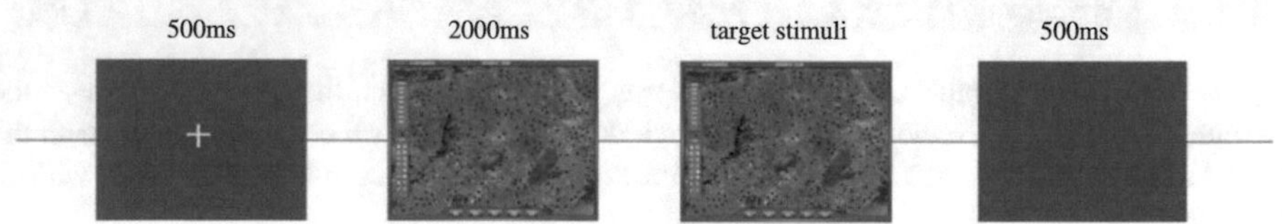

Fig. 2. Experimental pattern process

3.5 Data Analysis and Analysis Process

1. Analysis of behavioral data: E-prime2.0 software was used to collect and analyze the response time and accuracy of the experiment.
2. Analysis of EEG data: Analyzer data was used to analyze the experimental data. Firstly, the reference electrodes (TP9 and TP10) were converted. The ICA method was used to remove the electric eye artifacts (using Fp1 vertical eye as reference) and filtering. Artifacts less than 20% were removed, brainwave segmentation was performed and baseline correction and average superposition were completed. After all brain wave processing was completed, the brain waves of all subjects were overlaid and averaged for the same segment of brain waves to obtain the final brain wave waveform. According to the electroencephalogram, significant components in

brain waves, latencies, peaks, etc. were discovered. The ERP data were superimposed on the correct response EEG according to the level of the prompt range, and the target stimuli were superimposed as analysis points. The peak and latency of each component were measured, and two-factor repeated measures analysis of variance and paired T-test were performed using SPSS.

4 Analysis of Experimental Results

4.1 Behavioral Data

Among the experimental behavior data, the test number was 25, but the effective data were only 21 people. Behavioral data included the accuracy and reaction time of the recognition of the target stimuli. Through the analysis of the correct rate of the experiment, it was found that the correct rate in all the three cases were more than 99%, so the accuracy rate data did not have the significance of analysis. Therefore, this experiment only analyzed the reaction time. Under the stimulation of different presentation channels, the average values of the target stimuli RT were: visual channel color coding (496.63 ms) > visual-audio channel with size and Earcon dual coding (487.35 ms) > visual-audio channel with color and Earcon dual coding (454.1 ms).

4.2 ERP Data

1. Visual channel and visual-audio channels

Seven electrodes such as the central left-top region (CP5, CP1), the central right-top region (CP6, CP2), and the top region (P3, P4, PZ) were selected as the analysis electrodes. The three types of brainwaves were analyzed separately. From the beginning of target stimulus to 1000 ms was set as the EEG segmentation time, after the Grand Average, the brainwave waveforms of all the subjects were superimposed to obtain the EEG waveforms after the electrodes were superimposed. It was found that in the visual coding of the visual channel, N100 and P300 were more prominent in brain tissue components, and were more pronounced in the apical region, the central left apical region, and the central right apical region. In the double encoding of audio-visual channel size Earcons double coding and color Earcons, the N100, P200, and P300 were more obvious in the brain electricity component, and were more obvious in the top area, central left top area, and central right top area. Therefore, the P200 component was related to the auditory channel coding, which was the auditory-induced EEG component. In addition, from the EEG waveform, it could also be seen that for the incubation period of the N100 component simultaneously induced by the two types of presentation (visual, visual + audio), the latency of visual + audio presentation was significantly shorter than that of the visual presentation, and N100 was a kind of Pay attention to early components, which showed that visual + audio presentation could get people's attention faster. The peak value of the brain waves displayed by visual + audio was significantly higher than that of the visually presented brain waves, indicating that the visual and audio presentation had a higher degree of activation of the human brain. EEG conclusions were consistent with the conclusions of behavioral experiments.

2. Cognitive differences caused by differences in visual-audio channel coding

For the N100 component, statistical analysis was performed on the average peak value of EEG within −200 ms–300 ms of the target point. Do repeated analysis variance analysis of 2 (different coding changes: color change, size change) × 3 (area: central left top area, central right top area, and top area). Analysis showed that there was a significant main effect in the region ($F = 35.199$, $p = 0.000 < 0.05$) and there was no significant difference between coding stimuli ($F = 0.629$, $p = 0.629 > 0.05$). There was also no significant interaction between regions and coding stimuli ($F = 2.793$, $p = 0.085 > 0.05$).

In the visual + audio presentation mode, paired sample T-tests with different coding stimuli were performed on the central left top region, central right top region, and top region. The results showed that the color of the stimulus and the size of the stimulus, the absolute value of the central right top area of the absolute value were greater than the central left top and the top area (color change: 11.9115 uv > 11.0648 uv > 6.7692 uv, p are less than 0.05; scale changes: 12.3386 uv > 11.6397 uv > 6.8001 uv, p all less than 0.05). In both cases, the N100 was significantly more prominent in the central right apical area and the central left apical area than in the central apical area. The central right apical area was also significantly larger than the left apical area.

For the P200 component, the average peak value of EEG within 100 ms-400 ms of the target point was selected for statistical analysis. The repeated variance analysis of a 2 (different coding stimuli: color change, shape change) × 3 (region: central left top region, central right top region, top region) was conducted. The analysis obtained that the region had a significant main effect ($F = 5.024$, $p = 0.017 < 0.05$) and there was a significant difference between the different coding stimuli ($F = 5.758$, $p = 0.026 < 0.05$). There was also a significant interaction between coding stimuli ($F = 5.047$, $p = 0.017 < 0.05$).

In the visual + audio presentation mode, paired sample T-tests with different visual stimuli were performed on the central left top region, the central right top region, and the top region. The results showed that under visual + audio presentation, the color change stimulus and the shape change stimulus, the absolute value of the central right top region mean value were greater than the central left top and top regions. (Color change: 5.5352 uv > 5.3333 uv > 3.1486 uv; size change: 7.7961 uv > 7.4593 uv > 3.9533 uv). In addition to the difference between the central left apical region and the central right apical region, p value was not significantly greater than 0.05, and other comparisons were significant and p value was less than 0.05. In both cases, the difference between P200 in the central left apical region and the central apex region was significant. The difference between the central right apical region and the central apex region was also significant, but the difference between the central right apical region and the left apical region was not significant.

For the stimuli, there was a significant difference between the stimuli for color change and size change because of the significant difference between the stimuli

(F = 5.758, p = 0.026 < 0.05). According to the mean value of the regional T-test, the relationship among the mean value of the central right top zone, the central left top zone, and the top zone were: color change 5.5352 uv, 5.3333 uv, 3.1486 uv; size change 7.7961 uv, 7.4593 uv, 3.9533 uv. Therefore, M (color change) < M (size change), so the peak value of the size change was greater, and the degree of brain activation was higher.

For the P300 component, statistical analysis was performed on the average peak value of EEG within 100 ms–600 ms of the target point. Do repeated analysis variance analysis of 2 (different visual coding stimulus: color change, size change) × 3 (region: central left top region, central right top region, and top region). Analysis showed that there was no significant main effect in the region (F = 1.394, p = 0.271 > 0.05), and there was no significant difference between the different coding stimuli (F = 0.144, p = 0.708 > 0.05). There was also no significant interaction between regions and coding stimuli (F = 2.637, p = 0.096 > 0.05).

In addition, as can be seen from the EEG waveforms in Fig. 3, in the visual and audio presentation mode, the latency of the N100, P200, and P300 components caused by the color change and size change was not significant, which showed that these two coding stimuli were close to human stimulation. It could be seen from the EEG waveform that, in the visual-audio presentation mode, the peak value of the size factor was slightly higher than the color factor, indicating that the size change had a higher degree of activation of the human brain.

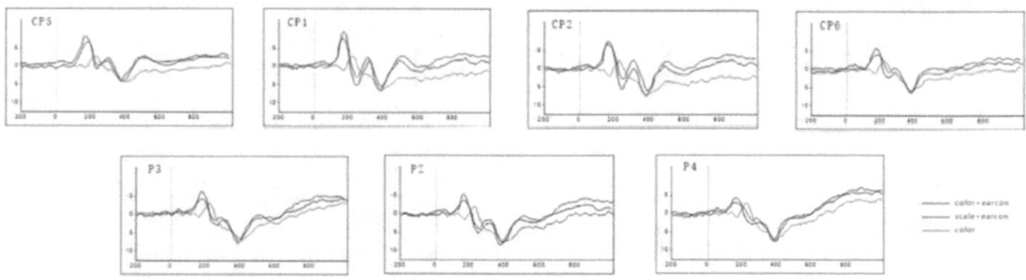

Fig. 3. ERP waves

5 Conclusion

Based on the analysis of the results of the above experiment, the conclusions are drawn as follows: For the presentation of digital interface alarm information, compered with visual coding, visual-audio dual channel is better performed. In highlighting fault information, people are more sensitive to color changes than size changes, but size changes have a higher degree of activation of the human brain, and in specific applications, sensitivity and activation are also related to the degree of information encoding changes, which needs further study.

Acknowledgments. This paper is supported by Science and Technology on Avionics Integration Laboratory and Aeronautical Science Fund (No. 20165569019).

References

1. Li, T., Wang, D., Peng, C., Yu, C., Zhang, Y.: Speed-accuracy tradeoff of fingertip force control with visual/audio/haptic feedback. Int. J. Hum. Comput. Stud. **110**, 33–44 (2018)
2. Grega, J., Christina, D., Jaka, S.: A user study of auditory, head-up and multi-modal displays in vehicles. Appl. Ergon., Part A **46**, 184–192 (2015)
3. Andrea, B., Ian, O., Dong, S.K.: Counting clicks and beeps: exploring numerosity based haptic and audio PIN entry. Interact. Comput. **24**(5), 409–422 (2012)
4. Hans-Jörg, S., Steffen, H.: Preset-based generation and exploration of visualization designs. J. Vis. Lang. Comput., Part A **31**, 9–29 (2015)
5. Abbate, A.J., Bass, E.J.: A formal methods approach to semiotic engineering. Int. J. Hum. Comput. Stud. **115**, 20–39 (2018)
6. Hooten, E.R., Hayes, S.T., Julie, A.A.: Communicative modalities for mobile device interaction. Int. J. Hum. Comput. Stud. **71**(10), 988–1002 (2013)
7. Antonio, M.R.: A multimedia ontology model based on linguistic properties and audio-visual features. Inf. Sci. **277**, 234–246 (2014)
8. Timothy, E.R., Ian, P., David, D.: Toward mobile entertainment: A paradigm for narrative-based audio only games. Sci. Comput. Program. **67**(1), 76–90 (2007)
9. Gabrielle, L.H., Eric, L., Cara, E.S.: Effects of augmentative visual training on audio-motor mapping. Hum. Mov. Sci. **35**, 145–155 (2014)
10. Jeffrey Jr., L.C., Robert, H.G., Brian, D.S.: The VERITAS facility: a virtual environment platform for human performance research. IFAC Proc. **46**(15), 357–362 (2013)

How Does the Mobile Phone PPI Design Affect the Visual Acuity with the Change of Viewing Distance?

Yunhong Zhang[1(✉)], Wei Li[1], Jinhong Ding[2], Anqi Jiao[2], Hongqing Cui[3], and Yilin Chen[3]

[1] AQSIQ Key Laboratory of Human Factor and Ergonomics, China National Institute of Standardization, Beijing, China
zhangyh@cnis.gov.cn

[2] School of Psychology, Capital Normal University, Beijing, China

[3] China Star Optoelectronics Technology Co., Ltd., Wuhan, China

Abstract. With the development of display manufacturing technology, the PPI of mobile phone screens has been upgrading. But how the visual acuity of the mobile phone has been changed with the increase of the mobile phone PPI has been the issues of concern among the manufacturers all the time. In this study, we use adaptive method to measure people's vision acuity level under different viewing distance conditions, so as to explore how mobile phone PPI affects human visual acuity under different distances. This study measured the visual acuity of 15 participants with different visual acuity level, and let them to watch different PPI mobile phones at different distance conditions. This study found that in general, with the increase of mobile phone PPI, the visual acuity of subjects increased. This indicates that the PPI of different mobile phones affects the visual acuity in the process of viewing the mobile phone. Under the view distance within 30 cm, the difference of visual perception threshold between different PPI mobile phones was significant, and the PPI effect of mobile phone disappeared after more than 30 cm. A group of participants whose visual acuity level was 5.0, for example, watch their mobile phone at close range (distance < 30 cm), PPI value fitting curve of mobile phone actual measurement angle is lower than the theoretical value curve; but watch their mobile phone at a far range (distance $\geq$ 30 cm), the actual PPI value curve approaches the theoretical value curve, showing the optimization effect of the phone PPI on the visual acuity.

Keywords: Intelligent mobile phone · PPI · The adaptive method
The visual acuity fitting curve

1 Introduction

With the rapid development of internet technology and mobile electronic products, people are getting more and more dependent on mobile phones in the process of obtaining external information and communicating with others. The good or bad of a mobile phone display is directly dependent on the resolution of the mobile phone. Cell phone resolution PPI (Pixels Per Inch) is also called pixel density, means that the

H. Ayaz and L. Mazur (Eds.): AHFE 2018, AISC 775, pp. 205–214, 2019.
https://doi.org/10.1007/978-3-319-94866-9_21

number of pixels per inch. It is mainly used to define the fine degree of the screen of a mobile phone. The higher the PPI value is, the higher the display displays the density of the image, and the higher the fidelity. As far as the senses are concerned, the visual effect of mobile phone screen imaging to consumers is more direct-viewing. Satoru Kubota and others [1] have found that the subjective experience of the participants will be better with the increase of the screen resolution of the mobile phone. The new PPI standard was put forward by the Apple Corp. When the Apple Corp released iPhone4 in 2010, it proposed a concept called "Retina", which says the highest pixel density that human eyes can distinguish is PPI 300. From 2012, PPI that was greater than 300 of the retina screen has become the standard of the big mobile phone manufacturers. But the standard of 300 PPI has also been questioned by many researchers, it is considered that there is no consideration of human retinal characteristics, viewing distance, and cell phone size and eyesight level of human eye. And there is no clear conclusion on how the screen resolution affects human screen fineness perception. Japanese scholar Kubota, S. used forced choice method to evaluate the perceived quality of different PPI phone screens on different fonts, and found that subjective image quality was improved when pixel density increased [1]. But when PPI is between 706 and 806, there is no significant difference in the quality of the subjective image; yet the perception of the edge of the serrated body character, even on the 806 PPI screen, can still be perceived; the visual acuity has no effect on the quality of the image, but it affects the perceptual on the edges characters of Serrated body. People with higher eyesight can better perceive the edge characters of serrated body. It is a subjective research. It is still essential to discuss whether the PPI of mobile phone is consistent with the vision curve of human being measured by actual measurement.

In this study, we used adaptive staircase method to verify the relationship between the change of cell phone PPI and human vision from the perspective of human visual perception threshold, and studied whether the level of human vision would increase with the increase of screen resolution of mobile phone. Adaptive staircase method is the classic method of threshold measurement, that is, the degree of the next stimulus is determined by the computer based on the response of the subject. Compared with the traditional fixed font size interval test method, this method can improve the efficiency of the experiment [2]. This study measured visual acuity using an adaptive staircase method, the four groups of subjects at different distances in different PPI screen on the threshold of discrimination, on the minimum size of subjects can correctly identify the international standard measurement character "E" direction, explore the visual level, viewing distance and mobile phone screen resolution, and fitting out of mobile phone the screen PPI, visual acuity and viewing distance visual acuity curve.

2 Method

2.1 Experimental Design

The experiment was designed with 3 factors (6 phones * 5 distances * 4 levels). The independent variable was different cell phone, different sight distance and different vision level. The visual level was 4 grades: 5.0 level of visual acuity, 4.5 level of visual acuity, 4.2 level of visual acuity, 3.8 level of visual acuity; The sight distance mainly refers to the distance between the eyes and the screen of the mobile phone, which was divided into 5 grades, and they are 10 cm, 20 cm, 30 cm, 40 cm, 50 cm; Mobile phone PPI are 200, 294, 400, 565 and 801, and the type and distance of the mobile phone were the variables in the subjects, and the visual acuity (levels) was the inter trial variable.

2.2 Participant

There were 15 participants, 9 women and 6 men, average age was 23.6 years old, and the average daily mobile phones use time was 4.50 h. The visual acuity range was from 0 to 800°, and the visual acuity of 5, 4.5 and 4.2 were 4 people, and 3.8 of the visual acuity were 3 people. All participants were healthy, normal visual acuities without color blind and ophthalmic diseases. And they were paid to participate in the experiment when the experiment was over.

2.3 Experimental Materials

The experimental material was converted to the mobile phone for the standard log remote visual acuity chart, as shown in Fig. 1.

2.4 Experiment Process

Experiments were conducted in a quiet laboratory that experimental environment illumination value was 500 lx, and the brightness of the mobile phone was controlled around 200 nits. During the whole experiment, the subjects were all sitting on the right side of the experimenter. Before the formal experiment, the experimenter first explained the instructions to the subjects, explained the experimental requirements, operation procedures, instructions for use and matters needing attention in the experimental operation, and then asked the subjects to fill out the user background information questionnaire. Then the subjects were given 2 groups of exercises in order to get the subjects to be familiar with the test process. In the formal experiment, the subjects sat in the natural state at first. According to the standard visual acuity test method, the visual acuity of subjects was measured at five levels of the subjects' eyes from the mobile phone 10 cm, 20 cm, 30 cm, 40 cm and 50 cm. The word "E" was first transferred to the size that the subjects could see. Then, when the participants correctly responded, the word "E" decreased correspondingly. When the participants chose the wrong direction of "E" for the two time, they ended the program. In the course of the experiment, the participants only need to answer the direction of "E", and the main test was to respond to the key. The whole experiment lasted about 1 h. The subjects got a certain reward at the end of the experiment.

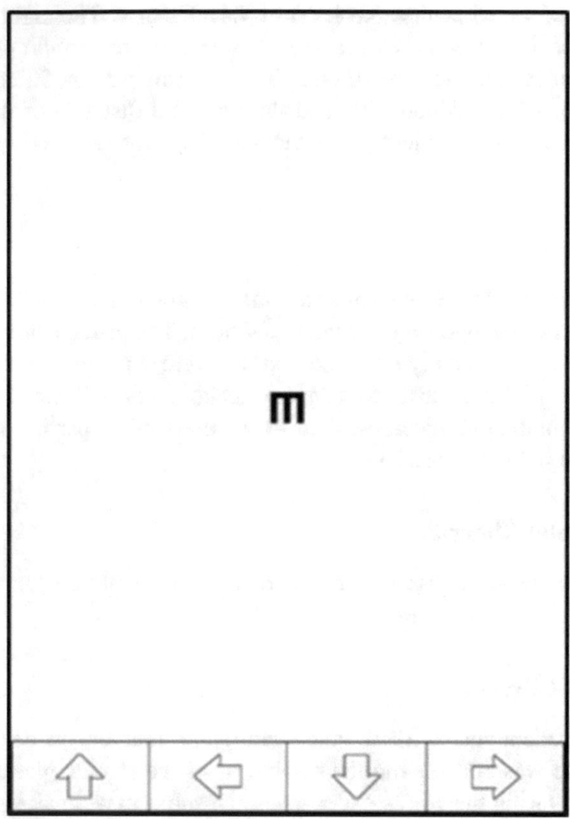

Fig. 1. Experimental materials

2.5 Data Analysis

In the process of data analysis, the minimum E size that the test can be identified was converted to the mobile phone screen resolution (PPI), and the conversion formula was as follows:

$$n \geq \frac{0.0635}{d * \tan \frac{1}{2} \alpha} \tag{1}$$

Among them, n is the value of mobile phone screen resolution that people need to see in the direction of minimum "E", alpha is the perspective of the word "E", and d is the distance between subjects from the screen. The higher the level of visual acuity measured on the cell phone, the higher the visual sensitivity of the subjects to be converted to the resolution, indicating the better the display effect of the mobile phone screen.

3 Experimental Result

3.1 Different Performance of Visual Acuity on Different Mobile Phones

The visual acuity was converted to the resolution value, and repeated measurement variance analysis was performed. Results were shown in Table 1. On the different PPI mobile phone screens, the discriminant threshold of subjects was significantly different ($F(5, 55) = 25.301$, $p < 0.001$). The results of the post analysis showed that there was no significant difference (M.D. + S.E. = 0.03 ± 0.02, $p = 0.200$) between the discrimination threshold (M. = 4.73) and the discrimination threshold (M. = 4.76) of the 200 PPI cell phone. The discrimination threshold of the 200 PPI mobile phone screen was significantly different from the discrimination threshold between the mobile phone screens above 400 PPI, respectively. There were significant differences in discrimination threshold of screen discrimination between 293 PPI and more than 400 PPI ($ps < 0.01$), and the discrimination threshold of other conditions was not significant. Figure 2 is the pixel density value of the mobile phone's visual acuity. It can be seen from the graph that with the increase of the screen resolution of the cell phone, the eyesight level of the subjects was also improved. Although the resolution of SONY mobile phone was very high, the 4 pixels in the display process represent one pixel, so the actual resolution was lower than 400 PPI. The results of this study divide the resolution of mobile phone into two categories: one was below 300 PPI, the other was at a lower level, and the other was 400 PPI with higher resolution.

Table 1. Variance analysis results for repeated measurements of different conditions

Effect	df	SS	F value	p value
Phone type	4	326860.8	36.9	0.000**
Visual range	4	2258281.7	59.5	0.000**
Phone type * Visual range	16	115247.3	17.0	0.000**

Note: df is the degree of freedom, SS is the sum of the total square,*p < 0.05, **p < 0.01

There were significant differences in discrimination threshold results under different present distance conditions, $F(4, 44) = 59.48$, $p < 0.001$. After the post test, it was found that with the narrowing of the sight distance, the vision level was getting better and better, and all conditions reached to significant levels ($ps < 0.001$). Figure 3 is the pixel density map of the subjects under different sight distance conditions, and as you can see from the picture, the farther the mobile phone is, the lower the vision level is.

There were interaction effects among different PPI presentation and different viewing distances, $F(16, 224) = 17.04$, $p < 0.001$. When the distance was 10 cm, there was significantly difference among the discrimination threshold results of different PPI screens, $F(4, 56) = 49.49$, $p < 0.001$; the difference in discrimination threshold results of different PPI screens was significant when the distance was 20 cm, $F(4, 56) = 9.64$, $p < 0.01$; the difference in discrimination threshold results of different PPI screens was significant when the distance was 30 cm, $F(4, 56) = 7.69$, $p < 0.01$. And when the

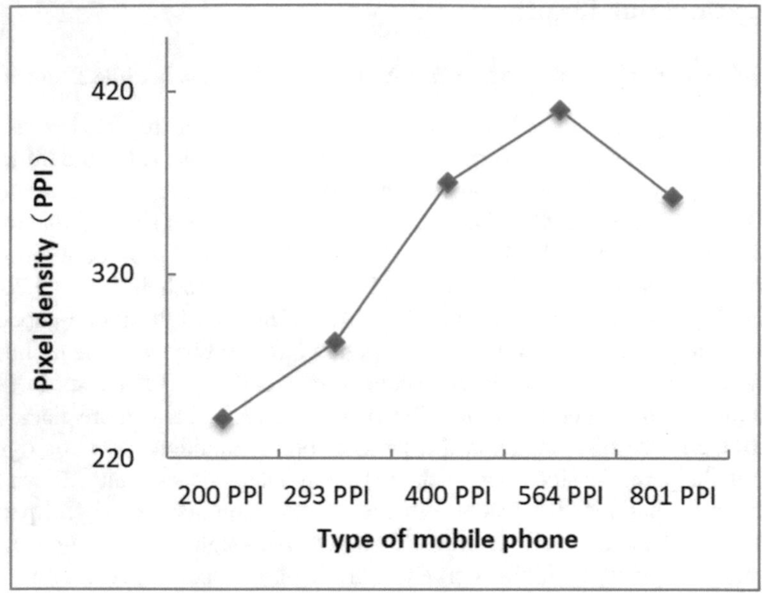

Fig. 2. Pixel density map of different types of mobile phones

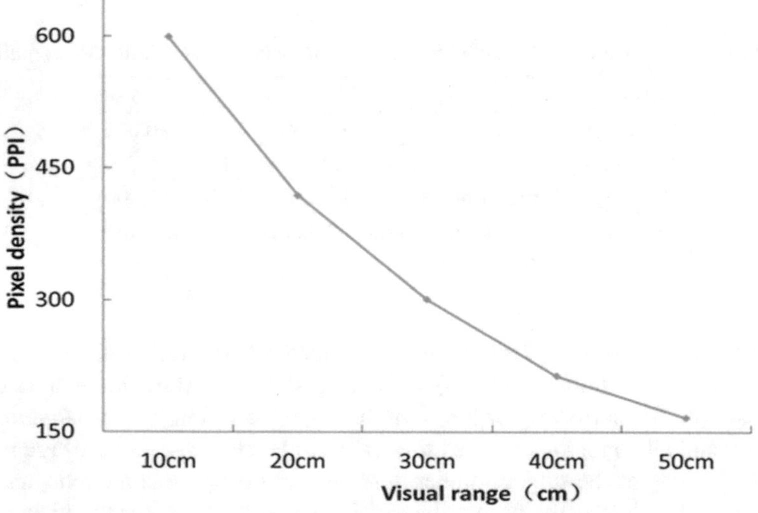

Fig. 3. Pixel density map of different sight distance

distance was 40 cm and 50 cm, there were no significant difference among the discrimination threshold results of different PPI screens (ps > 0.05) (See Fig. 4). From Fig. 3, we can see that at the close watch (<40 cm), the difference threshold of different PPI screens was different, and there was no significant difference in the threshold results between different PPI screens at the far position condition ($\geq$40cm).

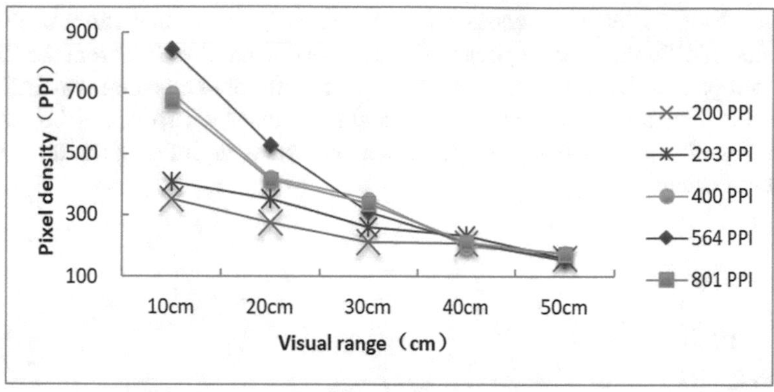

Fig. 4. The pixel density of different mobile phones under different sight distances

Repeated measures ANOVA analysis was performed on the results of discrimination threshold of subjects with different vision levels. The results showed that the discrimination threshold of subjects with different vision levels was significantly different ($F(3, 11) = 6.15$, $p < 0.05$). There was no obvious difference in the discrimination threshold results of normal visual acuity subjects and the degree myopia patients; the difference of the discrimination threshold result between 300° myopia and the 600° myopic patients was not obvious, but the discrimination threshold results of 300° myopia was significantly different from that of the 900° myopia patients; the discrimination threshold of 600° myopia patients was significantly different from that of 900° myopia.

There was no significant difference in discrimination threshold results between PPI presentation conditions and different visual acuity subjects conditions, $F(12, 44) = 1.50$, $p = 0.16$. For people with different visual acuity, different PPI screens had a great influence on discrimination threshold. The discrimination threshold of different PPI screens was significantly different in the normal visual acuity population. The differences in discrimination threshold results among different PPI screens were all significant ($F < 50° (4, 12) = 16.42$, $p < 0.01$; $F300°(4, 12) = 13.35$, $p < 0.01$; $F600°(4, 12) = 10.82$, $p < 0.01$; $F900°(4, 12) = 24.18$, $p < 0.01$).

3.2 The Fitting Curve of Visual Acuity Under Different PPI Conditions

The theoretical value that "E" corresponding to the minimum resolution threshold of different PPI mobile phone were calculated by visual acuity and the watching distance, and the actual PPI value were calculated as the smallest size conversion perspective E with different distances and different mobile phone condition. Figure 5 showed the PPI theoretical value and PPI actual value curves of subjects whose visual acuity was 5. From the graph, we can see that all cell phone curves are below the theoretical value at close distance, that is, the PPI value of mobile phones has an impact on the recognition threshold of E words; at long distance, all cell phone curves tend to the theoretical value curve, that is, when the screen has the ability to display "clear" E word, the vision level of the subjects will be restricted by their own vision level. Since the maximum

PPI value of the screen was required in the nearest location, this study uses the PPI actual value of different mobile phones at the nearest distance to represent the PPI value of each cell phone. However, the resolution of mobile phone screen was higher than that of its real distance from the nearest sight distance, which might be caused by different cognitive processing strategies when subjects were judging the direction of E at different distances.

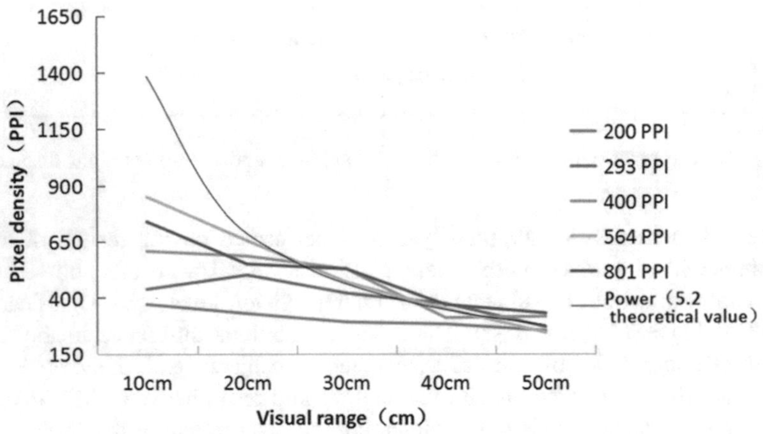

Fig. 5. The comparison between theoretical fitting curves and the mobile phones actual value of 5.0 visual acuity

4 Discussion

4.1 Differences of Visual Acuity Between Mobile Phones with Different Resolution

The results showed that the main effect of the type of mobile phone was significant, and the visual acuity of the subjects will be improved as the screen resolution of the mobile phone rises. The result is consistent with the hypothesis, and the mobile phone resolution increases, which will inevitably lead to a clearer screen. There are many factors that affect visual acuity. Brightness, contrast between objects and backgrounds, retinal adaptation and flash fusion will all affect our visual acuity [2]. The higher the resolution of mobile phone was to increase the contrast between visual identification object and its background, so it will increase people's recognition of objects. So the experimental results are consistent with expectations. The increase of contrast can indeed increase the level of human visual acuity. The main effect of subjects' visual acuity was significantly different at different distances, which indicates that the distance of vision was affected by distance. This was consistent with our experience in life. The closer we are, the better we can see.

This study found that there was interaction between distance and cell phone types. This means that the level of vision was not influenced by distance or cell phone resolution, but by the interaction between distance and cell phone resolution. It was found that the closer the distance is, the greater the change of visual acuity will be brought by resolution. When the sight distance reached a certain range (above 30 cm), the resolution of mobile phone will no longer affect our visual acuity. Many cognitive psychologists think that human information processing was not immutable and frozen, there was a lot of flexibility, information processing flexibility referred to the individual cues in different situations, the information processing mode of stimulation task (centralized processing, parallel processing) conversion in different task situations, individuals between centralized and parallel conversion processing method the [3]. The results of this study supported the view of cognitive processing flexibility. When people do the measurement of visual acuity at different distances, the difference of sight distance will lead people to produce different processing strategies, so as to ensure the best completion of cognitive activities. When the participants performed the visual acuity measurement at close distance, the detailed recognition of the symbol "E" resulted in the finer partial processing because of the clear details, so the screen resolution of the mobile phone had a more significant impact on the visual acuity of the subjects; when measuring the visual acuity at a long distance, because the retina itself does not achieve the fine processing of the recognition symbols, the subjects take the overall processing strategy, so the impact on the screen resolution of the cell phone was not so obvious.

4.2 Fitting Curves of Visual Acuity Under Different PPI Conditions

This study used 6 different resolutions of mobile phones to compare the theoretical values of visual acuity and cell phone resolution. It was found that the higher the screen resolution of the cell phone and the closer the distance theoretic curve was, the more clearly the different vision conditions were sensitive to the screen resolution of the cell phone at four kinds of visual acuity. At the same time, we found a strange phenomenon. When measuring the visual acuity of subjects at close range, the resolution of the folded cell phone screen was higher than that of the mobile phone screen resolution. This phenomenon disappeared in the long distance measurement, the reason may be that when the subjects in the near distance measuring visual acuity, and not by processing of the word "E" itself, but because of changes of identification symbols and interpretation of fine processing caused by judgment, so the judgment itself will increase our transformation when the value of PPI, so that at last the calculated value was greater than the mobile phone itself resolution; but in the long distance measurement, because of the limitation of the human retina to the resolution, people can only judge the whole process of E rather than the change of light and dark when they are judging, so it can represent the real resolution of the mobile phone at long distance. Though there was no way to represent the real resolution of mobile phones at close range, we still see the adjustment of different cell phone resolutions to the visual acuity. The higher the resolution of the mobile screen is, the better the recognition effect of the subjects is.

5 Conclusion

In this study, we compared the perception thresholds of four kinds of visually impaired people in 6 different PPI mobile phones at different sight distance, analyzed the relationship between PPI and visual threshold, and compared the PPI curves of measured data with theoretical curves. The results showed that with the increase of PPI, the visual acuity of subjects increased, indicating that the visual acuity of different cell phone screen resolutions affected the viewing process was significantly different within and outside 300 PPI. Meanwhile, the results of this study showed that within the range of 30 cm, there was a significant difference in visual perception thresholds between different PPI phones. There was no significant difference in visual perception thresholds between 30 cm and mobile phones. The visual acuity level of 5 groups of subjects as an example, the PPI value fitting curve of the actual measurement angle of mobile phone was lower than the theoretical value curve when the mobile phone was close to the mobile phone; remote viewing mobile phone (more than 30 cm), the actual PPI value curve approaches the theoretical value curve showed that the optimization effect of mobile phone resolution visual acuity. The results of this study also showed that the sensitivity of perception threshold depending on different distance of the mobile phone PPI, mobile phone PPI 300 as a watershed, viewing distance of 30 cm within and beyond the influence of model performance was different, the actual performance and theory of the concept of mobile phone PPI was not entirely consistent, moderated by people's cognitive strategy.

Acknowledgements. The authors would like to gratefully acknowledge the support of the National Key R&D Program of China (2016YFB0401203), and China National Institute of Standardization through the "special funds for the basic R&D undertakings by welfare research institutions" (522018Y-5942, 712016Y-4940).

References

1. Kubota, S., Hisatake, Y., Kawamura, T., Takemoto, M.: 5.1: Influence of pixel density on the image quality of smartphone displays. SID Symp. Digest Tech. Pap. **46**(1), 22–25 (2015)
2. Zhu, Y.: The Experimental Psychology, 2nd edn, p. 68. Peking University Press, Beijing (2000)
3. Miao, X., Li, Y., Wang, M., Zhang, Z.: The flexibility of information processing of compulsive individuals. Psychol. Sci. **38**(5), 1264–1271 (2015). (in Chinese with English abstract)

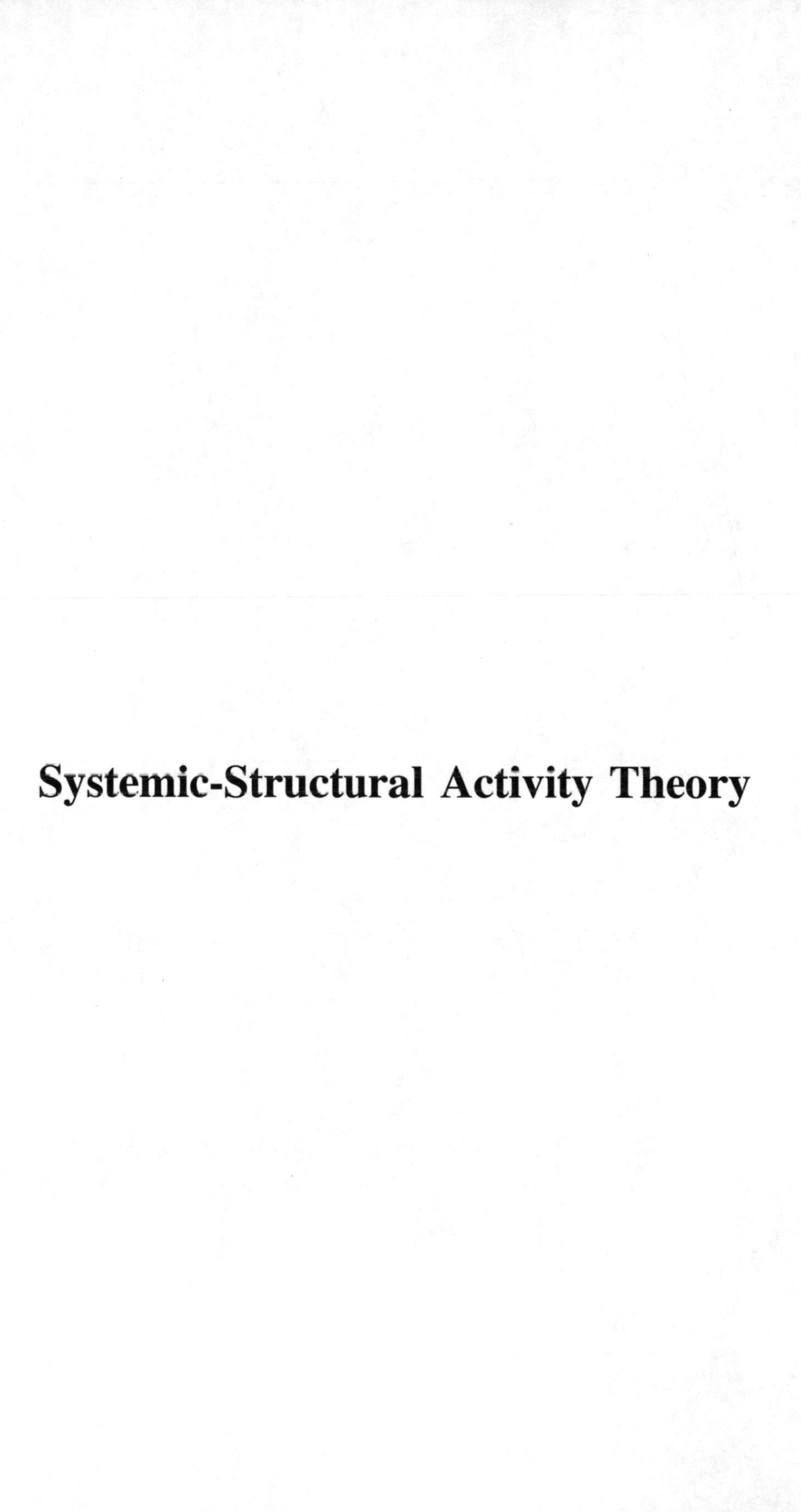

Systemic-Structural Activity Theory

Time Study in Ergonomics and Psychology

Gregory Bedny[1(✉)], Inna Bedny[1], and Waldemar Karwowski[2]

[1] Evolute, LLC, Wayne, NJ, USA
gbedny@optonline.net
[2] University of Central Florida, Orlando, FL, USA
wkar@ucf.edu

Abstract. Time studies are usually concerned with finding performance time for repetitive tasks that include elements that follow each other in the same sequence from cycle to cycle. The motor activity dominates in this type of tasks. The main purpose of a time study is to determine a standard time for the task performance for planning, control, measure of productivity, define wages, etc. The specialists utilize chronometrical methods of study, or normative data for performance of selected standardized elements of work that include selection of representative operators, evaluation of work pace, rating pace of performance, evaluation of efficiency of selected operators' performance, etc. An advanced method of time study is based on utilizing determined by the MTM-1 system motion times. This paper presents an overview of time study research applicable to ergonomics and psychology, and the general principles of the behavioral actions performance analysis

Keywords: Systemic-structural activity theory
Time standard for task performance · MTM-1 system · Pace of performance
Time structure of activity

1 Introduction

Time study has been originated by Taylor in the beginning of the twentieth century. This direction has been mainly used for determining the time standards for performance of various tasks. Around the same time, Gilbreths did their pioneering work on the motion study for analysis of the methods of task performance. Later these two directions have been integrated into motion and time study. In USA motion and time study have been conducted by industrial engineers and other specialists. Time study has been concerned with finding representative times for repetitive tasks' performance. Such tasks have different elements which follow each other in a regular manner from cycle to cycle. The motor activity dominate this type of tasks. The main purpose of time study is to determine the standard time for task performance for planning, controlling, measuring productivity, defining compensation, etc. Usually, in order to determine the standard performance time specialists utilize chronometrical methods of study, or use the normative data for performance of selected standardized elements of activity. Chronometrical methods include selection of representative performers, evaluation of pace, rating pace of performance, evaluation of efficiency of selected operators' performance, etc. All of this facilitates the basic chronometrical analysis. Currently, the

H. Ayaz and L. Mazur (Eds.): AHFE 2018, AISC 775, pp. 217–224, 2019.
https://doi.org/10.1007/978-3-319-94866-9_22

time study is an important area when assessing productivity in economics. The Gilbreth's method of motion study is not utilized now. The more efficient method of time and motion analysis has been developed by Maynard and his colleagues. This method is known as MTM–1system [1]. This system is effectively applied for analysis of repetitive work where motor components dominate. However, the specificity of the contemporary tasks changed significantly. They include a lot of cognitive components and are extremely variable. Traditional methods of time study are not adapted for analysis of such type of tasks. This explains the fact that traditional methods of time study are not efficiently used in ergonomics.

Ergonomists and engineering psychologists developed their own methods of study temporal parameters of activity. For example, in order to study skill-based behavior they measure various reaction times. Some interesting data that has been obtained through measurement of simple and choice reaction time has been used for applying human-information processing approach in cognitive psychology (Hick-Hyman Law). Another method of studying temporal characteristics of operators' performance is based on Fitts' Law. These two methods consider human activity as a reactive system that consists of a number of independent reactions, when an operator reacts with maximum speed. However, human activity should not be viewed as a set of independent reactions performed with maximum speed. When performing a task, an operator executes a goal directed, logically organized system of cognitive and behavioral actions that are directed to achieve a goal of a task. Elements of activity can be performed simultaneously and/or in sequence. The pace of performance of considered elements is significantly different from pace of reaction. Activity is not just a process as it is described in cognitive psychology but rather a complex structure that unfolds in time. Analysis of this structure and comparison of it with the structure of equipment is the major principle of design. If in the process of such comparison it would be discovered that the structure of activity is complex this means that usability of the equipment is low and the task is difficult for a performer. Thus, one needs to know the time structure of activity during task performance because the description of the logical organization of elements of activity and its time structure is prerequisite of quantitative evaluation of task complexity, reliability and others quantitative characteristics of the task performance.

In systemic-structural activity theory (SSAT) the time study includes new non-traditional methods of analysis of the temporal characteristics of activity. This approach allows to describe the time structure of variable activity during task performance that includes numerous cognitive components. This approach allows to introduce new methods of time study in the analysis of the operator performance in automated control systems, including computerized and computer based systems.

2 General Principles of Time Study of the Man-Machine Systems

Traditional methods of time study are applied for analysis of task performance of blue-collar jobs. Another direction of study is involved in temporal analysis of operators' activity in man-machine and human-computer interaction systems. Such systems can involve performance of computerized and computer based tasks. The time during which a system is transferred from the initial to the required state is called "cycle of time regulation". This time includes time for functioning of technical components of the system and time for the task performance by an operator. Task performance time very often constitutes a substantial part of the cycle of time regulation.

$$T0 = Tm + TH, \quad (1)$$

where T_0 – cycle of time regulation; Tm - time of functioning of technical components of the system that does not overlapped by time for human performance; T_H – operator's task performance time.

This is a simplified presentation of the cycle of time regulation because structural relationship between functioning of the considered components in time can be very complex. The substantial part of the cycle of time regulation often consists of operators' or users' task performance time. Hence, operator's task performance time is a critical factor for functioning of any system. This becomes clear when considering the concept of the reserve time. Reserve time is defined as the surplus of time over the minimum that is required for the operator to detect and correct any deviation of system parameters from allowable limits, and bring the system back to tolerance. Thus

$$Tres = T - T0, \quad (2)$$

where T – time that cannot be exceeded without peril to the system; T_0 – cycle of time regulation.

From the SSAT perspectives when activity is regarded as a self-regulative system, it is necessary to differentiate between objectively existing reserve time and the operator's subjective evaluation of that time. This is especially important when there is no externally presented information about how reserve time can be changed in a specific situation. This may lead to inadequate behavior of the operator in case of an accident. An operator often roughly evaluates the reserve time by making statements such as "I have plenty of time", or "I have little time", or "I have no time." Such statements may reflect a sharply changed activity strategy of task performance in emergency situations. We considered some aspects of time study at the system level for the man-machine functioning. Let us now consider some other aspects of time study.

The purpose of the time study is determining the standard time for performance of a specific task. In traditional time study, tasks are repetitive and elements of a task follow in the same order from cycle to cycle. The main units of analysis in such studies are motions or some elements of a task that erroneously are also labeled as motions. For example, in his "Time activity analysis" Rodgers [2] calls such elements of task as "lift table tape", "press tape on paper", etc. erroneously motions. However, these are not motions but rather the small elements of tasks that can include various motions. It can be seen that there is no standardized terminology for activity analysis in industrial engineering the same as in psychology and ergonomics.

The main idea of traditional time study is to find the best way of performing production operations or tasks to increase their efficiency and productivity. Such understanding of time study is not adequate for contemporary tasks analysis due to the fact that in the modern industry significant proportion of mental components of work make tasks extremely variable. Moreover, operators can utilize various strategies of task performance. This is especially relevant for operator of man-machine or human-computer interaction systems. Therefore, selection of adequate units of activity analysis, development of analytical methods for studying tasks at the design stage and for increasing efficiency of analysis of already existing tasks is especially important. As it is demonstrated in SSAT [3, 4], time study is a critical factor for the design of human activity in general. Comparison of the time structure of activity with configuration of equipment or human-computer interaction interface is the major principle of designing equipment and the efficient method of task performance. In time study the main units of analysis are not motions, but cognitive and behavioral actions. For determining duration of cognitive and behavioral actions developed in cognitive psychology chronometrical methods, and methods offered by applied and systemic-structural activity theory can be utilized.

3 Time Study of Cognitive Components of Activity During Task Performance

Let us consider the chronometrical analysis of the cognitive components of activity. For example, suggested by Donders (1862) method can be utilized for determining the duration of cognitive components of activity. This method also is used in contemporary cognitive psychology. A person performs two mental tasks, X and Y, where Y = X + K. Psychologists measure the time of performing X and Y, and subtract Tx from Ty to derive Tk. This procedure permits one to determine the duration of the mental process, even if this process cannot be directly observed and has very short duration. In the majority of performed by the operators task the visual information is the most important one. The developed in SSAT eye tracking method of interpreting this information is vital for determining duration of cognitive actions' performance. The traditional method of eye movement interpretation uses a cumulative scan for the entire task where scientists extract such data as scan path length (in pixels), cumulative dwell time or average fixation time, number of fixations, or number of saccades. Below we present an example of the cumulative scan path of the human-computer interface task (Fig. 1).

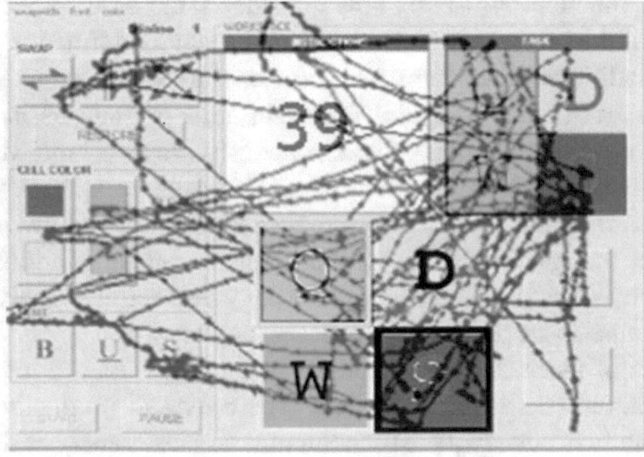

Fig. 1. Cumulative scan path of the human-computer interaction task.

Such data is not sufficiently informative. In SSAT we divide a task into a number of relatively independent fragments. Each fragment is associated with one or several images of eye movements. These images allowed us to trace eye movement from one element of task to the next with high precision and compare it with the mouse movements. Based on such data, rules of action extraction, and their standardize classification, we can accurately describe the logical organization of cognitive and motor actions when considering a particular fragment of task [5]. Each saccade and gaze or eye fixation that are associated with the conscious goal comprise a cognitive action. We present, as an example some, the criteria of actions extraction.

Actions are classified based on the following six criteria:

1. Dominance in a particular moment cognitive process;
2. Analysis of the action's purpose;
3. Relation of the gaze to visible elements on the screen;
4. Purpose of the following action and particularly the mouse clicks,
5. Duration of the gaze and its qualitative analyses;
6. Analysis of debriefing of the subjects and comparison of their reports.

Similar rules should be implied when operators receive information from an instrumental panel. Below we present Table 1 (fragment) which demonstrates

Table 1. An example of an eye movement and actions classification between two mouse clicks (fragment).

Eye Move And Final Position	Mental/ motor Actions Involved	Algorithm	Motor Action/ Motor Movement	Time/action (ms)	Path sequence generated/duration
				Move time to position	Dwell time at position
G_Q	Simultaneous Perceptual actions, with explorative thinking. Comparison of object and goal in relation to the program of performance.			5	6 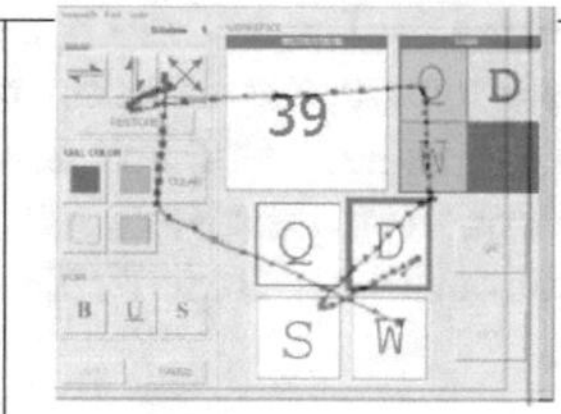
O_q				5	4 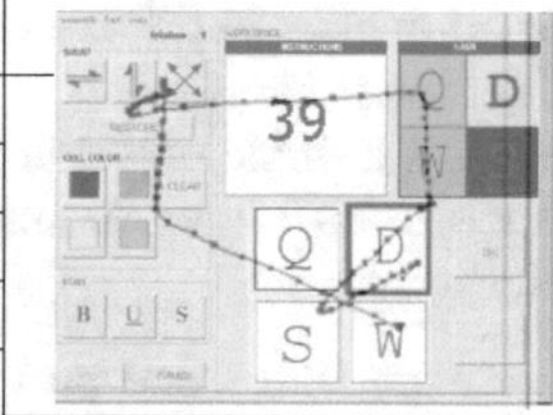
T_{CB}				6	5
G_S				6	4
G_D				5	3
O_q				7	4
O_W				5	11
O_D				5	3
G_D	Decision on program of performance and motor action based on decision.			7	3
O_D	Perceptual action with motor action and thinking based on program of performance		24	7	21
T_{PV}	Motor action of eye and mouse along with selection from choice			7	4

cognitive and associated with them motor actions and their duration during performance of the considered fragment of task.

This example demonstrates that the analysis of eye movements and associated with them motor activity are an important source of information for the extraction and classification of cognitive actions. There are some other recommendations for eye interpretations in SSAT that are out of the scope of this brief discussion.

4 General Principles of the Behavioral Actions Performance Analysis

In most situations motor actions' analysis can be performed without eye movement analysis. In this section we consider in an abbreviate manner the developed in SSAT method of motor actions analysis. One of the most advance systems of analysis of time and efficiency of performance of behavioral components of activity is Methods Time Measurement (MTM-1) [6]. Specialists in ergonomics and work psychology are usually not familiar with the full description of this system. This system is considerably complex and in order to use it one would have to invest some time and efforts to learn it. The given in various literature abbreviated versions of this system are rather misleading. Moreover, this system is not adapted for analysis of the present-day variable tasks where the substantial part of tasks includes cognitive components. The MTM-1 system has been developed for analysis of repetitive motor tasks in mass production. The main units of analysis in this system are human motions that are described in a standardized manner. Such motions are also called microelements. Depending on the specifics of motions' performance they have various predetermined time for their performance. The MTM-1 system has microelement EF (Eye Focus) which can be used to determining the duration of recognition of simple stimulus and performance of simple "yes-no," "if-then" decisions.

Here, in an abbreviate manner we describe, as an example, one standardize motion or microelement "Reach". There are five classes of Reach. We consider just three of them.

1. Reach to object at a fixed location, or to an object in other hand or an object, on which other hand rests;
2. Reach to a single object in a location, which may vary slightly from cycle to cycle;
3. Reach to an object jumbled with other objects in a group so that search and selection occur.

There are tables that provide performance time of different categories of Reach depending on the distance of the motions and some other factors. Other standardized motions are described similarly. The MTM-1 system is not adapted for analysis of tasks in man-machine system and specifically for analysis of tasks involving human-computer interaction. In SSAT this system is used in the enhanced manner. New standardized rules for application of the MTM-1 system were developed. The main units of analysis are not motions, but motor actions that can integrate motions into motor actions according to the goal of motor actions.

For example, the motor action "reach and grasp an object" includes two standardize motions "reach" (R) and "grasp" (G). SSAT contains the following hierarchically organized units of analysis of motor activity: motions, motor actions, and their combinations (members of the algorithm). Such members of the algorithm is depicted by symbol O_1^{ε} where the superscript designates the motor component of activity and subscript demonstrates a sequence number of the considered member of the algorithm. A member of the algorithm that describes motor activity can include one or several motor actions. According to the capacity of the short-term memory this can consist of 1–4 actions. Let us consider a simple example. A subject has to "move the left hand to the pin box and grasp one pin out of the bunch of other pins in pin box". Motor action can be described as follows (Table 2):

Below is another example that is a part of the description of the computerized task (Table 3).

Table 2. An example of description of the motor activity.

Member of algorithm	Verbal description of member of algorithm	Standardize description of motor action	Time
O_1^{ε}	Move left hand to the pin box and grasp pin	R32B + G1C1	1.23

Table 3. An example of description of the computerized task.

Member of algorithm	Verbal description of member of algorithm	Standardize description of two motor actions	Time
O_2^{ε}.	Type 1 and then press ENTER to choose ADD INVENTORY RECEIVING screen	(R50B + AP1) + (R30B + AP1)	1.68 x 1.2 = 2.01

The left column presents the standardized description of the member of the algorithm (psychological units of analysis), the second column gives a verbal description of the consider member of the algorithm, the third column present the standardized description of motor actions that are included in the content of the member of the algorithm. Columns one and three present psychological units of analysis because in these columns we utilize standardize description of activity elements. The right column lists the performance time of the considered member of the algorithm.

In the above examples R32B means performance of motion Reach, version R-B, distance of movement is 32 cm; R50B and R30B are the same type of motions with different distances. G1C1 means Grasp a nearly cylindrical object with the diameter larger than ½ where there is interference with Grasp. AP1 means Apply Pressure. Thus, in these two examples we presented three motor actions, each of which consists of two motions that are integrated by the goal of actions. Sometimes, a motor action can include more than two standardized motions.

The logic of the order of execution of the members of the human algorithm depends on the nature of the decisions or logical conditions. Some members of the algorithm can even be totally omitted. The performance time of the considered repetitive task vary. The reserve time for task performance is also variable.

5 Conclusion

Time is a critically important factor for the analysis of any task because activity unfolds in time as a process. Such methods as the reaction time measurement or similar methods are not adequate for the contemporary time studies in ergonomics. SSAT as a new framework suggests the totally new principles of time study. Usually, cognitive and motor actions are not extracted as basic units of activity analysis in cognitive psychology. This makes is very difficult to accurately determine the time required for the task performance. SSAT offers rigorous principles of cognitive and behavioral actions description and methods of determining time of their performance. The new method of application of the MTM-1 system is suggested by adapting it for the modern task analysis. The concept of the strategies of task performance, and the algorithmic analysis of human activity demonstrate that the idea of just having one best way of performing a task is not adequate. These days the purely manual work is significantly reduced, while the weight of the cognitive components of work is on the rise. The tasks now are mostly variable and they are performed in various ways. Such tasks have a logical, hierarchical, and probabilistic organization. SSAT suggest the new approach to the time study that presents an opportunity to not only determine the time of task performance, but also to apply various quantitative methods in task analysis that allow to enhance the efficiency of performance, reduce human errors, and decrease the complexity of the tasks' performance.

References

1. Maynard, H.B., Stegemerten, G.J., Schawab, J.L.: Method-Time Measurement. McGraw-Hill Book Co., New York (1948)
2. Rodgers, S.H., Eggleton, E.M. (eds.): Ergonomic Design for People at Work, vol. 2, 2nd edn, p. 120. Van Nostrand Reinnhold, New York (1986)
3. Bedny, G.Z.: Application of Systemic-Structural Activity Theory to Design and Training. CRC Press. Taylor & Francis Group, Boca Raton (2015)
4. Bedny, G.Z., Karwowski, W., Bedny, I.S.: Applying Systemic-Structural Activity Theory to Design of Human-Computer Interaction Systems. CRC Press. Taylor & Francis Group, Boca Raton (2015)
5. Bedny, G.Z., Karwowski, W., Sengupta, T.: Application of systemic-structural theory of activity in the development of predictive models of user performance. Int. J. Hum. Perform. **24**(3), 239–274 (2007)
6. Karger, B.: Engineering Work Measurement, 3rd edn. Industrial Press, New York (1977)

Decision Support of Mental Model Formation in the Self-regulation Process of Goal-Directed Decision-Making Under Risk and Uncertainty

Alexander Yemelyanov(✉)

Department of Computer Science, Georgia Southwestern State University, 800 Georgia Southwestern State University Drive, Americus, GA 31709, USA
Alexander.Yemelyanov@gsw.edu

Abstract. This paper considers the motivational approach to decision-making in problems of uncertainty from the position of the psychological theory of SSAT. Illustrated throughout the paper is the notion of how the use of the major provisions of this theory, related to the concepts of goals, self-regulation, positive feedback, etc., helps in understanding the deeper mechanisms of decision-making and in creating effective systems for their computer support. Instead of the traditionally used approach, the paper proposes that when the solution of the problem is preceded by the construction of its mathematical model, it is better to use a fundamentally different approach to modeling, such as when the mental decision-making model is constructed by the subject himself in the process of solving the problem under consideration. If the solution of the problem is determined by a previously-constructed mathematical model at the first approach, then at the second approach, the solution depends on a mental model, which may not even be fully acknowledged by the subject. This paper describes the Motivation Evaluation Process for solving a problem and the Performance Evaluation Process with decision support for making a decision.

Keywords: Systemic-structural activity theory
Decision-making under uncertainty · Self-regulation
Motivation Evaluation Process · Performance Evaluation Process
Mobile decision support

1 Decision-Making Under Risk and Uncertainty

Decision-making in situations of uncertainty is a complex problem, with the risk of potential losses of money, health, reputation, etc. always present. It is conventionally assumed that decision-making under uncertainty considers situations in which several outcomes are possible for each course of action, and the decision-maker cannot estimate the probability of occurrence of the possible outcomes; however, when the decision-maker is able to calculate these probabilities of occurrence, decision-making is considered to be performed under risk [1]. With this probabilistic approach of defining uncertainty, it relates directly to a lack of information on probabilities of outcomes; despite this, uncertainty is reduced when the probabilities of outcomes are specified – in other words, when a problem shifts from the category of uncertainty into the

H. Ayaz and L. Mazur (Eds.): AHFE 2018, AISC 775, pp. 225–236, 2019.
https://doi.org/10.1007/978-3-319-94866-9_23

category of risk. In the current work, the uncertainty of an outcome is viewed from a broader perspective and is associated both with a lack of information regarding the value of the outcome, and with a lack of information regarding the possibility of its occurrence. It is worth noting that an outcome's uncertainty is related to the uncertainty of the goal, while the uncertainty of its possibility is related to the uncertainty of the conditions of the problem. The necessity of considering possibility instead of probability is related to the fact that possibility is more broadly understood than probability and, apart from informational (statistical) characteristics reflected at the decision-maker's conscious level, possesses emotional (energy) characteristics, reflected at his/her unconscious level. Finally, emotional characteristics reflect the subjective complexity (or difficulty) of obtaining the considered outcome, which is an important characteristic of its uncertainty, directly influencing the decision that is made. Therefore, in our non-probabilistic approach, decision-making under uncertainty considers situations in which several outcomes are possible for each course of action, and in which the decision-maker cannot estimate the value of outcomes and the possibility of their occurrence; when the decision-maker is able to determine these values and possibilities, decision-making is considered to be performed under risk. It is important to consider that when evaluating uncertainty and risk, the analysis of subjective values and subjective possibilities of outcomes should not be quantitative, as is typically observed to be the case in the probabilistic approach, where these factors have relative monetary or statistical value. Instead, they should reflect a vagueness inherent in the decision-maker's perception of uncertainty and risk. As an alternative for representing uncertainty, it is suggested to implement possibility and fuzzy set theories, which both relate to vague linguistic variables, such as "high" or "often" [2].The presence of uncertainty in the occurrence of outcomes contributes a principle correction to the construction of the model of the decision-making problem. As will be demonstrated in the present paper, such a model can be constructed by an individual only in the process of decision-making. In addition, existing methods propose an initial construction of the model of the problem (primarily probabilistic or multi-criteria), followed by calculations of the solution to this problem with the help of quantitative methods of linear algebra, without any necessary psychological justification, which, as will be further established below, leads to an inadequacy in the obtained results. In the currently existing decision-making methods (Utility Theory, AHP, MAUT, etc.), the problem's model is constructed using mathematical methods, and traditionally, the process of constructing the model precedes its practical use. And although the goal itself is an important factor to consider in decision-making, it is not usually included in these decision-making models. However, the Analytic Hierarchy Process is an exception to this trend. Presented below is a brief description of the Analytic Hierarchy Process (AHP) with the goal of comparing it to the presently recommended Motivation Evaluation Process (MEP), along with its application for decision-making, the Performance Evaluation Process (PEP).

2 Analytic Hierarchy Process

AHP – is a multi-criteria decision theory that provides a general and systematic framework supporting complex decision-making situations with multiple and often conflicting objectives (*i.e.* criteria, attributes). The AHP method [3] includes four steps: problem modelling, weights valuation, weights aggregation and sensitivity analysis [4]. When using AHP, the problem is decomposed into a hierarchy of more easily comprehended sub-problems, each of which can be analyzed independently, according to a hierarchy in which the upper level is the goal of the decision. The second level of the hierarchy represents the criteria and sub criteria, and the lowest level represents the alternatives. As soon as the hierarchy is built, the decision makers systematically assess its various elements by comparing them to each other two at a time (pairwise comparison), with respect to their impact on an element above them in the hierarchy. In making the comparisons, the decision makers use their judgments regarding the outcomes' relative importance. For example, with the goal of *selecting the vehicle that best meets one's objectives*, the following factors could be considered as relevant criteria: initial cost, maintenance cost, prestige, and quality. The quality criterion could be divided into the sub-criteria of safety, frequency of breakdown, performance, and design. And, in turn, the design sub-criterion could be divided into the sub-sub-criteria of exterior design and interior design. Once the criteria, sub-criteria, and sub-sub-criteria have been determined, pairwise comparison of the factors will need to be made with respect to the goal. For example, the user should compare the relative importance of prestige and maintenance cost with respect to the goal. In order to make the pairwise comparisons, a 9-point linear scale was used. The scale allows to determine whether one criterion compared to another is equally important (1), slightly more important (3),..., or absolutely more important (9). Ultimately, each of the five criteria above, along with their sub- and sub-sub-criteria, will need to be compared to each other. When pairwise comparisons of all criteria with respect to the goal will be finished, the pairwise comparisons of all alternatives with respect to each criterion should be considered. All pairwise comparison numerical values are gathered in the comparison matrix. With the purpose of calculating the relative weight of each criterion or sub-criterion, linear algebra is used. This allows to determine the vector of weights as the normalized eigenvector of the matrix, associated with the largest (principle) eigenvalue that measures the degree of inconsistency. In the next step, AHP aggregates the local priorities across all criteria in order to determine the relative global priority of each alternative. AHP applies an additive aggregation with a normalization of the sum of the local priorities to unity. The AHP method is supported by Expert Choice software, which helps for the selected model to provide weights valuation (collecting the consistent pairwise comparisons), weights aggregation (determining the global priority), and sensitivity analysis (validation of the suggested model). A short overview of AHP is provided below, which is important to its subsequent consideration in the current work.

In this way, in AHP, the model of the problem must have independent criteria and cannot be changed within the process of making a decision. AHP solves the given problem by using quantitative methods of linear algebra, based on a previously-created multi-criteria evaluation model. After building evaluation model, this model remains unchanged at all subsequent steps of the AHP method: weights valuation, weights aggregation, and sensitivity analysis. In AHP, a rank reversal can occur due to a lack of sound psychological evidence for the data, which is collected from the user and then further extrapolated with the use of linear algebra models. In this case, the use of linear scales to measure human responses, as well as the use of linear algebra models for their subsequent transformation into the final decision regarding the relative priorities of the alternatives, are not based upon sound psychological evidence, which can thus result in rank reversal. AHP turns out to be more successful when applied to well-structured (or quantitatively described) problems, as opposed to ill-structured problems, which have unclear goals and incomplete information. AHP only determines the relative preferences of existing alternatives, without linking to their subsequent performance. This means that even the highest-ranked alternative, from the perspective of AHP, might not be selected for performance, perhaps because of a lack of necessary motivation for its execution on behalf of the decision-maker.

3 Self-regulation Model of Thinking Process

As it is presented to us, the description of the thinking process proposed by SSAT [5, 6] is by far the most productive approach in the study of decision-making processes. Below, we will illustrate the advantages of this approach, as well as describe its practical application to decision-making problems in situations of uncertainty. SSAT contributes key interpretations into the understanding of fundamental psychological concepts that describe human thinking, such as goals, motives, mental states, mental feedback, etc., and it considers thinking from a functional analysis perspective when it is presented as a self-regulation system with internal mental feedback. In SSAT, the *goal* is considered to be a basic concept of problem solving activity. At the start of making a decision, the goal may have a very general form. Only during the process of problem solving does the goal gradually become clearer and more specific, and it may even be corrected if necessary during the course of activity. This is why the application of AHP is limited to well-structured problems only, which have a fixed goal and criteria of selection. The problem solving process begins when the *initial mental model* of the problem is created. However, the initial mental model of the problem often is unable to facilitate the attainment of a desired result on its own. Therefore, after understanding the problem at hand, the subject divides the problem into *sub-problems*. He/she initiates the formation of sub-problems by formulating various *hypotheses*. Each hypothesis has its own potential goal. Based on the comparison and evaluation of such hypotheses, a subject selects one and formulates the first sub-goal associated with the selected hypothesis. Problem solving includes the continuous *reformulation (disaggregation*) of a problem and the development of its corresponding mental model. If the received data is evaluated *positively (positive feedback)*, the thinking process cycle is complete. However, if the result is evaluated *negatively (negative feedback)*, internal or external information should be added to continue this process on a lower level (or to search for other alternatives of solving the problem). In other words, in SSAT, the decision-maker is able to regulate his/her behavior not only externally, but also internally by using the inner mental plane. The self-regulation model of solving problem under uncertainty (Fig. 1) shows how the *initial mental model* of the problem was transformed into the *final mental model*. For this purpose, a problem should be disaggregated to such levels of sub-problems, which can be *positively evaluated.*

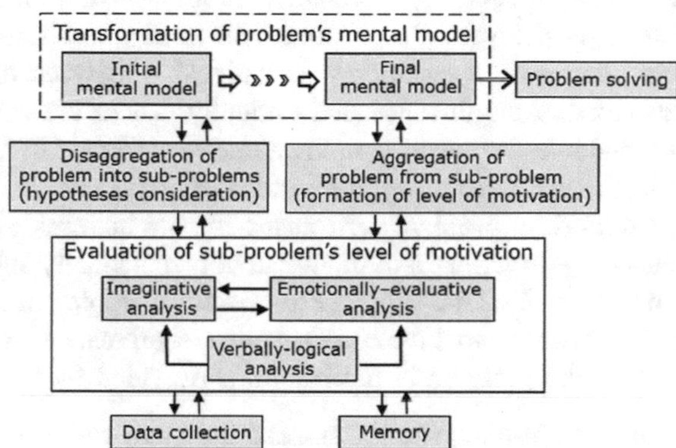

Fig. 1. Self-regulation model of decision-making under uncertainty.

These levels of motivation for solving problems are formed based upon the *imaginative, verbally-logical*, and *emotionally-evaluative analyses,* along with a collection of appropriate external and internal information. When the evaluation process for each positively evaluated sub-problem is finished, the aggregation process begins. Within this process, all determined levels of alternatives' motivations in sub-problems will be aggregated backward according to the previously created structural hierarchy (problem, sub-problems, sub-sub-problems, etc.) within the disaggregation process. The aggregation process results in the formation of the final mental model of the problem, with the levels of motivation available for all alternatives. This allows the individual to make a decision concerning whether the alternative with the highest level of motivation will be executed, or if its level of motivation is not high enough, which would then require new alternatives to be added for consideration. This may occur when the problem is too difficult or not significant enough. If the problem is complex, it will need to be reduced to sub-problems. The sub-problems, in turn, will be reduced to sub-sub-problems and so on, until some terminal level of specification (simplicity) will be reached, in which a separate level of motivation can be determined for each alternative. This terminal level contains the sub-problems which can be positively evaluated; in other words, these are sub-problems whose final mental model can be determined. When all sub-problems are evaluated, the aggregation process will begin; this aggregation process combines the collective motivational forces of all alternatives of the terminal level of consideration of sub-problems into a motivational force of each alternative for the given problem. After this process is complete, the final stage is decision-making regarding the alternative that will be selected for execution.

4 Motivation Evaluation Process

In the process of problem solving, the subject disaggregates a problem and assesses the obtained results according to the level of *significance* of the outcomes and the level of its *difficulty* (subjective complexity). The relationship between difficulty and significance determines the level of activity motivation. In the decision-making process, the alternative with the highest level of motivation will finally be chosen for execution. Therefore, we were interested to ascertain the relationship between the level of motivation and its indicators of significance and difficulty. There are several motivation rules that allow to evaluate the level of motivation. According to [5, 7], if a subject evaluates the problem as a personally significant and difficult one, the level of activity motivation that is required for problem performance would increase. At the same time, if a subject evaluates the problem as a difficult, but not a personally significant one, the level of motivation can drop, or the subject can reject the problem altogether. A more detailed relationship between positive and negative significance and adequate and inadequate difficulty of the problem in motivation is provided by Gregory Bedny:

1. If difficulty of task is not adequate for subject, then motivation is negative.
2. If difficulty of task is adequate for subject and there is only positive significance, then motivation is positive.
3. If difficulty is adequate for the subject and interaction between positive and negative significances is positive, then motivation is positive.
4. If difficulty of task is adequate for subject and interaction between positive and negative significances is negative, then motivation is negative.

Motivation Evaluation Process (MEP) for solving problems that we suggest in current paper, implements the motivation rules mentioned above and SD-Frame (Fig. 2) to evaluate the motivational levels of sub-problems in the process of solving problem.

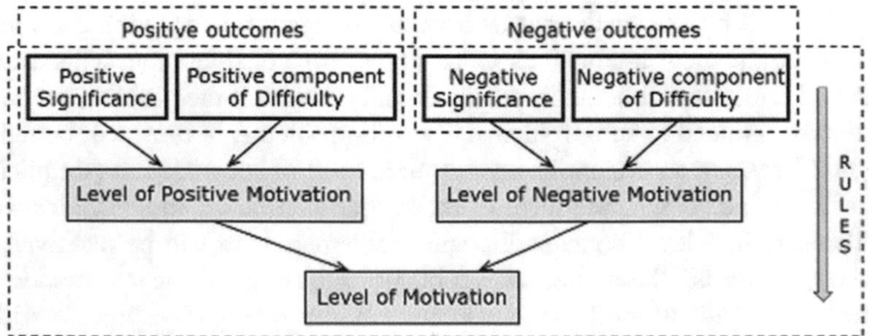

Fig. 2. SD-Frame of mental evaluation of motivational level.

The goal splits the outcomes of the alternative into positive and negative categories and thus determines the levels of their significances, respectively, as either positive or negative [8]. Positive significance and positive component of difficulty create a level of positive motivation, and negative significance and negative component of difficulty create a level of negative motivation. It should be noted that SSAT does not consider the different components of difficulty. Here, positive component of difficulty characterizes the difficulty associated with gaining positive outcomes, and the negative component of difficulty characterizes the difficulty associated with avoiding negative outcomes. We implemented this separation for difficulty in the event when rules that are more detailed are needed to describe the relationship between level of motivation and its indicators. Finally, the level of positive motivation and the level of negative motivation determine the overall level of motivation. SD-Frame is an abstract, schematic structure, which is more general than the actual model of motivation. It presents a mental frame or template that consists of a core (fixed information about the structure of formation of the level of motivation) and slots (containing variable information about the levels of positive and negative significances and difficulties). The possession of slots for significance and difficulty makes the SD-Frame adaptable and flexible to accommodate a formation of a motivation level at any intermediate mental state of a sub-problem. When the intermediate sub-problem, with its sub-goal, is adapted (positively evaluated), the corresponding levels of significances and difficulties can be assigned to the slots, according to the existing motivation rules, which allow the resultant levels of an alternative's positive, negative and integrated motivation to be determined. If the intermediate problem is so complex that the levels of significances and difficulties cannot be assigned, and the decision-maker is still motivated to continue the decision-making process, then the sub-problem, according to the motivation rules, should be further simplified into sub-sub-problems by using hypotheses to such an extent that the SD-Frame can be adapted. During the process of aggregation, the levels of motivation in the sub-problems, which are determined in their SD-Frames, are then combined into the final level of motivation of the problem.

In the Analytic Hierarchy Process theory analyzed above, the model of the decision-making problem was created before the individual solved the problem, so consequently, it did not consider such important factors as the potential correction of the mental model of the problem during the process of solving it. The inclusion of changes is regulated by internal mechanisms of self-regulation when the requested data are extracted from memory; these data are able to introduce fundamental changes into the results of solving a problem. Unfortunately, the cognitive approach, which recommends the creation of models before an individual has solved a problem, orients itself based only upon external feedback, which is essentially a link with the preliminary (and therefore, a largely inadequate) constructed model, but not with the individual (and his own mental model). Thus, the AHP – as all other multi-criteria decision theories, in which the criteria are not included in the model during the process of decision-making, but instead, before it – is limited only to the application of well-structured problems with previously determined criteria of selection. Another important concern of the implemented models is the fact that, within them, the factor of uncertainty in the occurring outcomes is largely described with the help of probabilities. Even though this is convenient for subsequent mathematical computations of the model, it is absolutely unfeasible when the guidelines of the problem possess an internal or emotional characteristic. In the mind of the decision-maker, the factor of uncertainty in the attainment of a goal, as well as the achievement of desired outcomes and the avoidance of undesired outcomes, do not possess a probabilistic character, but instead a problematic (related to difficulties) one.

5 Performance Evaluation Process

Motivation Evaluation Process allows the problem being solved to be divided into sub-problems, each possessing its own individual level of motivation for being solved. In this process, an important role is played by the rules that determine the formation of levels of motivation in sub-problems, which ultimately determine the final level of motivation in the problem being solved. The problem of decision-making in regards to selecting the best alternative is particular in that, in order to solve it, it is crucial to divide the problem into sub-problems related to measuring the level of motivation for each alternative. Only the alternative which, first of all, contains a sufficiently high level of motivation to perform and, second of all, contains a level that is higher than that of other alternatives, will be accepted for execution. In other words, the problem of making a decision requires measuring the level of motivation. In this case, the motivation rules described in the MEP are not sufficient. Therefore, for decision-making problems, it is suggested to use the Performance Evaluation Process (PEP) instead of the MEP. At the core of PEP lies the IL-Frame (Fig. 3), as opposed to the SD-Frame, which measures the *level of preference* in the problem being solved, instead of evaluating the *level of motivation*.

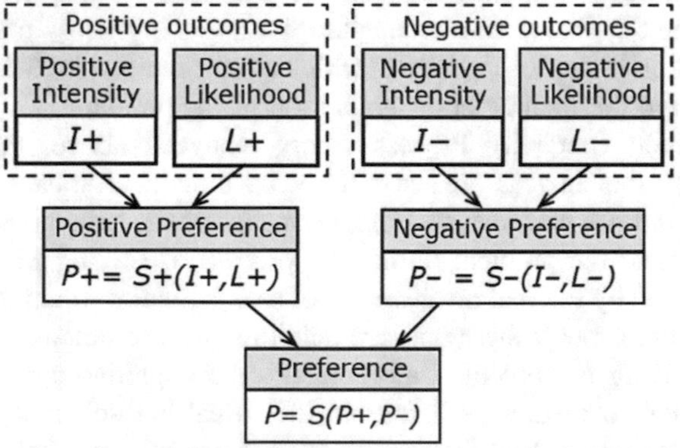

Fig. 3. IL-Frame of measuring the preference level.

For this purpose, using Multi-Attribute Utility Theory [9], special utility functions were built, which allow to measure the desirability or the preference of solving a problem. For a certain range of problems, these functions provide a fairly good approximation for measuring the level of motivation when making a decision. The problem of making a decision under conditions of uncertainty regarding the occurrence of two types of outcomes, positive (pros) and negative (cons), was considered. These outcomes were presented by their magnitude (intensity) and by their possibility of occurrence (likelihood). In other words, this problem had two criteria (*positive* and *negative*) and two sub-criteria (*intensity* and *likelihood*). The measurement of intensity (I) and likelihood (L) was provided on the verbal fuzzy scales of "weak-strong" and "seldom-often," respectively, and each of these scales contains nine levels [10]. In this way, the measured intensity and likelihood conveyed an emotional-evaluative aspect based on the expectation of occurrence of outcomes in conditions of uncertainty. *Positive preference (P+)* is considered as a composition of *positive intensity* (*I*+) (subjective value of the positive outcomes in the context of attaining the goal) and *positive likelihood* (*L*+) (subjective expectations of getting *I*+): $P+ = S+(I+, L+)$. *Negative preference (P–)* is considered as a composition of *negative intensity* (*I*–) (subjective value of the negative outcomes in the context of attaining the goal) and *negative likelihood* (*L*–) (subjective expectations of avoiding *I*–): $P- = S-(I-, L-)$. *Preference (P)* is considered as function of marginal utility functions *P*+ and *P*–: $P = S(P+, P-)$. The construction of such preference functions was possible as a result of numerous experimental studies conducted by Kotik [11]. Figure 4 demonstrates how the integral level of preference of 51% is formed based upon positive (55%) and negative (50%) levels of preference, determined by their positive ("strong") and negative ("weak") subjective intensities, as well as by their positive ("not seldom-not often") and negative ("often") subjective likelihoods, respectively.

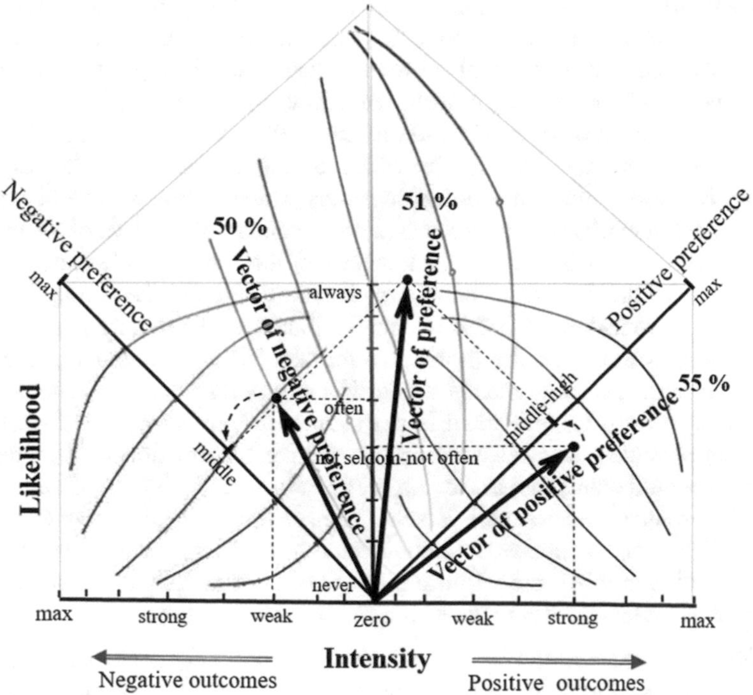

Fig. 4. Formation of the resulting vector of preference.

The distinctive features of these functions was that they were not, firstly, the numerical additive functions, which are traditionally used in MAUT and, secondly, they reflect the level of motivation to choose the most preferable alternative. At the same time, the positive and negative intensities represent, positive and negative subjective importance (significance) of outcomes respectively; while positive and negative likelihoods represent the possibility of getting positive outcomes (positive component of difficulty) and possibility of avoiding negative outcomes (negative component of difficulty) respectively. The validity of these functions when making decisions under uncertainty with the integrated positive and negative outcomes (pros and cons) was discussed in [12]: the preference function in 77% of cases, correctly determined the best alternative and in 66% of cases – priority ranking. This allows us to assume that the level of preference determined in this way reflects, to a certain extent, the motivational level in decision-making. This makes it possible to use the IL-Frame instead of the SD-Frame in decision-making problems, until more precise methods of measuring the level of preference appear, which reflect the corresponding level of motivation. In the future we will use Express Decision software for measuring these levels of preference. The Performance Evaluation Process allows to determine the preference of alternatives, and then to compare them and make a decision based on which alternative is more preferable. This method uses a *divide-and-concur algorithm* that implements a recursive breakdown approach in decision-making: decompose the problem into smaller sub-problems, solve them, and then recombine their results to solve the bigger problem. This division of the problem into sub-problems may span several levels deep until a *basic (ad hoc) level of certainty* will be reached, at which point the problem can be positively evaluated. In other words, the problem will contain only those outcomes, for which the decision-maker will be able to determine their positive (or negative) intensity and likelihood, which, in turn, will allow to determine the positive (or negative) preference level of each of these outcomes. PEP, unlike other decision-making methods, allows not only to decide which of the available alternatives is most preferable, but also to make a decision regarding which alternative can be chosen for execution. Such a motivational approach allows to discard the choice of an alternative, the implementation of which there is a low level of motivation (preference) that makes changes either to the goal or to the alternatives to determine the appropriate option. Another distinctive feature of PEP is that this method does not require a decision-maker to form a mental model of the problem before solving it, but instead, calls for the decision-maker to form this mental model in the process of arriving at the solution itself. In other words, an individual is not required to present a clear concept of the goal, nor to formulate any selection criteria for the construction of a model of the problem that would then assist in solving it. PEP allows an individual to build a mental model of a problem and to solve it using *self-regulation mechanisms* with *internal feedback* [6], which thus allows the individual to be led to the right choice through the construction of a problem model during the process of making a decision. The Performance Evaluation Process includes the following three stages:

1. *Problem decomposition*. The decomposition process starts from the evaluation of the decision-making problem by applying the IL-Frame. If for each alternative, the intensities and likelihoods of the positive and negative outcomes can be determined,

this means that the problem can be positively evaluated, does not need further decomposition of alternatives, and so the next stage, aggregation, will be applied. At the aggregation stage, the level of preference for each alternative will be measured with the IL-Frame, which allows to make a decision regarding which alternative is more preferable in the final, decision-making stage. If the problem is unable to be positively evaluated – for example, when the intensities of positive outcomes cannot be determined, the problem should then be divided into sub-problems with corresponding sub-goals. For a decision-making problem, this means that the decomposition process is required at the alternatives' levels with further evaluation and measuring of each alternative by using the IL-Frame. As a result of the decomposition, the problem should be divided into sub-problems, until the point that such a level of detail is reached that the positive and negative outcomes will be assessed by their levels of intensity and likelihood – so that in the aggregation stage, this will allow to determine (measure) the problem's preference level. For this purpose, hypotheses are used. Each hypothesis creates such a sub-problem that the alternatives are evaluated from the position of the outcomes (positive or negative), which are more specific (less uncertain) than the previous group of outcomes. If for such outcomes, their intensity and likelihood can be evaluated on the given scale, this signifies that further specification of the sub-problem and its outcomes is not required. Otherwise, a new hypothesis should be considered with even more specified outcomes. The hypothesis allows to assess the intensity(s) and likelihood(s) of the specific outcome(s), which in turn allow to determine the level of problem's preference, according to the IL-Frame. The choice of hypotheses is determined by the decision-maker and helps him to construct a mental model of the problem being solved. Hypotheses are usually considered for mutually exclusive situations, *(i.e.* the occurrence of some outcomes and their non-occurrence), in order to assess outcomes' degree of influence on the preference levels of sub-problems or alternatives. As a result of decomposition, a Decision Tree (DT) is determined for each alternative, and this reflects the functional structure of all the sub-problems examined. In this case, nonterminal vertices determine the sub-problems examined, and terminal vertices contain estimates of intensity and likelihood for positively evaluated sub-problems.

2. *Problem aggregation.* In the process of aggregation, a preference level will be determined (measured) for each alternative, which reflects the level of motivation for choosing a particular alternative. Aggregation includes three stages. In the first stage, by using the IL-Frame, the preferences levels of all positively evaluated sub-problems should be determined. In the second stage, it is necessary to separately determine the positive and negative preference levels for each alternative. In order to do this, we use the special rules for aggregation that present hierarchical and timing dependences of outcomes. The timing dependence of outcomes is determined by hypotheses that are usually considered for two mutually exclusive cases – the occurrence of an event (for example, an accident with severe consequences) and the non-occurrence of this event. In the third stage, by using the IL-Frame, the positive and negative preference levels of each alternative are combined into its cumulative level of preference. Thus, the result of problem aggregation is the construction of a Goal-Motivational Decision Tree (GMDT) for

each alternative. This tree differs from DT in that all sub-problems are assigned their measured levels of preference. The level of preference of the entire alternative aggregates the levels of preference of the sub-problems examined.

3. *Decision-making*. In this process, the preference levels of each of the alternatives are compared, and a decision is made regarding selecting an alternative for execution. An alternative with a sufficient level of preference for execution is selected. A low level of preference can occur when the problem is too complicated or not significant enough. The specification of the goal and the addition of new alternatives make it possible to arrive at an acceptable solution. If both alternatives have a sufficiently high level of preference, but their preference levels are equal or very close to each other, then positive and negative preference levels should both be considered.

Express Decision (ED), a mobile decision-making software application, is guided by PEP at its core to provide decision-making support. At the first step of this process, ED supports the decision-maker by transforming his/her mental model of the decision-making problem from the initial to the final state. By implementing the IL-Frame, ED helps to evaluate the problem, measure its positive, negative, and resulting levels of preference, and create DT and GMDT trees.

References

1. Knight, F.H.: Risk, Uncertainty, and Profit. Sentry Press, New York (1964)
2. Kotik M.A.: A method of diagnostics of a person's attitude towards an alarming event. In: Problems of Communication and Perception. Studies in Psychology VII. Scientific Notes of Tartu University, vol. 474, pp. 162–179. Tartu University Press, Tartu (1978)
3. Saati, T.: The Analytic Hierarchy Process. McGraw-Hill, New York (1980)
4. Ishizaka, A., Labib, A.: Review of the main developments in the analytic hierarchy process. Expert Syst. Appl. **38**(11), 14336–14345 (2011)
5. Bedny, G.: Application of Systemic-Structural Activity Theory to Design and Training. CRC Press (2015)
6. Bedny, G., Karwowski, W., Bedny, I.: Applying Systemic-Structural Activity Theory to Design of Human-Computer Interaction Systems. CRC Press, Boca Raton (2015)
7. Bedny, G., Karwowski, W.: Systemic-Structural Theory of Activity: Applications to Human Performance and Work Design. CRC Press, Boca Raton (2006)
8. Heath, C., Larrick, R.P., Wu, G.: Goals as reference points. Cogn. Psychol. **38**, 79–109 (1999)
9. Keeney, R., Raiffa, H.: Decisions with Multiple Objectives: Preferences and Value Trade-offs. Wiley, New York (1976)
10. Yemelyanov, A.M.: The model of the factor of significance of goal-directed decision making. In: Baldwin, C. (ed.) Advances in Neuroergonomics and Cognitive Engineering. Advances in Intelligent Systems and Computing, vol. 586, pp. 319–330. Springer International Publishing, Cham (2017)
11. Kotik, M.A.: Developing applications of "field theory" in mass studies. J. Russ. East Eur. Psychol. **2**(4), 38–52 (1994)
12. Kotik, M.A., Yemelyanov, A.M.: Emotions as an indicator of subjective preferences in decision making. Psychol. J. **13**(1), 118–125 (1992)

A Systemic Approach to Implementing Psychological Factors in Management

Fred Voskoboynikov(✉)

Baltic Academy of Education, Dekabristov 35, 190121 St.-Petersburg, Russia
fredvosko@hotmail.com

Abstract. In this brief presentation we will emphasize on what is essential for managerial activity from the psychological perspective. The influence of social environment on human productivity from the activity theory point of view is discussed in this work. Social interaction between the manager and subordinates is considered through the framework of the systemic-structural activity theory. Specific attention is paid to the way managers relate to subordinates and how this factor effects the group moral and psychological atmosphere in the workplace. We will dwell on the history of motivation in industry and analyze the concept of management and manager. We will present some important factors of the psychological nature which can be applied to the practice of management for achieving the team's objectives. Such factors as motivation, consideration for subordinates' personality features, the individual style of performance, the effect of a group environment on individual performance and the phenomenon of compatibility are considered in our work in utilizing the systemic approach.

Keywords: Management · Activity theory · Systemic-structural activity theory Self-regulation of activity · Motivation · Personality features Individual style of activity · Compatibility factor · Communicative abilities

1 Introduction

Systemic-structural activity theory (SSAT) considers human activity as a goal-directed self-regulative system [1]. It manifests itself in the way people through trials, errors, and feedback corrections create strategies of performance which are derived from the personality features. Activity can be defined as conscious, intentional, goal-oriented and socially formed behavior which is specific to humans. All actions are organized and directed toward the achievement of conscious goals of the task. Activity consists of actions that could be cognitive /or internal/ and behavioral /or external/ [2]. Activity theory distinguishes two types of activity: "object-oriented" and "subject-oriented". The former is referred to a subject using tools on material objects with the goal to complete the task and evaluate the results. The latter is referred to social interaction between people, which is the most important element in management.

Management is a multidimensional concept that includes technical, economic, psychological, and social factors. In management there are always two components - the subject, the one who provides managerial functions, and the object to whom the managerial actions are directed. The subject and the object of management form a

H. Ayaz and L. Mazur (Eds.): AHFE 2018, AISC 775, pp. 237–246, 2019.
https://doi.org/10.1007/978-3-319-94866-9_24

unified management system. Effective management is manifested in transformation of the system from one state to another, to the desired state. It is the interdisciplinary field that has an important meaning for the society. To effectively conduct multifaceted managerial activity requires a systemic approach.

Many factors influence on the psychological environment in the workplace, but the strongest one comes from the manager. The way managers relate to subordinates effects the group morale and psychological atmosphere in the workplace. If the manager does not project a positive image it automatically transmits into the relations between the team members. The working environment becomes stressful, people less incline to cooperate with each other, they feel uncomfortable and morally vulnerable. This factor affects the whole nature of business communication and largely determines the psychological environment in the workplace.

2 The Concept of Management and Manager

Historically, until the time of industrial revolution in the nineteenth century, the concept of management and managers did not exist. The society was the society of owners and their "helpers". In the twentieth century, our society became a society of organizations. The development of the society into the society of organizations led to a demand for people who were neither owners nor helpers. The new society needed people who practice professional management: planning, organizing, integrating and developing people etc. Thus, management as a specific activity and the area of study has been developed. Griffin [3] identifies the concept of management as the "set of activities directed at an organization's resources - physical, informational, financial and human - with the aim of achieving organizational goals in on *efficient* and *effective* manner". The efficient manner means using resources wisely and in a cost-effective way. The effective manner means making the right decisions and successfully implementing them.

The term *management* was first popularized by Frederick Winslow Taylor who is considered the earliest advocates of scientific management. In 1881 he proposed his scientific management theory as a way of making the conduct of work-related activities more efficient. He described his method in the book *The Principals of Scientific Management* [4]. The major postulate of the theory was an assumption that individual workers would be willing to work hard for monetary rewards. Taylor introduced wage incentives schemes so that workers could get paid more for the increased production. The book has inspired administrators to adopt productivity-enhancing and waste-reducing procedures and measures. Despite the fact that the wage is still considered the main motivating factor for increasing productivity, motivation by the only economic incentives works up to the certain point. People see more than just earnings in their work, they are filled with thoughts and ideas and want to see them implemented along with receiving monetary rewards.

Taylor believed that managers are required to do all the thinking related to the planning and design of work, leaving workers with the task of implementation. He writes, "In our scheme, we do not ask the initiative of our men. We do not want any initiative. All we want of them is to obey the orders we give them, do what we say, and do it quick." [5]. Scientific management boils down to five simple principles: 1. Shift all responsibility for the organization of the work from the worker to the manager 2. Use scientific methods to determine the most efficient methods for completing the work (while specifying the precise way in which the work is to be done) 3. Select the best person to perform the 'designed' job 4. Train the worker to do the work efficiently 5. Monitor worker performance to ensure the appropriate method is followed and that the appropriate results are achieved. The principles imply total visibility, strict accountability and absolute control [6]. In his endeavor to maximize manual efficiency Taylor abandoned the nuances and strengths of human nature and capability displaying psychological illiteracy. Indeed, a key criticism of Taylor's approach was that he treated people as machines. By concentrating on efficiency, he discounted and ignored human factors, addressing tasks and their execution without considering the human and their impact.

One of the earliest critics of Taylor's work was Lillian Gilbreth, an engineer and industrial psychologist. She was the one who instigated the psychological aspects of scientific management. Together with her husband Frank Gilbreth, a pioneer of motion study, they formed their own form of scientific management which focused on the human element in management. She was interested in exploring the psychological element within management in order to complement and augment the scientific management perspective. Her book *The Psychology of Management: The Function of the Mind in Determining, Teaching and Installing Methods of Least Waste* was published in 1914 [7]. Lillian Gilbreth incorporated concepts of human relations and worker individuality into management principles. It was a major early work in the field of industrial psychology. She expressed the view that scientific management should be built on the principle of recognition of the individual, stressing the importance of including the human element in management. Her contribution has been essential in positioning psychology in the context of management and emphasizing its role and value at a time when Taylor was still promoting the employee as a machine concept. Although her work has become less popular nowadays, Lillian Gilbreth is recognized as the 'first lady of management' [8].

Psychology has a lot to contribute to management in general, and more specifically to the management of projects. Projects are done in groups; they require team members to communicate, empathize, comprehend, influence and engage. Moreover, there is a crucial need to understand what motivates individuals to improve performance within teams, and to encourage the adoption of proposed change. Delivering successful projects requires an understanding of people and psychology. In our recent work [9] we were able to progress the agenda of the psychological aspects of management by encouraging a deeper consideration for the role of the human element in managing. The richness of working with people can therefore be used to excel, improve and grow performance through the use of human capabilities and relationships.

The concept of manager has to do with all people whose functions are to manage other people's activity: business owners and plant foremen, supervisors and heads of departments, commanders of military units and coaches of athletic teams, and many others fall into the category of managers. Regardless of the type of organization and the field of activity, management functions are essentially the same. In fact, management functions are considered to be universal. Managers plan and organize, coordinate and control, make decisions, motivate and reward. They handle physical, informational and financial resources. All these functions can be combined into four categories of resources: physical, informational, financial and human. The content of the first three categories is vary significantly depending on the specifics of the organization or business. Whereas for maintaining the fourth category of resources, the human resources, there is a lot of similarity regardless of the specifics of organizations.

3 Psychological Factors Applied to Management

Motivation is perhaps the most important factor, as it involves the action phase in human behavior, which has a direct impact on the outcome of the activity. Motivation is considered as source of energy to drive the activity. Motives are energetic components of activity, while goals are a cognitive element of activity. The interrelationships between goals and motives is dynamic and complex, and may vary over the course of activity. The critical motivational factor for working people is achievements and acknowledgment of achievements by their superiors. People desire recognition, they want to experience their importance. They want to have their ideas considered and want to feel a real sense of accomplishment. Everyone has a natural need for working activity and not just for generating income to meet consumer preferences, but for the work which brings satisfaction from the work process and from the achieved results.

To take an individual-focused approach in working with people is necessary for creating a cohesive team. From the social psychology perspective, the team is a small social group of persons who unite and cooperate for achieving the common goal. For the successful functioning of the team two factors are of the most importance: team members must possess the needed technical skills and experience in the field and they must complement each other. The first factor is usually well taken into account, but the second one is not always paid much attention to. Each person is different in his or her own unique way. Some people are quick and can easily adapt to the changing environment, others are slow and are not as dynamic. Some individuals can sustain tough impacts while others can't, but the latter are able to navigate in slightly noticeable changes of the surroundings, which enable them to react more keenly. Some feel comfortable in performing monotonous work while others are "falling asleep" in doing same. Some people are happy to work in a group environment, others prefer to work on individual assignments. To rely on people's strong qualities is more effective than to insist on fixing the weaker ones. As a result, managers will best benefit from what people are capable of and they will experience satisfaction by their performance. Hence, the golden rule in dealing with people is not to try to change people, but rather to build on what they are and compensate for what they are not. Respecting people's individuality and using it the best possible way will eventually benefit the team and the entire organization.

Individual characteristics of a person is a product of his or her heredity, physical being and the acquired experience of humankind. *Temperament* is an important characteristic of personality which manifests itself in activity. Temperament characterizes people's behavior from the position of force with which they respond to the same stimuli. It characterizes people only by the dynamic of their reaction on the impact and does not predetermine their mental ability or social significance. There are four known conditional temperament types. We will restrict ourselves to a very brief description of the types. People of *sanguine* temperament have a strong nervous system. They steady in their feelings and actions, they are sociable, talkative, easily converges with new people. *Choleric* type is individuals with a strong nervous system, quick-tempered, straight-forward and aggressive. They are characterized by stable aspirations and persistence in achieving their goals and are capable in overcoming great difficulties. People of choleric type are characterized by prevailing of excitation over inhibition. They are the ones who have "bad brakes" so-to-say. To put such individuals on the front line of communication with customers, where "customers are always right", is hardly a good idea. *Phlegmatic* type is individuals with a strong nervous system. They are balanced, diligent, patient, and peaceful; tend to be self-content and kind, relaxed and rational. People of *melancholic* type are individuals with a weak, easily vulnerable nervous system, capable of sustaining a short-term stress. Melancholic type persons are usually perfectionists, can sustain monotonous work, they possess an ability to pay close attention to details which is quite important in a number of professions.

The described four general types quite rare exist in a "pure" form. Indeed, many of us exhibit some mixture of temperament characteristics. This is how it can be explained. In any classification the type is characterized by the severity and the ratio of its constituent properties or other characteristics. Theoretically, the degree of severity of the properties may vary indefinitely thereby creating endless number of possible types. However, in reality there is no need for such theorization and the type-approach can be used for practical purposes. Singling out the most prominent feature of temperament attributes a person to one or another type. While one temperament type better relates to some kinds of activity, another type is good for some other kinds of activity. To take subordinates' temperament type and other personality features into account in the process of management is helpful for the purpose of achieving the team's objectives.

Extraversion – introversion and *neuroticism – stability* is another feature of personality which manifests itself in activity. Neuroticism (weak emotions, unstable) versus Stability (strong emotions). Extraversion is a characteristic of people as outgoing. Extraverts are in a constant need of "psychological food" from the social environment. They are characterized by high motor and speech activity, they are easy to respond to a variety of proposals and actively involve for their implementation. Introversion is understandably opposite to extraversion. Introverts are thoughtful, rational, inclined to planning their activity, inclined to self-analysis". Extraversion-introversion has a weaker effect on behavior, while neuroticism-stability has a stronger effect. For example, manager-introvert can learn to behave in an extraverted manner if circumstances require. Neuroticism can be described as an enduring tendency to experience negative emotional states, such as anxiety, guilt and depression. They draw some negative scenarios in their mind; their body language reflects their emotional state, their physiological reactions – blood pressure, pulse count, etc. evidence the same. Those, that score high on neuroticism tend to respond poorly to stress. People with high score on neuroticism are "emotion-focused" on stress and interpret situations as threatening, while people with low score in neuroticism (stable) are "problem-focused" and tend to ignore the source of stress. Low indicators on the scale stability–neuroticism say about emotional stability.

4 The Study of Individual Style of Activity in Management

There are two ways of ensuring the effectiveness of human performance. One is by professional selection, the so-called "screening out" individuals with specific attributes. The other one is through individual training methods directed towards the formation of individual strategies of activity based on personality features of the individual in the process of adaptation to the objective requirements of activity. The study of personality and individual differences is a critically important area of activity theory. The central notion in this area of study is the individual style of activity that connects features of personality with mechanisms of self-regulation and strategies of performance. According to the concept of individual style of activity people can efficiently adapt to the objective requirements of activity by utilizing the non-normatively prescribed methods of performance, but rather by their own individual style of activity. It should be understood however, that the use of the individual style of activity does not mean a violation of the standard job requirements, which exist in any type of working activity. Individualization in performance is restricted by a range of tolerance and by managers. Any performer can consciously or unconsciously develop efficient strategies of performance, which are adequate to her or his personality features, given they do not violate the objective requirements of working activity. Individual style of activity allows different individuals to rely on their personal strength as to compensate for their individual weaknesses. That is, people attempt to diminish the impact of their weaker features of personality in a given task situation utilizing their strong features for a more efficient performance. Individual style of performance can be developed based on the analysis of individual features of personality and on the obtained training methods that are directed towards the formation of individual strategies of performance [10].

The concept of individual style of activity was first introduced by the Soviet psychologist Merlin [11]. In SSAT individual style of activity is considered as strategies of performance deriving from the mechanism of self-regulation, which depend on personality features. Such strategy occurs at the conscious and unconscious levels and is based on the principles of self-regulation consciously or unconsciously. Both levels are tightly interconnected and transform from one to another. The process of self-regulation manifests itself in formation of desired goals, in developing of the program of actions, which correspondence with these goals, with conditions for achieving the goals and with persons' individual abilities. Other words, people through trials, errors and feedback corrections create strategies of performance suitable to their individuality. For example, people with inert nervous system develop a predisposition to organize and plan their work in advance and attempt to utilize stereotyped methods of performance. It should be distinguished however that the individual style of activity and methods of performance is not the same. The latter is not dependent upon individual personality features but rather upon organizational factors, imposed supervisory procedures, etc. Sometimes methods of performance that derive from organizational factors may contradict with the individual personality features. In cases of an inadequate training, which ignores individual features of personality, the subject may acquire methods of performance contradicting his/her individuality. It may negatively affect the performance level and the job satisfaction. Understanding of individual features of personality allows managers to assign the more adequate tasks for individual subordinates, and develop individual strategies of social interaction with subordinates.

Empirical facts and theoretical studies show that the adaptive mechanism of personality features can work up to the certain limit. In studying the individual style of activity, it is important to observe how people with different individual characteristics acquire the same knowledge and skills. On the other hand, it is as important to identify how subjects disintegrate into distinct groups with respect to their ability to skill acquisition. Such disintegration takes place as the capacity of some individuals to adjust to the requirements of the activity is reduced due to the increased task complexity. In the tasks of average difficulties all individuals exhibit similar level of achievement regardless of their abilities. In such situations it is hard to identify individual differences in performance. For example, when subjects take part in the training process, different curve of skill acquisition can be observed. Those trainees, whose individual properties more suited to the requirements of activity, adapt to the task more rapidly. As a result, individual curves of the skill acquisition have different positions and their difference are more noticeable at the beginning of the training process. Gradually differences in performance decreases and skill acquisition falls into a narrow acceptable area. At this stage individual differences of performance cannot be detected. At the final stages of the training almost all participants show approximately same results. This means that all trainees were able to adapt to the objective requirements of activity.

5 Communicative Abilities

Not all people easily get into contact with others. People's communicative abilities are different, and as a consequence, their preferences of communication with others are different too. Some people feel comfortable working shoulder to shoulder in groups, others prefer to work on individual assignments, some feel in "their shoes" when they lead others, while others feel comfortable being followers. The ability to communicate with people has its effect on both the team cohesiveness and on individual and group performance. A study conducted by psychologists at Leningrad University (Russia) in 1960s revealed four distinct types of communicative personality traits, which is manifested in ability for interpersonal communication: *leading, followers, closed, cooperating*. Leading type refers to people with a strong focus on power in the group. Such individuals can work more productively subject to the submission of other group members. Follower is the type with a pronounced orientation to the voluntary submission. Persons of this type can successfully participate in the solution of the group task only on condition of subordinating to a more confident, independent and competent member of the group. As a rule, such people are good performers. The relationship between the follower and the group is built on the basis of exposure to group influence. They are satisfied with such a position since they do not need to make decisions, which is not always an easy task. Closed type is the type with a pronounced individualistic orientation. They prefer working on individual assignments, they do not strive for leadership and cooperation. In this regard, it is advisable to use individuals of the leading type for carrying out organizational activities. People of the follower type, respectively, should be used performing tasks given them from “the above”. Closed type can successfully and efficiently work in performing independent tasks with a minimum degree of interaction with other group members. Cooperating type is good for working together with other members of the production group on an equal footing, without clearly singling out the leading position of any of them. They are striving to work together with others and follow them in case of reasonable decisions for the task solutions. As a rule, persons of this type are good partners in business. Managers should take into account the differences in people’s communicative characteristics in order to avoid a psychological mistake when completing the group for performance of the group tasks. In a group with two members of the leading type of communicative behavior and in a group that consists of followers and closed ones, the performance may be negatively affected. In the first case it will lead to conflicts, in the second - uncertainty and confusion. Thus, it’s important to take people’s communicative abilities into account in the process of management.

6 The Effect of Group Activity on Individual Performance

The increasing complexity of social structures and technological progress increases the share of collective knowledge-intensive sectors of industry which highlights the problem of increasing the effectiveness of individuals’ joint activities. In the study of the group activities an object of research becomes a group of people linked by a common purpose. Their activity is connected by the means of carrying out a common task. The concept of a group activity indicates multifaceted phenomenon, which must be differentiated.

It has long been observed that people behave differently in group settings as compared to the behavior in private. This is because the group is not the arithmetic sum of separate individuals and the result of the group performance is not always a positive sum of the results of individual performances by its members. Individual in a group appears in a new capacity – as a component of the system “individual – other individuals”. Groups have properties of their own; they are different from the properties of the individuals who form the group. Just as a combination of copper and tin results an alloy, which hardness is neither of each of them, people in the group act and behave differently. Representatives of various professions and other kinds of activity such as polar explorers, mountain climbers, commanders of aircrafts and ships’ crews experience in real life that not all people are equally fit for complex teamwork. For the effective execution of tasks in group environment not only specialists with their technical skills are needed, but also the degree of compatibility between them as well. Depending on the degree of compatibility the result of the group performance may either be equal to the sum of the results of individual performances, or greater or lower than that sum. Examples of the incompatibility can be observed in a working crew, where there is a significant difference in workers’ skills, or an athletic team formed by athletes of different skill levels. This kind of incompatibility is called *physiological*. In these examples such physical parameters as height, physical strength, motor skills etc.

are described. To note such differences in people is not that difficult and it's unlikely that anyone will instruct people with such differences to perform a task where these differences present a hindrance.

People always experience certain flow of feelings toward others. These feelings are based on the differences of psychological nature, such as temperament, character, social orientation, habits, amateur interests and so on. They may be positive or negative, or neutral, they can be mutual or non-mutual and therefore conflicting. The incompatibility by the described differences is called *psychological*. These differences are not always obvious and apparent, but precisely the differences of this kind have a decisive impact on compatibility and, in turn, on the group and individual performance. Psychological incompatibility has its negative influence not only on the group and individual performance, but on human health as well. Unfriendly uptight relationships between group members call up negative emotions. In mass professions, where there are no expressed extreme conditions, people can perform productively under the influence of negative emotions for a fairly long time. It's important to understand that all of this flows at the expense of the unnecessary stress. Working activity on the background of negative emotions for a long period of time may cause pathological developments in the central nervous system, which could lead to various diseases of a neurotic order. People become irritable, experience headaches, insomnia, blood pressure disorders, dysfunction of gastrointestinal tract, and other deviations in health condition. Typical medical approach for the treatment of such conditions does not always give positive results. There are statistical data in different countries on the loss of a large number of man-hours as a result of the nervous breakdown due to psychological incompatibility.

7 Summary

Activity theory distinguishes two types of activity: "object-oriented" and "subject-oriented". The former is referred to a subject using tools on material objects with the goal to complete the task and evaluate the results. The latter is referred to social interaction between people, which is the most important element in management. In this paper we demonstrated, that according to the systemic-structural activity theory, human activity is considered to be a goal-directed self-regulative system. In particular, individual style of activity allows the subject to adapt to the situation more efficiently because it connects features of personality with mechanisms of self-regulation and strategies of performance. We discussed the influence of the group environment on individuals' behavior and performance and the importance of taking the psychological compatibility into account. We analyzed some important psychological factors which should be applied into the process of management to create an optimum psychological atmosphere in the workplace.

References

1. Bedny, G.Z., Karwowski, W.: A Systemic-Structural Theory of Activity. Application to Human Performance and Work Design. Taylor & Francis Group, Boca Raton, FL (2007)
2. Bedny, G., Karwowski, W., Voskoboynikov, F.: The relationship between external and internal aspects in activity theory and its importance in the study of human work. In: Bedny, G., Karwowski, W. (eds.) Human-Computer Interaction and Operators' Performance, pp. 31–59. Taylor & Francis Group, Boca Raton, FL (2010)
3. Griffin, R.: Management. South Western Cengage Learning, Mason, Ohio (1999)
4. Taylor, F.: The principal of scientific management. Harper & Brothers, New York (1911)
5. Taylor, F.W.: Shop management; a paper read before the American society of mechanical engineers. Transactions of the American Society of Mechanical Engineers **24**, 1337–1480 (1903)
6. Morgan, G.: Images of Organization, 2nd edn, p. 23. Sage, Thousand Oaks, California (1997)
7. Gilbreth, L.M.: The Psychology of Management: The Function of the Mind in Determining, Teaching and Installing Methods of Least Waste. Sturgis and Walton, New York (1914)
8. Hindle, T.: Guide to management Ideas and Gurus. The Economist, London (2008)
9. Voskoboynikov, F.: The Psychology of Effective Management: Strategies for Relationship Building. Routledge, New York (2017)
10. Voskoboynikov, F.: The influence of personality features on performance in work, study and athletic activity. In: Marek, Tadeusz, et al. (eds.) Human Factor of a Global Society: A System of Systems Perspective, pp. 187–192. Taylor & Francis Group, Boca Raton, Fl (2014)
11. Merlin, W.S.: Outlines of Theory of Temperament. Perm Pedagogical University, Perm (1964)

Approach to Safety Analysis in the Framework of the Systemic-Structural Activity Theory

Gregory Bedny(✉) and Inna Bedny

Evolute, LLC, Wayne, NJ, USA
gbedny@optonline.net

Abstract. In this work we present safety analysis from the systemic-structural activity theory perspectives. This is an alternative framework to cognitive psychology. SSAT understands human activity as a goal-directed system that integrates cognition, external behavior and emotionally-motivational mechanisms. In contrast to cognitive psychology where goal integrates cognitive and motivational mechanisms, in SSAT goal is understood as a cognitive mechanism that includes some conscious elements. It is a conscious desired result of activity in a particular situation. The goal is always associated with motivational mechanisms and this creates a vector motives → goal. This vector is tightly interconnected with the emotionally-evaluative mechanisms through which a personal significance of information is determined which is specifically important in safety analysis. The prevention of errors requires determining the goal of the task, the actions required to achieve the goal, evaluation of significance of various elements of a task, and its complexity. The safety analysis should discover the preferable strategies of task performance. Considering activity from the functional analysis perspective when activity is understood as a complex self-regulative system is also an important aspect of safety analysis. In order to increase safety the design of equipment should consider the safety requirements. The first question is how to present information in an accidental situation and then to design controls and their layout to insure the adequate response. In this work we present qualitative analysis and morphological analysis of activity of an operator who works in a standing position with two arms controls and one pedal control. A special table on the left holds uncut pieces whereas one of the right holds finished pieces. This design, incorporating mutually exclusive modes of control, is well-justified from the ergonomics standpoint. However, developed in SSAT methods of morphological analysis of activity and method of task complexity evaluation demonstrate that such solution are not sufficient. This principle of safety analysis can be applied to the new equipment at the design stage. Suggested approach facilitates detection of potential dangerous points in the task performance.

Keywords: Systemic-Structural activity theory · Safety · Reliability
Complexity · Algorithmic analysis · Time-structure analysis

H. Ayaz and L. Mazur (Eds.): AHFE 2018, AISC 775, pp. 247–254, 2019.
https://doi.org/10.1007/978-3-319-94866-9_25

1 Introduction

One of the important purposes of human factor and ergonomic studies is improving operator's safety. The human-technology relationship is a critical factor in this field of study where the considerable research efforts have been expended. However, the existing methods of safety analysis in ergonomics have various limitations. Currently, most of the safety analysis are based on cognitive psychology [1, 2]. However, systemic-structural activity theory [3, 4] presents an alternative psychological approach of safety analysis. The SSAT approach utilizes such methods as qualitative, algorithmic, time structure, and complexity evaluation to enhance the safety analysis. SSAT has also adapted the MTM-1 system for description of motor actions in order to analyze the safety of performance of motor components of activity.

Let us consider as an example safety analysis method that has been utilized for analysis of a press operator's transition from leg to two-hand control. The modern mechanical presses used for serial and small-serial production processes typically are built for either two-hand or one-leg control. For example, a press operator in the standing position can start the ram moving down by simultaneously pressing two buttons, one with each hand. Alternatively, the ram can be set in motion by pressing a foot-pedal. The operator can switch between these modes of control by manipulating a two-position switch with the right hand. Safety rules determine which mode of control should be employed: the leg control is permissible when the metal sheet feed and removal of blanks and finished parts are mechanized, as such a set-up does not require the operator to place his/her hands into the danger zone. An additional safety requirement for the leg control mode is the use of a guard, which prevents inadvertent entry of the operator's hands into the danger zone. Guard removal is permitted only when hand-control mode is selected via the two-position switch. Once the press is in hand-control mode, the movement of the ram is possible only when both left and right hands are pressing their respective buttons, which must be held down until the ram movement is completed. Any break in contact with either button immediately interrupts the movement of the ram or slide.

However, a closer consideration reveals some drawbacks of such safety method. Neither mode of control automatically deploys or withdraws the protective guard; the operator may ignore safety procedures and remove the protective guard while in the leg-control mode, raising the possibility of inadvertently operating the press while the hands are in the danger zone. The risk is heightened by the fact that switching between control modes may not necessarily follow any regular pattern, the operator's choice of the control mode being dependent on the technical requirements of the specific work process and the operator's judgment as to the best way to tackle it. This suggests that the press design should be modified so that whenever the operator removes the protection guard the machine should be automatically switched from the leg control mode to the two-handed control.

At the first glance the foregoing conclusion may seem self-evident and the design flaw easily detected. However, experience shows that such insights are not always nearly so obvious at the equipment design stage; rather, there is often a mismatch between the designer's understanding of how the equipment will be used and what operators will choose to do in practice.

Below we present an example of the analytical description of a production operation involving the manual loading of a blank. The first and second stages of task analysis are qualitative and algorithmic safety analysis.

2 Qualitative and Algorithmic Methods of Safety Analysis

Qualitative and algorithmic methods of study can be utilized at the analytical stage of the analysis of the safety of the task performance.

Prior to any design innovation the performance of this production operation can be described as follows: the press operators work in a standing position; a special table to their left holds uncut metal pieces, while a similar table on their right holds finished pieces. An operator can switch from arm control to the leg control by manipulating a two-position switch with the right hand. Safety rules determine which mode of control should be employed: leg control is permissible when the metal sheet feed and removal of blanks and finished parts are mechanized. This eliminates the need for the operator to move his or her hands into the danger zone. An additional safety requirement for a leg control mode is the use of a guard, which prevents inadvertent entry of the operator's hand into the danger zone. In the absence of a device preventing the movement of hands into the danger zone, movement of a ram is only possible when both left and right hands are pressing their respective buttons, which must be held down until a ram movement is completed.

The uncut blanks weigh 10 kg. In order to take a blank from the left table or deposit a finished piece on the right the operator must make a body rotation of 45°; taking a blank from the left requires a hand movement of 80 cm, while depositing a finished piece on the right table requires a hand movement of 50 cm. The hand movement to reach the two-position switch (as described above) is 30 cm. The calculated distance of the hand movements includes some body motion in the same direction; the effect of the body movement is to diminish the magnitude of the hand movement distance. Other distances of movement are considered during the described below more detailed time-structure analysis, which combines SSAT and MTM-1. At this early stage of analysis a simple narrative description of the work process suffices.

An operator selects two-hands or leg control by turning the two-position switch to the required position with his/her right hand, then takes one blank from the left table with both hands, moves it onto the work surface of the press, pushes it against the stop, and then activates the press by simultaneously pressing the left and right buttons. When the cutting process is completed the operator releases the buttons, takes the finished piece with both hands, and deposits it on the table on the right. In cases, an operator forgets, or chooses to ignore the safety regulations, it is possible to use the leg control without the protection of the safety guard. Table 1 presents an algorithmic description of the production process as it is outlined above.

Table 1. Algorithmic description of the production operation performed by a press operator involving transfer from one mode of control to another without automatic switching to the guard protection.

Member of algorithm	Description of algorithm member
$\boldsymbol{O_1^{\mu}}$	Recall safety rules or forget/ignore them intentionally
1(1-2) $\boldsymbol{l_1^{\mu}}\uparrow$	If safety rules are forgotten or ignored decide to perform O_3^{ε}; if recalled decide to perform operator O_2^{ε}
1(1) $\downarrow\boldsymbol{O_2^{\varepsilon}}$	Move two-position switch to the required position with the right hand (for two-hand control of the press) and remove protection guard
1(2) $\downarrow O_3^{\varepsilon}$	Take a blank from the left table with both hands and put it on the work surface of the press
O_4^{ε}	Push the blank to the stopper
2 (1-2) $\boldsymbol{l_2^{\mu}}\uparrow$	If safety rule is performed (O_2^{ε} is performed) decide to turn on the press with two-hand control (go to O_5^{ε}). If O_2^{ε} is not performed decide to use leg control (perform O_6^{ε}) even when protection guard is removed
2 (1) $\downarrow O_6^{\varepsilon}$	Turn on the press with two-handed control when protection guard is removed and go to ω_1
ω *$\omega_1\uparrow$	Always-false logical condition (go to $O_6^{\alpha w}$)
2 (2) $\downarrow \boldsymbol{O_6^{\varepsilon}}$	Turn on the press with leg control even when protection guard is removed; then go to $O_7^{\alpha w}$
ω $\downarrow \boldsymbol{O_7^{\alpha w}}$	Wait based on visual control until ram completes its working movement
O_8^{ε}	Release the two buttons or pedal and move the finished piece to the right-hand position table

*The 'always false' logical condition is a syntactical device used to indicate the transition from one member of the algorithm to another (go to $O_6^{\alpha w}$). It does not represent any actual actions or operations during task performance.
**Symbols in bold designate danger points during the production process.
***Logical condition l_2^{μ} (decision making) performs checking functions, in the case.

As it can be seen from Table 1, the individual members of a human algorithm are designated by special symbols. Each member consists of a set of qualitatively different actions integrated by a common goal, where the possible combinations of actions are constrained by the specificity of their logical organization and the capacity of the operator's short-term memory. The arrows associated with the members of the algorithm in Table 1 indicate the transition from one member of the algorithm to another.

Symbolic description of the individual member of algorithm in the left column of the table is an example of utilizing the psychological units of analysis. On the right side of the table we present technological units of analysis which are just verbal description of members of a human algorithm. An analysis of the algorithm discloses its potential danger points, understood as the cognitive or behavioral actions (or their combination), whose execution could lead to injuries. In Table 1 such members of the algorithm are designated by bold type, and comprise of $O_1^{\mu}, O_2^{\varepsilon}, O_6^{\varepsilon}, l_1^{\mu}, l_2^{\mu},$ and $O_7^{\alpha w}$. The 'active waiting period' of task performance ($O_7^{\alpha w}$) is the time during which an operator observes the press in operation - that is, the downward movement of the ram - after having pressed the two-button or leg control. Although the operator does not perform any motor actions during this period, these actions are still required to actively focus an operator's attention on the machines of the operation. Risks can emerge during this period, especially if an operator has ignored the safety instructions and is working in leg/pedal control mode. If this is the case, a distraction can lead to injury, a higher level of focused attention is required, increasing the complexity of the task.

In order to illustrate the design potential of the algorithmic analysis of operator's performance strategies under differing conditions, Table 2 shows the same production operation after the implementation of a design innovation. In this alternative design, the switch from leg to two-hand control mode is carried out automatically whenever the protective guard is removed. This change means that it is no longer possible to carry out the production operation in violation of safety requirements.

Table 2. Algorithmic description of a production operation performed by a press operator involving transfer from one mode of control to another with the automatic switching to the guard protection.

Member of algorithm	Description of algorithm member
O_1^ε	Take a blank from the left table with both hands and put it on the work surface of the press
O_2^ε	Push the blank to the stopper
O_3^μ	Recall safety rules or forget/ignore them intentionally
$l_1^\mu \overset{1}{\uparrow}$	If safety rules are recalled to perform O_4^ε. If safety rules are forgotten or ignored perform O_4^ε
$\overset{2}{\downarrow} O_4^\varepsilon$	Move two-position switch to the required position with the right hand for two-hand control
O_5^ε	Turn on the press with two-handed control and go to ω_1
$^{*}\omega_1 \overset{\omega}{\uparrow}$	Always-false logical condition (go to $O_7^{\alpha w}$)
$\overset{1}{\downarrow} O_6^\varepsilon$	Turn on the press with leg control even when protection guard is removed and go to l_2
$l_2 \overset{2}{\uparrow}{}^{*}$	If leg control does not work then return to O_4^ε
$\overset{\omega}{\downarrow} O_7^{\alpha w}$	Wait based on visual control until ram completes its working movement
O_8^ε	Release the two buttons or pedal and move the finished piece to the right-hand position table

*Typically, a logical condition has two or more outputs; in this case one output option is set to 0.

A comparison of the algorithmic descriptions of the task performance before (Table 1) and after (Table 2) the design innovation demonstrates the removal of all dangerous points of the production operation. The decision-making associated with $l_2\mu$ in Table 1 is eliminated and thus there is no need for the later checks. As erroneous operation of the leg control is no longer possible, member $O_7^{\alpha w}$ is no longer a potential danger point.

3 Design of the Activity Time Structure During the Task Performance and Safety Analysis

In cases where operator's task performance strategies are difficult to predict, or where a more in-depth assessment of possible risks and their associated prevention costs is required, further, more detailed stages of SSAT task analysis can be carried out [5]. In such situations the time-structure analysis is utilized. It involves describing the time-structure of activity in the tabular or graphic form. When the time structure of activity includes elements that can be performed only sequentially, the tabular form of presentation is usually applied. The discussed here example includes relatively simple cognitive actions. The time allotted for the performance of these actions is based on data presented in the engineering psychology handbook [6, 7]. The duration of the basic motions is based on data taken from the MTM-1 system manual [8]. The MTM-1 system manual contains a full description of the system and comprehensive tables of motion data (the abbreviated description given in the time-study handbooks is inadequate for practical use). The SSAT methodology recommends that the MTM-1 analysis should be performed only after studying the possible strategies of task performance and the algorithmic description of the task. SSAT has developed new principles for the use of MTM-1 system for the tasks analysis where there is a complex combination of cognitive and motor components of activity and the performed tasks have very flexible strategies of performance. Table 3 shows the time structure of activity during performance of the production operation depicted in Table 1.

Table 3. Time-structure analysis of the production operation depicted in Table 1

Members of algorithm	Description				Mental components of activity
	Right hand or leg	Time in sec	Left hand	Time in sec	Time in sec
O^{μ}_1	Extraction information from memory				1.2
l_1^{μ}					0.3
	Decision making (yes/no type)				
O^{ε}_2		0.98			
	R30A+G1A+M2.5+RL1+R30E				
O^{ε}_3		1.86	*R50ABA+G1B+M50B10/2+RL1 (AS30) (AS30)*		
	R80ABA+G1B+M80B10/2BA+RL1				
O^{ε}_4	*(AS30) (AS30)*				
		0.7			
l_2^{μ}	*R5A+G5+M35A(10/2)x0.4+RL2*				
O^{ε}_5	Decision making (yes/no type)				0.3
		0.79	The same		
ω_1	*R40A+G5+AP2*			--	
		--			
O^{ε}_6	No time				
		1.47			
$O^{\alpha w}_7$	One step and pressing pedal				
		3.0			
O^{ε}_8	Active waiting period			1.54	
	RL2+R30A+G1B+M50B10/2BA+RL1 (AS30)		*RL2+R30A+G1B+M80B10/2BA+RL1 (AS30)*		

Members O_1^{μ}, l_1^{μ} and l_2^{μ} of the algorithm in Table 3 include one cognitive action; operator O_3^{ε} includes two motor actions combined with the body movement. For the right hand, the first action is *R80ABA* + *G1B* and the second one is *M80B10/2BA* + *RL1*. Both hand actions are combined with the assistance of the body movement *AS30*, where 30 indicates the distance the body moves; the operator reaches the metal blank and moves it the distance of 80 + 30 = 110 cm. Each motor action includes motions; the first motor action includes three motions integrated by one action goal: motion *R80A BA* is accompanied by the assisting body movement and by grasping the blank (*G1B* – 'grasp object lying close on a flat surface'). O_6^{ε} (move leg and press pedal) includes one motor action which is usually performed by the right leg. So, it can be seen that SSAT uses MTM-1 system differently from other approaches.

The last stage of safety analysis includes quantitative evaluation of task complexity. In quantitative safety analysis the common practice is to assess the probability of errors leading to injury; this approach can be greatly enhanced by applying the developed within SSAT quantitative methods of task complexity evaluation. The combination of probability and complexity evaluation methods arguably provides a more comprehensive safety assessment, allowing improved quantification of the degree of danger and thus more accurately targeted risk-reduction measures. Recognizing that non-subjective methods of obtaining the probabilistic characteristics of human performance are difficult and also tend to be insufficiently accurate, the presented here method combines derived from subjective expert judgment probabilistic measures with objective measurement procedures. It should be noted that what is being assessed using these methods is not the probability of an accident *per se* but rather the probability of potential danger points in the performance process. Accidents will occur at such danger points only under certain combination of conditions. Quantitative safety measures are useful tools for identifying, and thus better avoiding such conditions.

In this work we did not consider complexity evaluation method in safety analysis. However, obtained data demonstrates that complexity measures can be used not only for discovering and assessing the potentially dangerous points of task performance but also for making recommendations for their reduction.

4 Conclusion

In is very difficult to predict human error at the design stage, when real equipment does not exists yet. That is why the development of analytical methods for analysis of human erroneous actions and related to them possible accidents is an important aspect of safety analysis. Presented in this work material demonstrates the new method of safety analysis that can be used at the analytical stage of equipment development. Quantitative analysis of task complexity is particularly important for such approach. SSAT developed a unified and standardized approach to the analysis of task performance and to safety analysis. The main units of analysis at the stage of formalized and quantitative task assessment are cognitive and motor actions. Motor actions can be described in terms of standardized motions. SSAT developed the new principles of utilizing the MTM-1 system for analysis of variable tasks where cognition and motor components of activity can be combined in numerous manners.

The described here studies demonstrate that an algorithmic description of activity, followed by a time-structure analysis and the quantitative evaluation of task complexity allow the detection of potential danger points in the tasks' performance. The value of this approach is that these points are detected not by observation or experiment, but by building models of human activity. This facilitates an effective safety analysis and problem-solving at the early design stage.

References

1. Reason, J.T.: Human Errors. Cambridge University Press, New York (1990)
2. Senders, J.W., Moray, N.P.: Human Errors: Cause, Prediction, and Reduction. Lawrence Erlbaum Associates, Publishers, Hillsdale, NJ (1991)
3. Bedny, G. Z.: Application of Systemic-Structural Activity Theory to Design and Training. CRC Press. Taylor & Francis Group (2015)
4. Bedny, G. Z., Karwowski, W. Bedny, I. S.: Applying Systemic-Structural Activity Theory to Design of Human-Computer Interaction Systems. CRC Press. Taylor & Francis Group (2015)
5. Bedny, I.: On systemic-structural analysis of reliability of computer based tasks. Sci. Educ. Odessa, Ukraine **1–2**(# 7–8), 58–60 (2006)
6. Myasnikov, V.A., Petrov, V.P. (eds.): Aircraft Digital Monitoring and Control Systems. Manufacturing Publishers, Leningrad (1976)
7. Lomov, B.F. (ed.): Handbook of Engineering Psychology. Manufacturing Publishers, Moscow (1982)
8. UK MTMA: MTM-1. Analyst manual. London: The UK MTM Association (2000)

Express Decision – Mobile Application for Quick Decisions: An Overview of Implementations

Alexander Yemelyanov(✉), Simon Baev, and Alla Yemelyanov

Department of Computer Science, Georgia Southwestern State University,
800 Georgia Southwestern State University Drive, Americus, GA 31709, USA
{Alexander.Yemelyanov, Simon.Baev,
Alla.Yemelyanov}@gsw.edu

Abstract. Express Decision (ED), a mobile software application, was developed in order to offer decision-making support in situations of uncertainty. The current work presents a description of the Preference Evaluation Process (PEP), which was created based on Systemic-Structural Activity Theory with the goal of conducting thorough analysis of self-regulation processes that determine the selection of the most preferable alternative, a process in which ED plays a key role. PEP, along with ED, helps to divide a problem into sub-problems, determine the preference for solving each of them, and, as a result, determine the preference of the best alternative for execution. PEP, when splitting problems into sub-problems, implements a divide-and-concur algorithm, in which ED serves as a template (frame) for determining the terminal (ad hoc) level of sub-problems under consideration. The current work demonstrates the implementation of PEP with ED with the use of a real-life example based on solving the problem of selecting the best type of car insurance policy. For this particular problem, decision- and goal-motivation trees are constructed.

Keywords: Systemic-structural activity theory · Decision making under uncertainty · Performance evaluation process · Decision tree · Goal-motivational decision tree · Mobile decision support · Car insurance

1 Introduction

Express Decision (ED) is a mobile decision-making software application, which was developed with the purpose of supporting individuals who need to make quick decisions in situations of both uncertainty and risk. An important characteristic of ED is that it is based upon a theoretical framework that reflects current psychological knowledge regarding processes of thinking and decision-making [1], which helps set ED as a state-of-the-art-product in decision-making. In problems of uncertainty, regarding the given goal, conditions, and criteria, in which the classical models of risk largely do not function, ED likely remains the sole means of support; the only feasible alternative to it still remains human intuition. It is worth noting that uncertainty while solving some problems is determined by the uncertainty of its mental model in the head

H. Ayaz and L. Mazur (Eds.): AHFE 2018, AISC 775, pp. 255–264, 2019.
https://doi.org/10.1007/978-3-319-94866-9_26

of the individual. ED, while it does not replace the human mind, allows to enter certain specificities into the individual's mental model, which then allows for problem solving. Because the existing methods are aimed at a comparison of available options of decision making without considering their motives, while developing ED, we were going by the notion that the decision being made should not necessarily be relatively better, as much as it should be motivated by its execution. The solution may be relatively better, but in light of difficulties regarding its execution, it might also possess low levels of motivation [2]. In order to alter an individual's motivation (to raise or lower its level) towards selecting an alternative in light of the same conditions, an individual's mental model of the problem must change in his head, first [1]. This can happen, for example, due to a specification of a goal (which can lead to a change in the division of outcomes into positive and negative) and/or due to the implementation of new criteria into the analysis (which will lead to a change of weights of these outcomes in the decision that's being made). ED assists an individual in managing the level of his motivation and in making reasonable decisions. With this in mind, ED only strengthens an individual's opportunities for decision making, while not replacing him, altogether.

ED was used in supporting the following types of decisions: everyday decisions ("Help a student make the decision whether or not to accept a friend request on Facebook from boss/colleague/teammate"); life decisions in healthcare ("Help a disabled person make the decision regarding where to live: stay in her current residence, move in with her relative's family, or move into an assisted-living facility"); human-operator decisions in aviation ("Help the investigator in selecting the best option for landing an aircraft based on both the results of simulating various options and on the results of comparative analysis of the anticipated consequences in each of these options").

In 83% of all considered cases ED selected the same best alternatives as it was provided by AHP [3], which is a recognized leader of solving multi-criteria decision problems.

The inclusion of ED in the Preference Evaluation Process is acknowledged to be a significant influence on the quality of decision-making. Below is presented a description of the Preference Evaluation Process with its subsequent use together with ED when making a decision regarding the selection of the best car insurance policy.

2 Performance Evaluation Process with Express Decision

At the core of PEP lies the IL-Frame (Fig. 1), which measures the level of preference in the problem being solved.

For this purpose, using Multi-Attribute Utility Theory [4], special utility functions were built, which allow to measure the desirability or the preference of solving a problem. For a certain range of problems, these functions provide a fairly good approximation for measuring the level of motivation when making a decision. The problem of making a decision under conditions of uncertainty regarding the occurrence of two types of outcomes, positive (pros) and negative (cons), was considered. These outcomes were presented by their magnitude (intensity) and by their possibility of occurrence (likelihood). In other words, this problem had two criteria (*positive* and *negative*) and two sub-criteria (*intensity* and *likelihood*).

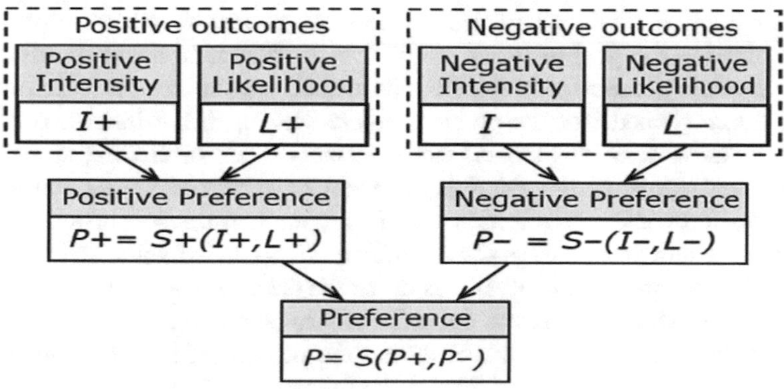

Fig. 1. IL-frame of measuring the preference level.

The measurement of intensity (*I*) and likelihood (*L*) was provided on the verbal fuzzy scales of "weak-strong" and "seldom-often," respectively [5]. In this way, the measured intensity and likelihood conveyed an emotional-evaluative aspect based on the expectation of occurrence of outcomes in conditions of uncertainty. *Positive preference (P+)* is considered as a composition of *positive intensity* (*I*+) (subjective value of the positive outcomes in the context of attaining the goal) and *positive likelihood* (*L*+) (subjective expectations of getting *I*+): $P+ = S+(I+, L+)$. *Negative preference (P−)* is considered as a composition of *negative intensity* (*I*−) (subjective value of the negative outcomes in the context of attaining the goal) and *negative likelihood* (*L*−) (subjective expectations of avoiding *I*−): $P- = S-(I-, L-)$. *Preference (P)* is considered as function of marginal utility functions *P*+ and *P*−: $P = S(P+, P-)$. *Where*

$I^+, I^- \in$ {extremely weak (*I1*), very weak (*I2*), weak (*I3*), not weak-not strong (*I4*), strong (*I5*), very strong (*I6*), extremely strong (*I7*)};

$L^+, L^- \in$ {extremely seldom (*L1*), very seldom (*L2*), seldom (*L3*), not seldom-not often (*L4*), often (*L5*), very often (*L6*), extremely often (*L7*)}.

The construction of such preference functions was possible as a result of numerous experimental studies conducted by Kotik [6, 7].

The Performance Evaluation Process allows to determine the preference of alternatives, and then to compare them and make a decision based on which alternative is more preferable. This method uses a *divide-and-concur algorithm* that implements a recursive breakdown approach in decision-making: decompose the problem into smaller sub-problems, solve them, and then recombine their results to solve the bigger problem. This division of the problem into sub-problems may span several levels deep until a *basic (ad hoc) level of certainty* will be reached, at which point the problem can be positively evaluated [8]. In other words, the problem will contain only those outcomes, for which the decision-maker will be able to determine their positive (or negative) intensity and likelihood, which, in turn, will allow to determine the positive (or negative) preference level of each of these outcomes. PEP, unlike other decision-making methods, allows not only to decide which of the available alternatives is most preferable, but also to make a decision regarding, which alternative can be chosen for execution. Such a motivational approach allows to discard the choice of an alternative, the implementation of which there is a low level of motivation (preference) that makes changes either to the goal or to the alternatives to determine the appropriate option. Another distinctive feature of PEP is that this method does not require a decision-maker to form a mental model of the problem before solving it, but instead, calls for the decision-maker to form this mental model in the process of arriving at the solution itself. In other words, an individual is not required to present a clear concept of the goal, nor to formulate any selection criteria for the construction of a model of the problem that would then assist in solving it. PEP allows an individual to build a mental model of a problem and to solve it using *self-regulation mechanisms* with *internal feedback* [1, 8], which thus allows the individual to be led to the right choice through the construction of a problem model during the process of making a decision. The Performance Evaluation Process includes the following three stages:

1. *Problem decomposition*. The decomposition process starts from the evaluation of the decision-making problem by applying the IL-Frame. If for each alternative, the intensities and likelihoods of the positive and negative outcomes can be determined, this means that the problem can be positively evaluated, does not need further decomposition of alternatives, and so the next stage, aggregation, will be applied. If the problem is unable to be positively evaluated – for example, when the intensities of positive outcomes cannot be determined, the problem should then be divided into sub-problems with corresponding sub-goals. For a decision-making problem, this means that the decomposition process is required at the alternatives' levels with further evaluation and measuring of each alternative by using the IL-Frame. As a result of the decomposition, the problem should be divided into sub-problems, until the point that such a level of detail is reached that the positive and negative outcomes will be assessed by their levels of intensity and likelihood – so that in the aggregation stage, this will allow to determine (measure) the problem's preference level. For this purpose, hypotheses are used. Each hypothesis creates such a sub-problem that the alternatives are evaluated from the position of the outcomes (positive or negative), which are more specific (less uncertain) than the previous group of outcomes. As a result of decomposition, a Decision Tree (DT) is determined for each alternative, and this reflects the functional structure of all the sub-problems examined. In this case, nonterminal vertices determine the sub-problems examined, and terminal vertices contain estimates of intensity and likelihood for positively evaluated sub-problems.
2. *Problem aggregation*. In the process of aggregation, a preference level will be determined (measured) for each alternative, which reflects the level of motivation for choosing a particular alternative. Aggregation includes three stages. In the first stage, by using the IL-Frame, the preferences levels of all positively evaluated sub-problems should be determined. In the second stage, it is necessary to separately determine the positive and negative preference levels for each alternative. In order to do this, we use the special rules for aggregation that present hierarchical and timing dependences of outcomes. In the third stage, by using the IL-Frame, the positive and negative preference levels of each alternative are combined into its cumulative level of preference. Thus, the result of problem aggregation is the construction of a Goal-Motivational Decision Tree (GMDT) for each alternative.

This tree differs from DT in that all sub-problems are assigned their measured levels of preference. The level of preference of the entire alternative aggregates the levels of preference of the sub-problems examined.

3. *Decision-making*. In this process, the preference levels of each of the alternatives are compared, and a decision is made regarding selecting an alternative for execution. An alternative with a sufficient level of preference for execution is selected. A low level of preference can occur when the problem is too complicated or not significant enough. The specification of the goal and the addition of new alternatives make it possible to arrive at an acceptable solution. If both alternatives have a sufficiently high level of preference, but their preference levels are equal or very close to each other, then positive and negative preference levels should both be considered.

Express Decision (ED), a mobile decision-making software application, is guided by PEP at its core to provide decision-making support. At the first step of this process, ED supports the decision-maker by transforming his/her mental model of the decision-making problem from the initial to the final state. By implementing the IL-Frame, ED helps to evaluate the problem, measure its positive, negative, and resulting levels of preference, and create DT and GMDT trees.

3 Express Decision Application in Selecting Car Insurance

When John, who lives in Richmond, Virginia, acquired a vehicle (2010 Toyota 4Runner SR5 with 95 k miles), the first problem that he faced was whether he needs to buy insurance coverage for his car. Virginia's insurance laws are somewhat different from other states'; the state does not require its drivers to maintain a car insurance policy, but those who do not are required to pay an uninsured motor vehicle fee. John used Express Decision (ED) with the goal of making a decision regarding whether *he needs to buy car coverage or pay an uninsured motor vehicle fee*. Making this decision was a problem for him, and we will denote this for brevity as *Problem 1 (Decision-making: Insurance vs. Fee)*. According to the IL-Frame used in ED, John must name the alternatives, and then for each of them, he must input four verbal characteristics by using the provided scales for subjective intensity of positive and subjective intensity of negative outcomes, as well as for the subjective likelihood for each of them. Assume that these characteristics were determined by the following reasoning for each alternative:

Alternative 1 (buy car coverage)

Pros (+): Purchase of car insurance is associated with strong (I5) protection on which John can often (L5) rely.

Cons (−): At the same time, in the past, possessing car insurance seldom (L3) elicited in John a negative reaction of average (I4) strength.

Alternative 2 (pay an uninsured motor vehicle fee)

Pros (+): Paying an uninsured motor vehicle fee very strongly (I6) allows John to save money, but this happens seldom (L3).

Cons (−): On the other hand, extremely strong (I7) risks are possible, even though their occurrence takes place very seldom (L2).

John was able to provide these characteristics because he had an understanding (although perhaps not a full awareness) of the extent of the Pros and Cons of the outcomes, and regarding the likelihood of their occurrence. Since all slots in the IL-Frame were filled, Problem 1 is considered to be positively evaluated. This allowed for ED to conduct further analyses with the obtained verbal characteristics, as well as to determine the following levels of preference (motivation) for choosing each alternative.

Level of preference for alternative 1 (*P1*) was 60% ("very high"); *P1 = S(S+(I5, L5), S−(I4, L3)) = 0.60*

Level of preference for alternative 2 (*P2*) was 51% ("high"); *P2 = S(S+(I6, L3), S−(I7, L2)) = 0.51.*

Considering the fact that John was highly (60%) motivated to buy insurance coverage for his car, he was ready to spend his time and energy on this issue. John started searching for insurance companies that could provide him with the best terms of coverage. From John's perspective, the best coverage reflected an inexpensive car insurance policy that would allow him to legally drive on the road. This could either be liability-only coverage or full coverage. In the case of full coverage, John was ready to pay from $500 to $1000 for a deductible to reduce the car insurance premium (a car insurance deductible is the amount of money you have to pay toward repairs before your insurance covers the rest). Since John already had experience with acquiring various types of car insurance in the past, he decided to visit the insurance companies and to choose the one that would offer him the best terms for a car insurance policy. Below, as an example of appraising one of these hypothetical companies (let us call it *Car-Ins*), the type of support that ED can offer John in solving his problem will be illustrated.

Therefore, John needs to solve the problem regarding selecting the car insurance company Car-Ins for one of two insurance options: liability-only coverage or full coverage. Let us call this decision-making problem as *Problem 2 (Decision-making: Liability-only coverage vs. Full coverage, Car-Ins).* If John is well-versed with the terms offered by Car-Ins, his problem will be positively evaluated, and ED will determine levels of preferences in the selection of each of the alternatives. If the level of preference in favor of selecting the better alternative (for example, liability-only coverage) will be sufficiently high, then John may conclude his selection with this alternative. In the example being considered, let's assume that John is not too familiar with the terms of coverage offered by Car-Ins. In this case, both positive and negative outcomes are too uncertain for him, which causes him to experience difficulties with the input of the required characteristics of intensity and likelihood into the IL-Frame. If this does happen to be the case, Problem 2 is considered to be negatively evaluated and should be divided into simpler sub-problems. In the given situation, this means that John must individually solve two evaluation sub-problems: *Problem 2.1 (Evaluation: Liability-only coverage, Car-Ins)* and *Problem 2.2 (Evaluation: Full coverage, Car-Ins).* John decides to begin with *Problem 2.1*, since he is expecting that liability-only coverage would be more suitable for him. But even this problem turns out to be too uncertain and thus too difficult for him, and he struggles with the selection of coordinating values for IL-Frame. This is why *Problem 2.1* is also considered to be negatively evaluated. This means that John needs to continue to simplify the problem and to lower its uncertainty. Since considering real-life scenarios involved with driving a vehicle with car insurance, such as the occurrence of a "car accident" or its absence, clarifies the problem and simplifies its subsequent evaluation and solution, John divided *Problem 2.1* into two sub-sub-problems: *Problem 2.1.1 (Evaluation: Liability-only coverage, car accident, Car-Ins)* and *Problem 2.1.2 (Evaluation: Liability-only coverage, no car accident, Car-Ins)*. Each of these evaluation problems should be separately assessed on IL-Frame by using corresponding verbal characteristics.

Because the occurrence of a car accident possesses an incidental nature, each of the abovementioned problems occurs in situations of accompanying hypothetical scenarios (occurrence of a car accident or its absence). According to car insurance industry estimates, an average driver files a claim for a collision about once every 17.9 years. So the probability of a "car accident" in the current year: p = 0.056. Let's consider that for John, this signifies a frequent or "often" probability of getting into a car accident, since he has already been driving for 15 years without any incidents, but could now end up experiencing one with great likelihood. The fact that a car accident might not occur ("no car accident") he considered to be "very seldom." Keeping this in mind, both *Problem 2.1.1* and *Problem 2.1.2* were positively evaluated.

Problem 2.1.1 *(Evaluation: liability-only coverage, car accident, Car-Ins)*: John *often (L5)* bears *extremely strong (I7)* responsibility for the cost of repairs and has *extremely weak (I1)* savings in case his car will need to be fixed (or even upgraded).

Problem 2.1.2 *(Evaluation: liability-only coverage, no car accident, Car-Ins):* John has *very seldom (L2) extremely strong (I7)* savings for not paying for full coverage, as well as a *very weak (I2)* peace of mind that he is not protected in the case of a car accident.

Here, it is necessary to note that the advantage of the considered model is related to its use of fuzzy verbal characteristics. When John determined the likelihood of getting into a car accident as "often," this characteristic is not reflected his subjective probability of getting into a car accident, rather indicated his perceived difficulty in avoiding it. The factor of difficulty [1], specifically – not the factor of probability – ultimately determined John's decision. However, it is important to note that the replacement of a decision-making model with uncertainties with a model with probabilities prevents the consideration of the factor of difficulty.

Further on, John switched to solving *Problem 2.2*, regarding the evaluation of full car insurance coverage. This problem, as well as *Problem 2.1*, which was analyzed above, is difficult for John, and so it received a negative evaluation and led him to consider two sub-problems: *Problem 2.2.1 (Evaluation: full coverage, car accident, Car-Ins)* and *Problem 2.2.2 (Evaluation: full coverage, no car accident, Car-Ins)* in terms of the occurrence of a "car accident" and "no car accident."

Problem 2.2.1 *(Evaluation: full coverage, car accident, Car-Ins)* was negatively evaluated, because the "car accident" needed specification regarding the severity of its consequences.

The issue is that John has not yet made a decision regarding the amount of deductible he would want to pay, so he now needs to relate his spending (deductible + insurance premium) and the benefits involved to whether he can obtain full coverage in the case of car accident consequences of varying degrees of severity. John divided the consequences of a car accident into those that cause more than $1,000 worth of damage to his car versus those that cause no more than $1,000 worth of damage to his car.

It is worth noting here that at the given stage, John may consider that the problem is becoming so difficult that it is simply easier to reject further analysis. However, as was indicated above, John was highly motivated to buy insurance coverage for his car, so it is easy to presume that the importance of the problem being solved determined the necessity to continue further with the analysis [2]. John considered the possibility of getting into a car accident with more than $1,000 worth of damage as *very often (L6)* for him and thought that the occurrence of a car accident with no more than $1,000 worth of damage was *seldom (L3)*. As a result, *Problem 2.2.1*, in relation to the severity of the car accident, was divided into two sub-problems: *Problem 2.2.1.1* and *Problem 2.2.1.2*, which were positively evaluated in the following way.

Problem 2.2.1.1 *(Evaluation: full coverage, car accident, > $1000, Car-Ins). Very often (L6)*, car insurance provides *very strong (I6)* compensation for the expense of repairs with an *average level* (*I4*) of spending on the part of the policy-holder.

Problem 2.2.1.2 *(Evaluation: full coverage, car accident, ≤ $1000, Car-Ins). Seldom (L3)* does car insurance provide *very weak (I2)* compensation for the expense of repairs with *extremely weak (I1)* spending on the part of the policy-holder.

The final problem that remains to be solved is *Problem 2.2.2*: the evaluation of full coverage in the case of “no car accident”. This problem is positively evaluated with the following assessments:

Problem 2.2.2 *(Evaluation: full coverage, no car accident, Car-Ins). Very seldom (L2)*, the car owner pays a *not weak-not strong (I4)* full price for coverage, while having a *weak (I3)* peace of mind that he is protected in the case of a car accident.

In this way, all sub-problems of *Problem 2* turned out to be positively evaluated, or in other words, they were solutions. This allowed to finalize the conclusion of the solution of *Problem 2* with the help of Express Decision, which provided the *following results:* the level of preference for choosing full coverage was 47%, whereas the level of preference for choosing liability-only coverage was 35%.

Thus, although the conducted analysis indicated that John preferred the option of full coverage to the option of liability-only coverage, the level of preference in selecting full coverage was not high (47%). This made it easy to anticipate that John would not select this option of car insurance from the company Car-Ins.

In order to raise the level of preference to select each of the considered alternatives, John had the opportunity to clarify his goal, or to make changes to the conditions of the car insurance options. For example, raising the deductible would lead to the lowering of the premium and would then likely make the full coverage option more appealing. If this does not work to increase the level of preference in selecting one of the insurance policies of Car-Ins, it is necessary to consider the alternatives of the insurance policy in other companies, with potentially more appealing terms for full and liability-only coverages.

As we could see, ED provides decision support by means of dividing the problem into sub-problems, which then simplifies their evaluation and final decision-making. With the help of ED, the process of forming the model of the problem, as well as its solution, occurs simultaneously with the use of internal feedback. This allows to correct the model of the problem in the process of its solution by means of specification of the goal, conditions, and criteria for outcomes. In the process of solving the problem, ED constructs a Decision Tree for the problem (Fig. 2), which reflects the hierarchy of the problem and its sub-problems. For the problem, ED also creates a Goal Motivation Decision Tree in which levels of preference are assigned for all sub-problems.

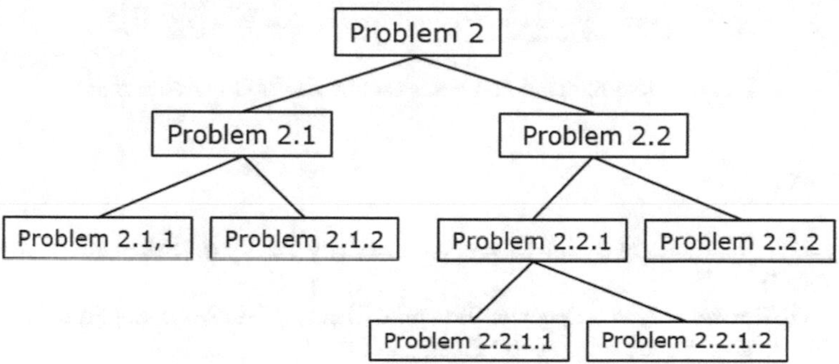

Fig. 2. Decision tree for decision-making problem 2

For decision-making *Problem 2 (Liability-only coverage vs. Full coverage),* Fig. 1 demonstrates the Decision Tree; Figs. 3 and 4 demonstrate the Goal-Motivational Decision Trees for "Liability-only car coverage" and "Full car coverage" respectively, with levels of preferences for sub-problems and initially used data defined as human's verbal characteristics.

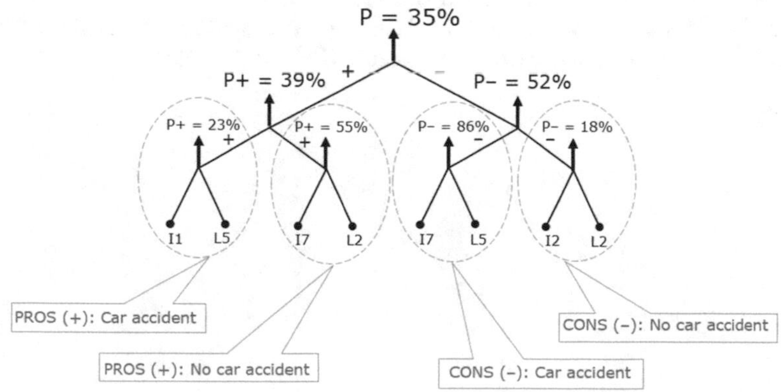

Fig. 3. Goal-motivational decision tree for "liability-only car coverage"

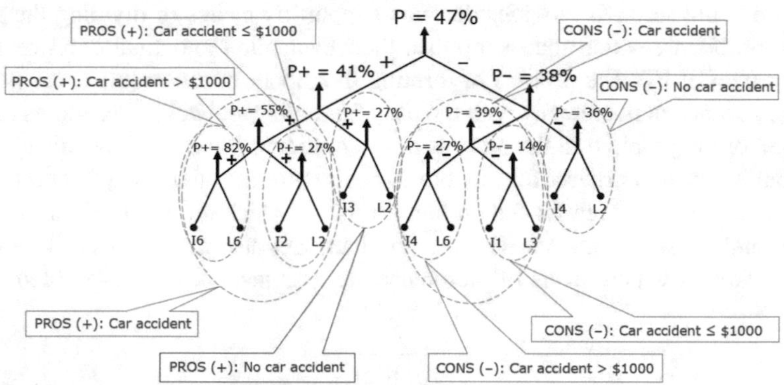

Fig. 4. Goal-motivational decision tree for "full car coverage"

References

1. Bedny, G.: Application of Systemic-Structural Activity Theory to Design and Training. CRC Press (2015)
2. Bedny, G., Karwowski, W.: Systemic-Structural Theory of Activity: Applications to Human Performance and Work Design. CRC Press (2006)
3. Saati, T.: The Analytic Hierarchy Process. McGraw-Hill, New York (1980)
4. Keeney, R., Raiffa, H.: Decisions with Multiple Objectives: Preferences and Value Trade-offs. John Wiley & Sons, Inc., New York (1976)
5. Yemelyanov, A.M.: The model of the factor of significance of goal-directed decision making. In: Baldwin, C. (ed.) Advances in Neuroergonomics and Cognitive Engineering (Advances in Intelligent Systems and Computing), vol. 586, pp. 319–330. Springer International Publishing (2017)
6. Kotik, M.A.: Developing applications of "field theory" in mass studies. J. Russ. East Eur. Psychol. **2**(4), 38–52 (1994)
7. Kotik, M.A., Yemelyanov, A.M.: Emotions as an indicator of subjective preferences in decision making. Psychol. J. **13**(1), 118–125 (1992)
8. Bedny, G., Karwowski, W., Bedny, I.: Applying Systemic-Structural Activity Theory to Design of Human-Computer Interaction Systems. CRC Press (2015)

Dichotomy of Historicity and Subjective Perception of Complexity in Individuals' Activity Goal Formation and Decision Outcomes

Mohammed-Aminu Sanda[1,2(✉)]

[1] University of Ghana Business School, P. O. Box LG 78, Legon, Accra, Ghana
masanda@ug.edu.gh
[2] Luleå University of Technology, SE-97187 Luleå, Sweden
mohami@ltu.se

Abstract. This study examined how the historicity of individuals' organizational activity and subjective perception of task complexity influence individuals' activity goal formations, strategies and decision outcomes considerations. The findings showed that conscious goal-directed processes of individuals are influenced by the dichotomy of historicity and their subjective perceptions of task complexity. It is concluded that; when "highest goal" is desired for an impending task, the dichotomous relationship of historicity of the individual's self-regulation activity and his/her subjective perception of task complexity will have direct influence on both his/her Activity Goal Formation and decision outcome processes, while his/her activity strategies consideration will be directly influenced by only the historicity of his/her self-regulation activity. Also, when "best goal is desired, the dichotomous relationship will directly influence only the individual's consideration of decision outcomes, while both his/her Activity Goal Formation and strategies consideration will be directly influenced by only the historicity of his/her self-regulation activity.

Keywords: Activity historicity · Activity complexity · Activities dichotomy Subjective perception · Activity goal formation · Individual decision outcomes

1 Introduction

The survival of organizations depends on their ability to adapt to their external environment [1]. Organizations thus, require monitoring and feedback mechanisms to enable them sense changes in their work environments, and for them to be able to develop the requisite capacity that can enable them to make responsive adjustments [2]. This therefore, calls for the need to determine a concept of context which captures the teleological content of a person's action [3], just as the understanding of social situation of development entailed the formation of a situational concept, which captures the way in which the situation determines social interactions and psychological development [3]. Actors at the workplace engage in activities that constitute practices and whose performances leads to the actors attaining habitual accomplishment of specific tasks. Practices that actors engage in, are influenced by both the social history and rationality

H. Ayaz and L. Mazur (Eds.): AHFE 2018, AISC 775, pp. 265–277, 2019.
https://doi.org/10.1007/978-3-319-94866-9_27

of the practices, even though the former was viewed by [4] to be of higher significance than the latter. As such, actors interact with the social, psychological and physical features of context in the activities they are to engage in. This implies that the activities of actors in organization occur both in the macro contexts that provide commonalities of action, and the micro contexts in which action is highly localized [5]. It is the interaction between these contexts that provides an opportunity for adaptive practice [4]. Thus, in the approach towards identifying factors that influence practices evolving from actors' activities in organizations, it is important to understand the practices that are entrenched in an actor's conscious goal-directed processes that leads to the emergence of thoughtfully mastered learning activity [2]. This is due to the growing interest in the use of the 'practice' approach to overcome concerns on the gap between the realistic understanding of what actors do at the workplace, and the theoretical understanding of what actors are expected to do at the workplace [4, 5].

The purpose of this study, therefore was to establish how the historicity of an individual's self-regulation system relates with the individual's subjective perception of complexity to influence the individual's activity goal formation, consideration of activity strategies and decision outcomes.

2 Dichotomy of Historicity and Complexity in Actor's Activities

Using the postulations of [6] as point of departure, and arguing from organizational activity theoretical perspective, the concept of self-regulation provides context for capturing the historical content of an actor's action in an organizational activity [2]. As it is noted by [6], the major problem facing the self-regulation system is the process of continuing reconsideration of activity strategies when internal and external conditions or situations have changed. This process sometimes leads to changes in the goal as well as the methods of achieving the goal [2, 6]. Thus, the program of self-regulation, its criteria of evaluation, and its goal can change during an actor's self-regulation process [7]. This is because, self-regulation is time-oriented, and is based on the evaluation of experience, present situation, and anticipated future scenarios [7]. This implies that internal changes in the psychological aspects of an actor's self-regulation may not emerge from experiences only, but also from memories that can be adapted to a situation [7]. Such memory epitomizes historical recollections of adaptable practices that are manifestations of declarative knowledge [2].

History, according to [8], is made in future-oriented situated actions and it is important to look for ways of capturing how actors discursively create new forms of activity and organization and the uncertainty associated with such organizational activity [2]. Uncertainty in a task is dependent on both the objective and subjective characteristics of an organizational activity [7, 9]. Therefore, it is important to understand an actor's activity from the perspectives of the history of its local organization and against the more global history of the organizational concepts, procedures and tools employed and accumulated in the local activity [2, 10]. This is because, history-making is embedded in an actor's everyday activities [11]. It is also important to identify and distinguish the complexities associated with the activity's cognitive attributions, as informed by the specificity of its information processing, and its emotional-motivational attribution, as informed by its energetic aspects [2, 12]. Making visible these attributions of complexity will enable the designers of actors' activities to understand the practice enhancing strategies used by actors to mediate the cognitive difficulties and the emotional-motivational challenges inherent in their designed activities [2, 12]. Hence, in understanding the dialectical complexities of an activity, there is the need to consider the existence of multiple goals [2]. This is because, in most cases, an actor's activity cannot be adequately interpreted upon the assumption that it is organized around a single, neatly identifiable goal [2].

The complexity of an actor's activity can be viewed as the number of rules inherent in the cognitive and motivational-emotional attribution of the activity [12, 13]. This is because, actors use their generic skills, such as flexibility, technical intelligence, perceptive ability, technical sensibility, a sense of responsibility, trustworthiness, and independence [14] to mediate complexities arising from their organizational activities [12]. This makes it important to understand ways actors engage in continual reconsideration of activity strategies, since it sometimes results, not only in changes in the methods of achieving the goal, but a change in the goal itself [2]. Arguing from the perspective of [7, 15], the conscious goal-directed processes of an actor preparing for an organizational activity is deemed to be influenced by the history of an actor's task engagements on his/her subjective perception of new task. This is based on the premise that the discovery of goals is essential to true activity [17], and the goals, being discrete elements of activities, can be transformed into contradictions, which may demand creative solutions, as well as expanded and generalized into a qualitatively new activity structure [15].

In this regard, it is proposed that; the goal formation, activity strategies and decision outcome associated with the emergence of an actor's thoughtfully mastered learning activity are directly influenced by the dichotomy of the actor's historicity and subjective perception of task complexity.

3 Methodology

Various attempts (e.g., [16, 17]) have been made to develop suitable methods for task complexity evaluation, including the use of various units of measure, such as the number of controls and indicators, or the number of actions. But there is the argument that task complexity cannot be successfully evaluated by such methods, principally because they employ incommensurable units of measure [7]. This therefore suggests that while task complexity can be evaluated both experimentally and theoretically [7], expert judgments, such as the use of a five-point scale can also be used for the subjective evaluation of the task's complexity [2, 7], as well as the historicity, goal formation, strategy formulation and outcome decision-making associated with the task [2]. In this study, the actors' historicity, complexities, goal formation, strategy and outcome-decision formulations in examination writing activity were measured using a five-point scale quantitative measure and expert subjective judgements.

3.1 Data Collection Procedure

Based on the well-established knowledge (e.g. [7, 14]) that activities of individuals are realized by goal-directed actions, informed either by mental or motor conscious processes [7], as objects of the cognitive psychology of skills and performances [7, 14], the individual differences of maximization in examination activity goal formation and decision outcomes was measured using the maximization measurement scale [18]. Guided by the scale, data was collected from 338 Graduate Students, comprising of I86 (55%) females and 152 (45%) males, using a self-administering questionnaire. All the study participants are pursuing master's degree programmes at the University of Ghana Business School, in Accra, Ghana. Being Graduate students, they all have first university degrees, and as such, were highly educated and have written examinations severally over the years as they progress on the education ladder. At the time of data collection in the month of October 2017, the respondents were preparing to write their end-of-second semester examinations between November–December 2017.

3.2 Data Analysis Procedure

Using the systemic analytical approach [7, 12]), the proposition (Sect. 2 above) that "goal formation, activity strategies and decision outcome associated with the emergence of an actor's thoughtfully mastered learning activity are directly influenced by the dichotomy of the actor's historicity and subjective perception of task complexity" relative to examination writing activity is tested. The collated data was analyzed stepwise, using structural equation modelling (SEM) with the analysis of moment structures (AMOS) as a modeling technique [19]. This procedure has the advantage of maximizing the validity of the estimates [20]. The AMOS software also gives the opportunity to conduct analyses for multiple levels of variables using a range of in-built statistical techniques [19]. In the first step, path analysis was conducted to test the indicator predictiveness of actors' Historicity, Subjective Perception of Complexity, Activity Goal Formation, Consideration of Activity Strategies, and Consideration of Decision Outcomes).

The path analysis was followed by a confirmatory structural analysis of the conceptual model underlined by the proposition (Sect. 2 above). The AMOS graphics statistical software is used as the analytical tool. The indicator fit is interpreted from the perspective of [21] that estimated indicator loadings on the latent variable (neuro-activities) must be 0.7 or higher. This allowed for model modification towards attaining superior goodness of fit by rejecting indicators whose loadings fall below the baseline of 0.7. The criteria used to establish model fit include Chi Square (CMIN), which is the absolute test of model fit and Comparative Fit Index (CFI). For the CMIN, a probability value below 0.05 implies model acceptance. The CFI value close to 1.0 also indicates a very good model fit.

4 Results and Discussion

4.1 Analysis of Indicators' Predictiveness of Latent Variables in Model

Seven (7) predictive indicators for five (5) latent variables (neuro-activities) were tested per the data collected in this study, and minimum was achieved for the model. A summary of the indicator variables and their factor loadings or standardized regression weights (*R*) for the respective latent variables (neuro-activities) are shown in Table 1 below. By convention, the indicators should have loadings of 0.7 or higher [21] on the latent variable for them to be considered significant predictors.

Table 1. Standardized regression weights of indicators on neuro-activities

Neuro-activities	Indicators	Label in model	R	R^2
Historicity of self-regulation activity	Decides on best outcome first	RSTO	0.71	0.50
Subjective perception of task complexity	Not lowering self-confidence, no matter how difficult the task	PTC	0.71	0.50
Activity goal formation	Form best goal	RATP	0.71	0.50
	Set highest goals	GPT	0.69	0.49
Consideration of activity strategies	Look for best option of performing task, no matter how long it takes	PATP	0.71	0.50
Consideration of decision outcomes	Expect activity outcome to be good	PSTO	0.71	0.50

It is observable from the results in Table 1 above that, almost all the indicator variables have factor loadings greater than or equal to 0.7. It is only the setting of highest goals for the task (GPT) that has a factor loading of 0.69 which is approximated to 0.7. Therefore, all the indicators have factor loadings or standardized regression estimates (*R*) of approximately 0.7 and higher, and as such are predictive of their various latent variables (neuro-activities) as recommended by [21]. The implications being that by considering the writing of examination as an activity, a good measure of an individual's Historicity of Self-Regulation Activity (neuro-activity) is characterized by the individual deciding on the best outcome of the activity first ($R = 0.71$, $R^2 = 0.50$), while that of the individual's Subjective Perception of Task Complexity (neuro-activity) is characterized by the individual trying not to lower his/her confidence on the expected outcome, no matter how difficult he/she perceive the tasks ($R = 0.71$, $R^2 = 0.50$).

Similarly, a good measure of the individual's Goal Formation for the Activity (neuro-activity) are characterized by (i) the individual forming the best goal to be achieved, no matter what it takes ($R = 0.71$, $R^2 = 0.50$), and (ii) the individual setting the highest goals for himself/herself ($R = 0.70$, $R^2 = 0.49$). Also, a good measure of the individual's Consideration of Activity Strategies (neuro-activity) is characterized by the individual looking for the best option of performing the task, no matter how long it takes ($R = 0.71$, $R^2 = 0.50$), while that of the individual's Consideration of Decision Outcomes (neuro-activity) is characterized by the individual always wanting the outcome to be good ($R = 0.71$, $R^2 = 0.50$).

As it is highlighted Table 1 above, the individual's Activity Goal Formation (neuro-activity) entailed two distinct predicative indicators, that is, (i) the individual forming the best goal to be achieved (RATP), and (ii) the individual setting the highest goals (GPT) for himself/herself. As such the correlations among the neuro-activities were tested by performing confirmatory structural analysis, varying the character of the individual's Activity Goal Formation.

4.2 Confirmatory Structural Analysis of Model with GPT-Influenced Activity Goal Formation

In this confirmatory structural analysis, the individual's Activity Goal Formation is predicted by the individual's formation of the best goal to be achieved, no matter what it takes (GPT). The AMOS-generated standardized path diagram with the standardized indicator loadings of the respective latent variables, as well as loadings between the various latent variables (neuro-activities) in the structural model is shown in Fig. 1 below. The goodness of fit statistics showed that the overall model-fit is quite good. This is because the estimated χ^2 of 0.00 (df = 0) has probability level of 0.000 which is smaller than the 0.05 used by convention. Thus, the null hypothesis that the model fits the data is accepted. Also, the estimate for the Comparative Fit Index (CFI) of 1.0 indicates an acceptance of the null hypothesis of a very good fit for the tested model.

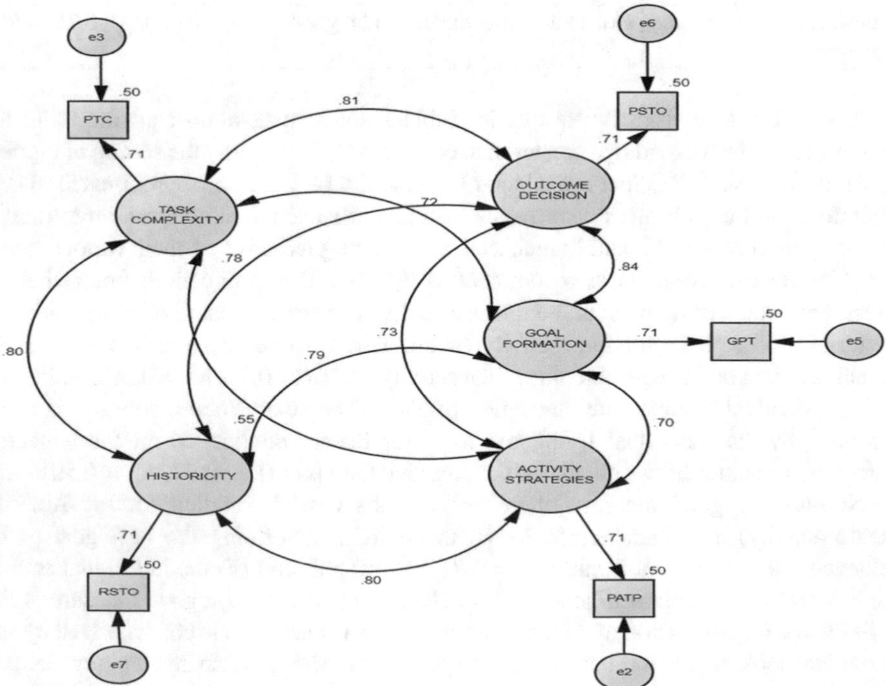

Fig. 1. AMOS-generated path diagram showing standardized loadings in structural model with GPT-influenced activity goal formation.

As it is observable in Fig. 1 above, the standardized correlation weights (α) for most of the association between the various latent variables (neuro-activities) are also above the threshold value of 0.7 recommended by [21] and thus signify strong association. It is inferred from Fig. 1 that in the Conscious goal-directed process of the individual, a strong bond exists between the individual's Historicity of Self-Regulation Activity and his/her Subjective Perception of Task Complexity ($\alpha = 0.80$). Similarly, in the emergence of the individual's thoughtfully mastered learning activity, influencing associations are found to exist between the individual's Activity Goal Formation and his/her Consideration of Activity Strategies ($\alpha = 0.70$), as well as, his/her Consideration of Decision Outcomes ($\alpha = 0.84$), and also, between the individual's Consideration of Activity Strategies and his/her Consideration of Decision Outcomes ($\alpha = 0.73$).

The two bonded elements of the individual's Conscious goal-directed process (i.e. individual's Historicity of Self-Regulation Activity and his/her Subjective Perception of Task Complexity) significantly influence the individual's' Consideration of Decision Outcomes which is reflective of the emergence of the individual's thoughtfully mastered learning activity. The strength of the influencing association between the individual's Consideration of Decision Outcomes and his/her Historicity of Self-Regulation Activity is $\alpha = 0.78$, while that with his/her Subjective Perception of Task Complexity $\alpha = 0.81$. Additionally, the individual's Historicity of Self-Regulation Activity, as an element of his/her Conscious goal-directed process, is found to influence the emergence of the individual's thoughtfully mastered learning activity from the perspective its relational effect on the individual's Activity Goal Formation ($\alpha = 0.79$), and his/her Consideration of Activity Strategies ($\alpha = 0.80$). There is also a significant influencing association between the individual's Subjective Perception of Task Complexity and his/her Activity Goal Formation ($\alpha = 0.78$). On the contrary, there is no significant influencing association between the individual's Subjective Perception of Task Complexity and his/her Consideration of Activity Strategies ($\alpha = 0.55$), which correlation weight is less than the threshold value of 0.7. This therefore, indicates that an individual's Subjective Perception of Task Complexity, as an element of his/her Conscious goal-directed process does not influence his/her emergence of thoughtfully mastered learning activity, informed by his/her Consideration of Activity Strategies. Thus, the derived model shown in Fig. 2 below shows the dichotomous influence of historicity and complexity in individuals' activity goal formation and consideration of decision outcomes, when the individual set for himself/herself highest goals for an activity.

The model highlighted in Fig. 2 below shows the existence of the dichotomous interrelationship between the historicity of an individual's self-regulation activity and the individual's subjective perception of task complexity engraved in the conscious goal-directed processes of the individual. Similarly, the emergence of thoughtfully mastered learning activity of the individual entails dichotomous interrelationships among the individual's activity goal formation, and consideration of activity strategies, as well the individual's considerations of decision outcomes. But in the transition of the individual's conscious goal-directed processes to the emergence of thoughtfully mastered learning activity of the individual, the dichotomous interrelationship of the historicity of the individual's self-regulation activity and his/her subjective perception of

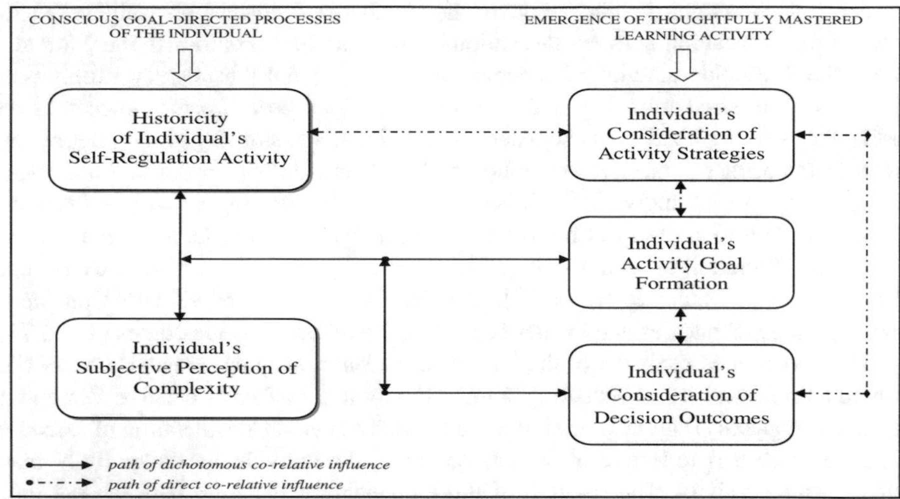

Fig. 2. Dichotomous historicity and subjective perception of complexity in individuals' activity goal formation and consideration of decision outcomes when highest goals are set.

task complexity influences both the individuals Activity Goal Formation and his/her consideration of decision outcomes for the task to be performed (writing examination in this study). The dichotomous interrelationship of historicity of the individual's self-regulation activity and his/her subjective perception of task complexity has no effect on the individual's consideration of activity strategies. Rather, it is the historicity of the individual's self-regulation activity, as a separate entity in the individual's conscious goal-directed processes, which has direct influence on the individual's consideration of activity strategies in the emergence of thoughtfully mastered learning activity of the individual. The individual's subjective perception of task complexity does not influence his/her consideration of activity strategies.

4.3 Confirmatory Structural Analysis of Model with RATP-Influenced Activity Goal Formation

In this confirmatory structural analysis, the individual's Activity Goal Formation is predicted by the individual's setting of the highest goals for himself/herself. The AMOS-generated standardized path diagram showing the standardized indicator loadings of the respective latent variables (neuro-activities) in the re-tested model-fit structural model is shown in Fig. 3 below. The goodness of fit statistics showed that the overall model-fit is quite good. This is because the estimated χ^2 of 0.00 (df = 0) has probability level of 0.000 which is smaller than the 0.05 used by convention. Thus, the null hypothesis that the model fits the data is accepted. The estimate for the Comparative Fit Index (CFI) is 1.0, indicating an acceptance of the null hypothesis of a very good fit for the tested model.

As it is highlighted in Fig. 3 below, the standardized correlation weights (α) for the association between the all latent variables (neuro-activities) are above the threshold value of 0.7 recommended by [21], signifying strong associations. It is thus inferred that in the Conscious goal-directed process of the individual, a strong bond exists between the individual's Historicity of Self-Regulation Activity and his/her Subjective Perception of Task Complexity ($\alpha = 0.80$). Similarly, in the emergence of the individual's thoughtfully mastered learning activity, influencing associations are found to exist between the individual's Activity Goal Formation and his/her Consideration of Activity Strategies ($\alpha = 0.71$) and, also between the individual's Consideration of Activity Strategies and his/her Consideration of Decision Outcomes ($\alpha = 0.73$). No influencing association exist between the individual's Activity Goal Formation and his/her Consideration of Decision Outcomes ($\alpha = 0.59$).

The two bonded elements of the individual's Conscious goal-directed process (i.e. individual's Historicity of Self-Regulation Activity and his/her Subjective Perception of Task Complexity) significantly influence the individual's' Consideration of Decision Outcomes which is reflective of the emergence of the individual's thoughtfully mastered learning activity. The strength of the influencing association between the individual's Consideration of Decision Outcomes and his/her Historicity of Self-Regulation Activity is $\alpha = 0.78$, while that with his/her Subjective Perception of Task Complexity is $\alpha = 0.81$. Additionally, the individual's Historicity of Self-Regulation Activity, as an element of his/her Conscious goal-directed process, is found to influence the emergence of the individual's thoughtfully mastered learning activity from the perspective its relational effect on the individual's Activity Goal Formation ($\alpha = 0.70$), and his/her Consideration of Activity Strategies ($\alpha = 0.80$). On the contrary, there is no significant influencing association between the individual's Subjective Perception of Task Complexity and his/her Activity Goal Formation ($\alpha = 0.44$), on the one hand, and his/her Consideration of Activity Strategies ($\alpha = 0.56$), on the other, whose respective correlation weights are less than the threshold value of 0.7. This therefore, indicates that an individual's Subjective Perception of Task Complexity, as an element of his/her Conscious goal-directed process does not influence his/her emergence of thoughtfully mastered learning activity, informed by his/her Activity Goal Formation or his/her Consideration of Activity Strategies. Furthermore, if the individual refrains from setting highest goals for himself/herself, his/her Consideration of Decision Outcomes will not be influenced by his/her Activity Goal Formation.

Thus, the model shown in Fig. 4 below is derived for understanding the dichotomy of historicity and subjective perception of complexity in individuals' activity goal formation and decision outcomes (without the individual setting highest goals for himself/herself). The model shows that by when the individual set the best goals for himself/herself as an indicative measure of the individual's Goal Formation for the Activity, a dichotomous interrelationship still exists between the historicity of the individual's self-regulation activity and the individual's subjective perception of task complexity, both of which are engraved in the conscious goal-directed processes of the individual. The exception in this instance is that, the emergence of thoughtfully

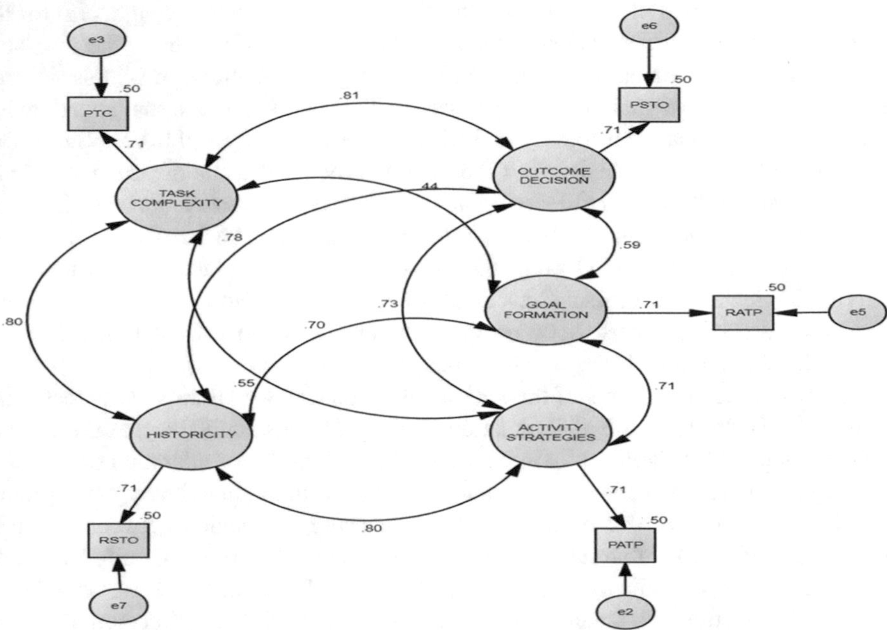

Fig. 3. AMOS-generated path diagram showing standardized indicator loadings in the structural model with RATP-influenced activity goal formation

mastered learning activity of the individual entails dichotomous interrelationships only between the individual's activity goal formation, and individual's consideration of activity strategies, as well as between individual's consideration of activity strategies and the individual's considerations of decision outcomes.

As it is highlighted in Fig. 4 below, the individual's activity goal formation does not relate with the individual's considerations of decision outcomes. But as it is shown in Fig. 4 below, in the transition of the individual's conscious goal-directed processes to the emergence of his/her thoughtfully mastered learning activity, the dichotomous interrelationship of the historicity of the individual's self-regulation activity and his/her subjective perception of task complexity influences only the individual's consideration of decision outcomes for the task to be performed (i.e. writing examination in this study). The dichotomous interrelationship of historicity of the individual's self-regulation activity and his/her subjective perception of task complexity has no effect on the individual's activity goal formation, as well the individual's consideration of activity strategies.

In the emergence of thoughtfully mastered learning activity of the individual, it is rather the historicity of the individual's self-regulation activity, as a separate entity in the individual's conscious goal-directed processes, which has directly influences both the individual's activity goal formation, and the individual's consideration of activity strategies. The individual's subjective perception of task complexity does not influence his/her activity goal formation, as well as his/her consideration of activity strategies.

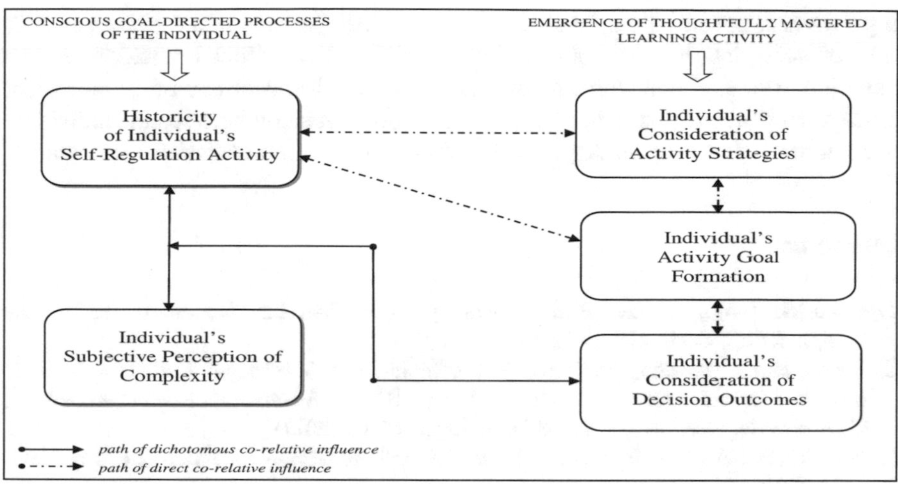

Fig. 4. Dichotomous influence of historicity and subjective perception of complexity in individuals' consideration of decision outcomes when best goals are set.

5 Conclusion

This study has shown that the conscious goal-directed processes of an actor are influenced by the history of the actor's previous task engagements and his/her subjective perception of complexity in an impending task. The study has also established that when an actor is preparing to engage in a pending organizational activity (writing examination in this study), different neuro-oriented activities occur in the transition of the individual's conscious goal-directed processes to the emergence of thoughtfully mastered learning activity of the individual. The transitional dynamics in such neuro-oriented activities is found to be shaped by the psycho-characteristics of the activity goal formation, in terms of the individual "aiming for a best goal" or "setting the highest goal". It is therefore concluded that; in the transition of the individual's conscious goal-directed processes to the emergence of his/her thoughtfully mastered learning activity, in which the "highest goal is anticipated", the dichotomous relationship of the historicity of the individual's self-regulation activity and his/her subjective perception of task complexity directly influences both the individuals Activity Goal Formation and his/her consideration of decision outcomes for the activity to be performed, but not the individual's consideration of strategies. The individual's consideration of strategies is directly influenced by only the historicity of the individual's self-regulation activity. In this vein, the individual's consideration of activity strategies and outcome decisions also have direct influence on each other. It is also concluded that in the transition of the individual's conscious goal-directed processes to the emergence of thoughtfully mastered learning activity of the individual, in which the "best goal is set", the dichotomous interrelationship of the historicity of the individual's self-regulation activity and his/her subjective perception of task complexity directly influences only the individual's consideration of decision outcomes for the activity to be performed, but not the individual's Activity Goal Formation and his/her consideration of strategies. Rather, both the individual's Activity Goal Formation and consideration of strategies are directly influenced by only the historicity of the individual's self-regulation activity. In this vein, there is no direct relation between the individual's consideration of activity strategies and his/her consideration of decision outcome.

References

1. Hendrick, H.W., Kleiner, B.M.: Macroergonomics: An Introduction to Work System Design. HFES, Santa Monica (2001)
2. Sanda, M.A.: Mediating Subjective task complexity in job design: a critical reflection of historicity in self-regulatory activity. In: Carryl, B. (ed.) Advances in Neuroergonomics and Cognitive Engineering, pp. 340–350. Springer, Cham (2017)
3. Blunden, A.: An Interdisciplinary Theory of Activity. Koninklijke Brill, Leijden, Netherlands (2010)
4. Jarzabkowski, P.: Strategy as social practice: an activity theory perspective on continuity and change. J. Manage. Stud. **40**(1), 23–55 (2003)
5. Sanda, M.A.: Cognitive and emotional-motivational implications in the job design of digitized production drilling in deep mines. In: Hale, K.S., Stanney, K.M. (eds.) Advances in Neuroergonomics and Cognitive Engineering, Advances in Intelligent Systems and Computing, vol. 488, pp. 211–222. Springer, Cham (2016)
6. Bedny, G.Z., Karwowski, W., Bedny, I.S.: Complexity evaluation of computer-based tasks. Int. J. Hum.-Comput. Int. **28**(4), 236–257 (2012)
7. Bedny, G.Z., Karwowski, W.: A Systemic-Structural Theory of Activity: Applications to Human Performance and Work Design. Taylor and Francis, Boca Raton, FL (2007)
8. Engeström, Y.: Managing as argumentative history-making. In: Boland, R.J., Collopy, F. (eds.) Managing as Designing, pp. 96–101. Stanford University Press, Stanford (2004)
9. Sanda, M.A., Johansson, J., Johansson, B., Abrahamsson, L.: Using systemic structural activity approach in identifying strategies enhancing human performance in mining production drilling activity. Theor. Issues Ergonom. Sci. **15**(3), 262–282 (2014)
10. Engeström, Y.: Expansive learning at work: toward an activity theoretical reconceptualization. J. Educ. Work **14**(1), 133–156 (2001)
11. Engeström, Y., Middleton, D.: Introduction: studying work as mindful practice. In: Engeström, Y., Middleton, D. (eds.) Cognition and Communication at Work, pp. 1–14. Cambridge University Press, Cambridge (1996)
12. Sanda, M.A., Johansson, J., Johansson, B., Abrahamsson, L.: Using systemic approach to identify performance enhancing strategies of rock drilling activity in deep mines. In: Hale, K. S., Stanney, K.M. (eds.) Advances in Neuroergonomics and Cognitive Engineering, Advances in Intelligent Systems and Computing, vol. 488, pp. 135–144. CRC Press, Boca Raton, FL (2012)
13. Kieras, D., Polson, P.G.: An Approach to the formal analysis of user complexity. Int. J. Man-Mach. Stud. **22**, 365–394 (1985)
14. Abrahamsson, L., Johansson, J.: From grounded skills to sky qualifications: a study of workers creating and recreating qualifications, identity and gender at an underground iron ore mine in Sweden. J. Ind. Rel. **48**(5), 657–676 (2006)
15. Engeström, Y.: Learning by Expanding: An Activity-Theoretical Approach to Developmental Research. Orienta-Konsultit, Helsinki (1987)

16. Mirabella, A., Wheaton, G.R.: Effects of task index variations on transfer of training criteria. Report NAVTRAEQUIPEN 72-C-0126–1. Naval Training Equipment Center, Orlando, Florida. January (AD 773-047/7GA) (1974)
17. Venda, V.F.: Engineering Psychology and Synthesis of Informational Sources. Manufacturing Publishers, Moscow (1975)
18. Dalal, D.K., Diab, D.L., Zhu, X.S., Hwang, T.: Understanding the construct of maximizing tendency: a theoretical and empirical evaluation. J. Behav. Decis. Mak. **28**, 437–450 (2015)
19. Sanda, M.A., Kuada, J.: Influencing dynamics of culture and employee factors on retail banks' performances in a developing country context. Manage Res. Rev. **39**(5), 599–628 (2016)
20. Di Stefano, C., Zhu, M., Mîndrilă, D.: Understanding and using factor scores: considerations for the applied researcher. Pract. Assess. Res. Eval. **14**(20), 1–9 (2009). http://pareonline.net/getvn.asp?v=14&n=20
21. Schumacker, R.E., Lomax, R.G.: A Beginner's Guide to Structural Equation Modeling. Lawrence Erlbaum, Mahwah (2004)

Diagnosing a Concussion by Testing Horizontal Saccades Using an Eye-Tracker

Atefeh Katrahmani(✉) and Matthew Romoser

Western New England University, Springfield, MA, USA
atefeh.katrahmani@gmail.com, matthew.romoser@wne.edu

Abstract. It is difficult to diagnose a concussion because neuro-images of the brain usually do not show signs of any physical brain damage. The primary purpose of the present research was to investigate the eye movement patterns of patients with concussion symptoms and differences between those patterns of healthy patients. Twenty-two participants were recruited for this study in two groups: the concussed teen group and the non-concussed teen group. They were asked to wear an eye-tracker and complete a cognitive test. The results show significant differences ($p < 0.05$) in the eye movement patterns between the two groups from different aspects: (1) The concussed group showed more chaotic eye movements while the non-concussed participants had smoother eye movement patterns with less saccadic jumping. (2) The concussed patients tended to have more random eye glances on the areas around the targets. In general, the eye-tracking methods can be useful in concussions diagnosis.

Keywords: Concussions · Eye-tracker · Human factors · Saccades
Concussion diagnosis

1 Introduction

While there are no precise statistics about the number of sports-related concussions, emergency rooms report 1.4 to 3.8 million sports-related mTBI (mild Traumatic Brain Injury) patients in the United States annually [1–4]. Diagnosing concussions can be difficult because many concussion assessment tools are subjective and their results are open to interpretation. Currently, self-reporting is the only way to evaluate concussions in many cases [5]. Patients do not necessarily show all the symptoms, and most of the diagnoses are dependent on patients' self-reports rather than on having a measurable and reliable diagnosis tool. Studies showed that this evaluation technique is not an accurate method of concussion assessment [6].

There is a serious lack of knowledge about the definition of concussions even among athletes. McDonald and colleagues state that only 66% of student athletes receive prior concussion education [7]. Research shows 43% of athletes with a history of concussions hide symptoms in order to be able to return to playing [4, 8]. The cognitive tests that are commonly used for concussion assessment have some issues. One of the main issues is the lack of a baseline test for some test methods.

Since every person's brain function are unique, scientists are not able to use a baseline test comparing one patient to another. In addition, the brain's functionality

H. Ayaz and L. Mazur (Eds.): AHFE 2018, AISC 775, pp. 278–283, 2019.
https://doi.org/10.1007/978-3-319-94866-9_28

changes over time [9, 10]. Therefore, before using the baseline, there is a need to ensure that the existing baseline test for the individual is valid.

Suh and colleagues [11] have found that people with mTBI show symptoms such as decreased target prediction, greater eye position error, and variability of eye position in a circular tracking test. In addition to causing a lack of smooth pursuit, concussions cause other visual impairment symptoms such as convergence abnormalities which occur in 47% to 64% of concussed patients [12]. Concentration of both eyes on a single point in space requires complete convergence and research shows that up to 90% of concussed patients show symptoms of convergence disorders [13–16]. Almost 30% of mTBI impaired cases show saccadic dysfunction [17].

Diagnosing concussions based on eye movements is critical and clinicians need to add it to their skill set of clinical expertise [18]. Requiring eye-tracking for concussion diagnoses might limit usage of sideline tests, where athletic trainers are responsible for baseline concussion assessment. Therefore, the investigation of eye movements can be useful in diagnosing concussions. Eye movements can also be used as a measurement tool for assessing the healing process, evaluating fitness to drive, and assessing whether or not a patient's healing has progressed enough to be able to go back to daily life.

2 Method

The present research seeks to provide a method to diagnose a concussion in patients with concussion symptoms. A cognitive horizontal saccades test was developed to assess whether or not a participant is concussed. More than diagnosing a concussion, these tests provided a means of assessing the healing progress and determining if the patient has recovered enough to return to daily life activities, like driving.

2.1 Participants

A total of twenty-two participants was recruited. Group 1 consisted of 11 concussed teen participants and group 2 consisted of 11 non-concussed teen participants. Participants with a history of natural visual impairment, neurological or eye disorders like Parkinson's or colorblindness were excluded from the study. Concussed cohort participants were among the patients who have been referred to Connecticut Children's Medical Center (CCMC) in Farmington, CT.

2.2 Eye-Tracker

Each participant's gaze point was recorded using a Tobii Pro-II mobile eye-tracker (Fig. 1). The eye-tracker is a pair of glasses with the scene camera and eye cameras included in the frame. The system was calibrated via a wireless connection to the analysis laptop and video is recorded by a recording unit clipped to the participant's belt. The point of view of the participants, represented as a red circle, was covered up upon the scene camera video image.

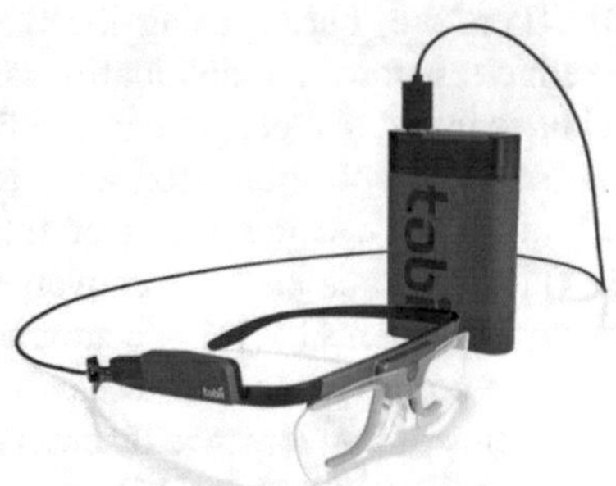

Fig. 1. Tobii Pro-II mobile eye-tracker. Reprinted from Tobbipro.com, from http://www.tobiipro.com/fields-of-use/psychology-and-neuroscience/customer-cases/audi-attitudes/. Copyright © 2017 Tobii AB. Reprinted with permission.

2.3 Experimental Procedure

This study was a single session study, which occurred in a lab at Connecticut Children's Medical Center (CCMC) in Farmington, CT. During this session, participants met the study administrator, signed the informed consent and had the opportunity to ask any questions.

Participants were asked to sit at a desk in front of a monitor. Two single solid circles appeared horizontally, one on the left side and the other one on the right side of the monitor. The participant needed to move their eyes quickly from one to another for ten seconds (Fig. 2).

Fig. 2. Horizontal saccades test

3 Results

Using SPSS, the results show a significant difference between the eye movement patterns of the concussed versus non-concussed participants. The non-concussed participants had significantly more accurate eye movements from one point to another point. Their fixation points were closer to the target points (A & B). The vertical (X), horizontal (Y) and total distances from the target points (D) were calculated. Significant differences between the fixation points' distances – in horizontal, vertical and total distance – were seen between two groups. Table 1 shows the statistical analysis results.

Not only were the concussed group less able to fixate at the target points accurately, they showed more undeliberate fixations and gaze points. Figure 3 illustrates the

Table 1. ANOVA statistical analysis

		df	Mean square	F	Sig.
X	Between groups	1	6407007.129	65.198	0.000
	Within groups	36978	98269.934		
	Total	36979			
Y	Between groups	1	7619476.077	1157.462	0.000
	Within groups	36978	6582.915		
	Total	36979			
D	Between groups	1	9638673.466	117.471	0.000
	Within groups	36978	82051.300		
	Total	36979			

difference in convergence of the fixation points of the concussed and non-concussed participants. As the figure shows, the non-concussed participants' fixation points showed more condensed fixation points while the results express more spread and undeliberate fixation points in concussed participants.

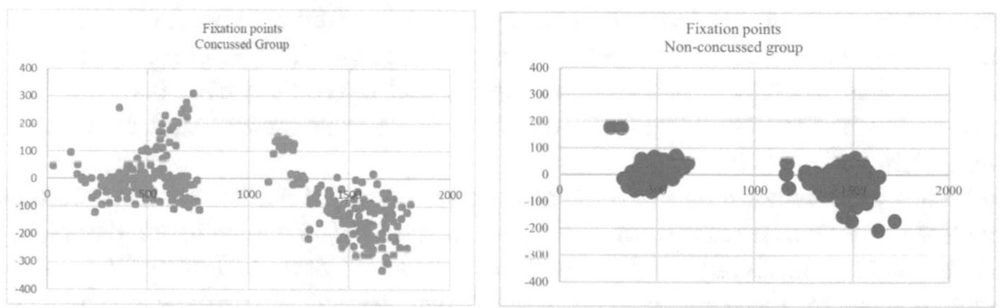

Fig. 3. Fixation points of concussed (left) vs. non-concussed (right) participants

In addition to more accurate and more deliberate eye movements, the non-concussed participants showed significantly faster eye movements when moving their eyes back and forth between two target points. The amount of time in which the non-concussed participants were able to look back and forth between two target points in ten seconds was significantly greater in comparison with same amount in concussed patients. This shows a faster cognitive processing in non-concussed participants.

According to Fitt's Law [19], when moving back and forth between two points with your fingers or looking back and forth, the faster the finger or the eyes move, the less accurate movement is expected. It means that, when one group shows more accurate eye movements it is expected to have a fewer number of unfocused jumps in a specific period of time. However, while the non-concussed group showed less accurate gaze points, they had fewer target hits. This can be a result of their injuries.

In the case of assessing the cognitive test data from the participants, the eye-tracking visual analysis may provide a better insight into the information that patients were processing and how they implemented what they were asked to do. Figure 4 shows a sample heat map of a concussed and a non-concussed participant.

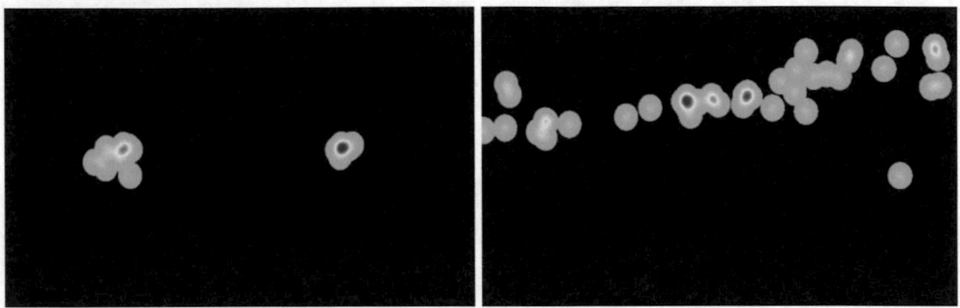

Fig. 4. Heat map of the non-concussed (left) and concussed (right) participants

4 Conclusion

In general, the results showed that using an eye-tracker provides a helpful concussion assessment tool. Patients with concussion symptoms showed a higher level of saccadic impairments in comparison with the non-concussed people. The concussed participants were less able to control their eye movements and move their eyes back and forth between two targets. They had lesser ability to fixate at the targets, and their fixation points were less accurate.

Beside the less accurate eye movements, the concussed participants showed more undeliberate eye movements. In a fixed time duration, the concussed participants showed more unnecessary fixations which randomly occurred, mostly far from the target points. In the case of eye movements' speed, in comparison with the non-concussed participants, the concussed participants had issue moving their eyes rapidly while trying to fixate at targets. They showed slower eye movements and the number of the times they hit the targets in a specific duration was significantly less than healthy participants.

This study suggested using an eye-tracker as an assessment tool for concussion diagnosis via implementing cognitive tests. This study is meant to contribute to doctors and physical therapists who diagnose concussions. Currently, concussion diagnosis is very subjective and open to interpretations. This method provides a more reliable way to diagnose a concussion and track healing progress by implementing follow-up tests.

References

1. Bazarian, J.J., Veazie, P., Mookerjee, S., Lerner, E.B.: Accuracy of mild traumatic brain injury case ascertainment using ICD-9 codes. Acad. Emerg. Med. **13**, 31–38 (2006)
2. Halstead, M.E., Walter, K.D., et al.: Sport-related concussion in children and adolescents. Pediatrics **126**, 597–615 (2010)

3. Langlois, J.A., Rutland-Brown, W., Thomas, K.E.: Traumatic brain injury in the United States: emergency department visits, hospitalizations, and deaths. Department of Health and Human Services, Centers for Disease Control and Prevention, Division of Acute Care, Rehabilitation Research and Disability Prevention, National Center for Injury Prevention and Control (2004)
4. Torres, D.M., Galetta, K.M., Phillips, H.W., Dziemianowicz, E.M.S., Wilson, J.A., Dorman, E.S., Laudano, E., Galetta, S.L., Balcer, L.J.: Sports-related concussion anonymous survey of a collegiate cohort. Neurol. Clin. Pract. **3**, 279–287 (2013)
5. Daneshvar, D.H., Picano, J.D., David, O., McKee, A.C.: Self-reported concussion history: impact of providing a definition of concussion. Ann. C McKee1 **2**, 6–8 (2014)
6. Kerr, Z.Y., Marshall, S.W., Guskiewicz, K.M.: Reliability of concussion history in former professional football players. Med. Sci. Sports Exerc. **44**, 377–382 (2012)
7. McDonald, T., Burghart, M.A., Nazir, N.: Underreporting of concussions and concussion-like symptoms in female high school athletes. J. Trauma Nurs. **23**, 241–246 (2016)
8. Register-Mihalik, J.K., Guskiewicz, K.M., McLeod, T.C.V., Linnan, L.A., Mueller, F.O., Marshall, S.W.: Knowledge, attitude, and concussion-reporting behaviors among high school athletes: a preliminary study. J. Athl. Train. **48**, 645–653 (2013)
9. Cartensen, L.L.: Growing old or living long: a new perspective on the aging brain. Public Policy Aging Rep. **17**, 13–17 (2007)
10. Kerr, Z.Y., Zuckerman, S.L., Wasserman, E.B., Covassin, T., Djoko, A., Dompier, T.P.: Concussion symptoms and return to play time in youth, high school, and college American football athletes. JAMA Pediatr. **170**, 647–653 (2016)
11. Suh, M., Basu, S., Kolster, R., Sarkar, R., McCandliss, B., Ghajar, J., Cognitive and Neurobiological Research Consortium., et al.: Increased oculomotor deficits during target blanking as an indicator of mild traumatic brain injury. Neurosci. Lett. **410**, 203–207 (2006)
12. Brahm, K.D., Wilgenburg, H.M., Kirby, J., Ingalla, S., Chang, C.-Y., Goodrich, G.L.: Visual impairment and dysfunction in combat-injured servicemembers with traumatic brain injury. Optom. Vis. Sci. **86**, 817–825 (2009)
13. Ciuffreda, K.J., Kapoor, N., Rutner, D., Suchoff, I.B., Han, M.E., Craig, S.: Occurrence of oculomotor dysfunctions in acquired brain injury: a retrospective analysis. Optom.-J. Am. Optom. Assoc. **78**, 155–161 (2007)
14. Szymanowicz, D., Ciuffreda, K.J., Thiagarajan, P., Ludlam, D.P., Green, W., Kapoor, N.: Vergence in mild traumatic brain injury: a pilot study. J. Rehabil. Res. Dev. **49**, 1083 (2012)
15. Kapoor, N., Ciuffreda, K.J.: Vision disturbances following traumatic brain injury. Curr. Treat. Option Neurol. **4**, 271–280 (2002)
16. Thiagarajan, P., Ciuffreda, K.J., Ludlam, D.P.: Vergence dysfunction in mild traumatic brain injury (mTBI): a review. Ophthalmic Physiol. Opt. **31**, 456–468 (2011)
17. Capó-Aponte, J.E., Urosevich, T.G., Temme, L.A., Tarbett, A.K., Sanghera, N.K.: Visual dysfunctions and symptoms during the subacute stage of blast-induced mild traumatic brain injury. Mil. Med. **177**, 804–813 (2012)
18. Ventura, R.E., Jancuska, J.M., Balcer, L.J., Galetta, S.L.: Diagnostic tests for concussion: is vision part of the puzzle? J. Neuroophthalmol. **35**, 73–81 (2015)
19. Fitts, P.M.: The information capacity of the human motor system in controlling the amplitude of movement. J. Exp. Psychol. **47**, 381 (1954)

Cognitive Computing and Internet of Things

Runtime Generation and Delivery of Guidance for Smart Object Ensembles

Daniel Burmeister(✉) and Andreas Schrader

Insitute for Telematics, University of Lübeck, Ratzeburger Allee 160, 23562 Lübeck, Germany
{burmeister, schrader}@itm.uni-luebeck.de

Abstract. Driven by the current developments in the area of the Internet of Things, the number of devices and the variety of (natural) interaction modalities which users of smart environments are confronted with are increasing. However, this trend can mentally overwhelm users due to a multitude of challenges in usability and the novel circumstances in ubiquitous surroundings. In order to reduce this cognitive load for users, guidance needs to be generated on the basis of interconnection of the devices and runtime, since the a-priori creation is not possible due to the heterogeneity of devices and their unpredictability of interconnections. Within this contribution, we present a framework which is able to generate and deliver guidance at runtime on the basis of self-descriptions of smart devices. At the same time, an API for developers is offered to extend the framework in a simple way to provide additional generators for guidance. By means of a generator for tutorials the effectiveness of this approach was evaluated with real users in a smart light scenario.

Keywords: Smart object guidance · Self-reflection · Guidance generation

1 Introduction

Due to the increasing availability and the growing market readiness of smart objects, such devices are entering everyday life of many users and shape existing households into so-called smart homes. Although such devices have the purpose of facilitating the daily routines of users due to the richness of new functions, they also entail new difficulties in usability [2, 13]. Numerous devices are developed with a dedicated use and offer their functionality without an inherent control.

To the contrary, devices - primarily so-called natural user interfaces (NUI) - are solely developed to control other devices or applications. In a smart home context, these devices form arbitrarily device ensembles via network communication. Due to the unpredictability and the possibility to connect these devices as desired, manufacturers are not able to adapt comprehensive manuals to the devices or to adapt the appearance of an interaction device to the functionality and thus design signifiers [9].

Hence, users need to be aware of the current (wireless) interconnection state of devices in order to correctly control planned actions. Furthermore, the interconnections of devices in ubiquitous environments may also change automatically depending on the current context, e.g., location or time. In addition, functionality and controls are

H. Ayaz and L. Mazur (Eds.): AHFE 2018, AISC 775, pp. 287–296, 2019.
https://doi.org/10.1007/978-3-319-94866-9_29

embedded into everyday objects or even the environment itself and are thus no longer recognizable for the user.

Although several approaches exist to explain users interactions using guidance, such as the learning of gesture control by projecting onto the hand [11], the representation of the required gestures by displaying shadow cast by a hand within a GUI [5], projected help to learn playing piano [11], or to help with interaction with devices via augmented reality [7], such approaches represent a static and not dynamically responsive solution to the circumstances of the environment. Within this contribution, we present the *Ambient Reflection*-approach in order to generate and deliver instructions of interaction capabilities of smart device ensembles based on a self-description of devices.

Thus, we enable the smart objects to explain their interaction capabilities and functionality themselves to the user, independently of their outer appearance and output capabilities. Using simple interfaces, developers can easily develop new instruction generators and integrate them seamlessly into the framework.

2 Self-reflection

Based on the self-x or self-* principle, in which applications should be enabled to organize their whole life cycle autonomously, we have developed the principle of self-reflection [1]. For this, the paradigm of reflection, known from the programming, in which programs are enabled to read out their structure as well as their behavior at run-time, can be adapted to the level of functionality and interactivity of smart objects.

By means of a structured self-description, such devices are enabled to receive knowledge about their functionality, current system states as well as possible interactions and generate guidance for users of smart environments. Device ensembles are thereby given the capabilities to explain themselves entirely to the user.

A self-reflection within ubiquitous environments is therefore understood as an ambient reflection.

However, in order to maintain this information and to process it automatically, it must be available as a structured data type. In the following, we present a self-description language developed for this purpose.

3 Smart Object Description Language

In order to describe smart objects in their functionality and their interaction capabilities, we have developed the Smart Object Description Language (SODL) on the basis of an in-depth analysis of existing description languages and smart devices manuals as an XML-Schema [3].

In addition to a general specification of the devices, their components and related state groups can be described in a structured manner. An interaction is understood in this context as a change in a state on a component which may have been caused by a third device. In order to obtain a suitable description of an interaction and to illustrate the sequence of these, the Virtual Protocol Model (VPM) of Jakob Nielsen and the technique of Hierarchical Task Analysis (HTA) were combined [8, 12]. While the Hierarchical Task Analysis gives the syntactic order, the Virtual Protocol Model describes the complete course of an interaction in seven different degrees of detail, which are also called levels, starting at the user's goals, the necessary partial tasks with relevant devices and components as well as the change of states down to the physical movements of the user's body parts. On the basis of the principle of task decomposition in HTA, tasks are broken down into ever smaller tasks, until they are atomic. The conjunction of these two approaches allows the complete description of Human Computer Interaction (HCI) with smart objects including Natural User Interfaces.

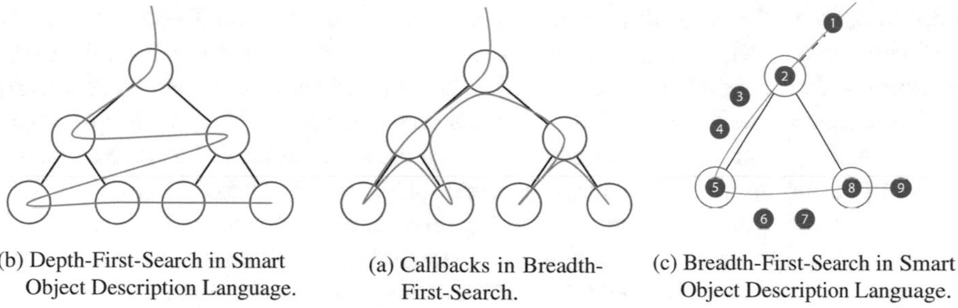

(b) Depth-First-Search in Smart Object Description Language.

(a) Callbacks in Breadth-First-Search.

(c) Breadth-First-Search in Smart Object Description Language.

Fig. 1. Traversing and callbacks to generate instructions.

4 Ambient Reflection Framework

To enable the principle of self-reflection with smart objects in ubiquitous environments, we have developed the Ambient Reflection Framework as a middleware approach, essentially consisting of the following two components [2]. First, the Smart Object Library can be integrated into the firmware of objects and is responsible for reading out the self-descriptions of the smart objects. Based on such information, relevant parameters are set and the semantical monitoring of system states is possible. At the same time, suitable network interfaces are opened which, on the one hand, provide the self-descriptions for third-party, and, on the other hand, provide available output modalities of a device for displaying any guidance. Second, the Description Mediator is responsible to collect the self-descriptions of all devices available in the network, and to combine these to new self-descriptions of device ensembles on the basis of their current interconnection state. This layout-neutral form of the description of interactions across device boundaries can be used to generate and deliver guidance when needed for users at runtime.

4.1 Guidance Generation

In order to facilitate the generation of instructions, simple interfaces have been developed for the framework, which provide call-back functions and thus a simple way of generating guidance. Since the Smart Object Description Language makes use of a tree structure because of its separation on different information levels by Nielsen using HTA, it is possible to obtain the information of the respective level in two different ways of traversing. A *Breadth-First Search* (BFS) (see Fig. 1(a)) can be used to obtain and process information level-wise [4]. This is particularly suitable for instructions which, for example, include a running text. By means of a *Depth-First Search* (DFS) (see Fig. 1(b)), each path of the tree structure is traversed in complete depth [4]. This procedure could be used for instructions which, at any time, should have the full amount of information of a single interaction but need not necessarily represent all details at once. E.g., collapsible views in a web page might be a conceivable application for this approach. As exemplarily presented for breadth-first search (see Fig. 1(c)), developers can use callback functions at different times of the traversal in order to build on the basis of the currently transmitted information. Before (1) and after (9) the complete visit of a level of the VPM, developers are able to define the overall structure of their support structure (e.g., the content structure of the guidance). During the run through a leaf or a node of an interaction description (2, 5, 8), information of the respective level can be processed. In addition, individual nodes can also be transformed appropriately into single guidance steps by the callbacks before (4, 7) and after (3, 6) visiting a node. Analogous to Fig. 1(c), callback functions are also provided for the depth-first approach. Within the Ambient Reflection framework, guidance generators are registered by their media type. Depending on the users preferences and needs, the best suitable generator is selected and started once a new device ensemble is formed. Based on the interconnection, all guidance media is stored in a cache to deliver them immediately when needed.

4.2 Guidance Delivery

As previously described, suitable devices with output modality offer these by means of the smart object library on the basis of started network interfaces for displaying guidance. Such information is also available in the form of MIME-types within the self-description of the smart objects. In order to provide optimal support based on the users needs, a user profile is used to generate the most appropriate guidance by selecting the best matching out of all registered guidance generators. Since these use the identical callback functions, it is also possible to cross-link generators. The delivery process can take place in two different stages of initiative:

- **Reactive:** The user explicitly requests guidance during or before an interaction with a device ensemble.
- **Proactive:** The device ensemble recognizes the need for a guidance autonomously and requests it for the current interaction directly at the Description Mediator.

Due to the variety of devices, we offer an interface for the proactive initiative to request of guidance during device operation via an eventing channel. The recognition of the time of the need is, however, application and therefore also manufacturer-specific. Once requested (independent of the initiative), suitable devices are sought to present the generated guidance. Since there often might be no output capability within the requesting device ensemble, other devices in the room are also considered for the delivery process. However, if this should be possible, the generated guidance is presented within an ensemble.

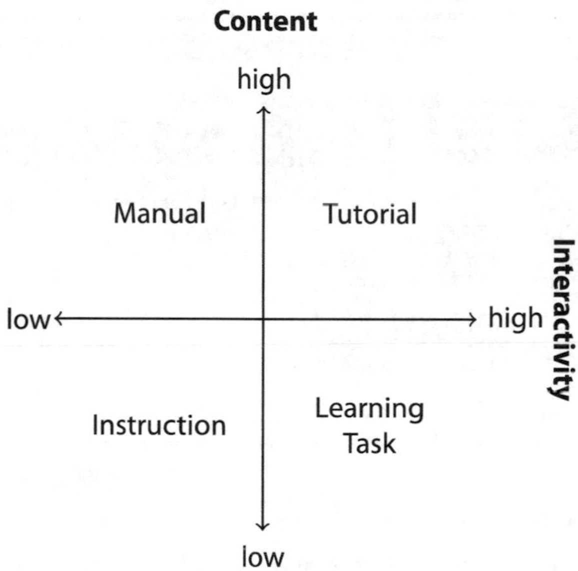

Fig. 2. Types of guidances depending on the dimensions content and interactivity.

5 Guidance Types

In order to classify types of guidance in a ubiquitous environment, we distinguish them by two dimensions: the completeness or density of content of a device ensemble and the degree of interactivity of the presented guidance. The presentation of guidance in general can also be carried out depending on the context or a previous (multiple) incorrect operation by the user. As shown in Fig. 2, we distinguish between

- **Instructions** are understood as single descriptions of individual interaction capabilities or functionalities of an entire system or application.
- **Manuals** are a compilation of instructions from a complete system. Therefore, a manual is a complete description of system functionality as well as necessary interactions. In addition to the description of functionality, a manual can also contain specifications and safety instructions of devices. Such manuals are frequently included in printed form of conventional electrical devices packaging.
- **Learning Tasks** provide a user to perform a function and give appropriate guidance and explanation. During and after execution, a user receives feedback on the success of his action.

- **Tutorials** are a compilation of learning tasks to provide a complete system functionality to a user in the form of interactive tasks. In a secure environment, the user has the possibility to form a complete mental model of an application by trying to learn interactions and exploring the functionality of a system.

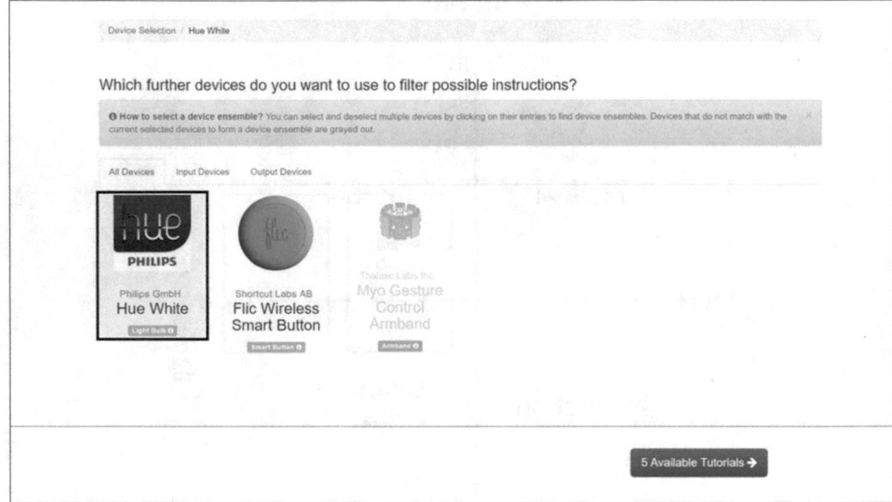

Fig. 3. Overview of smart devices within the network.

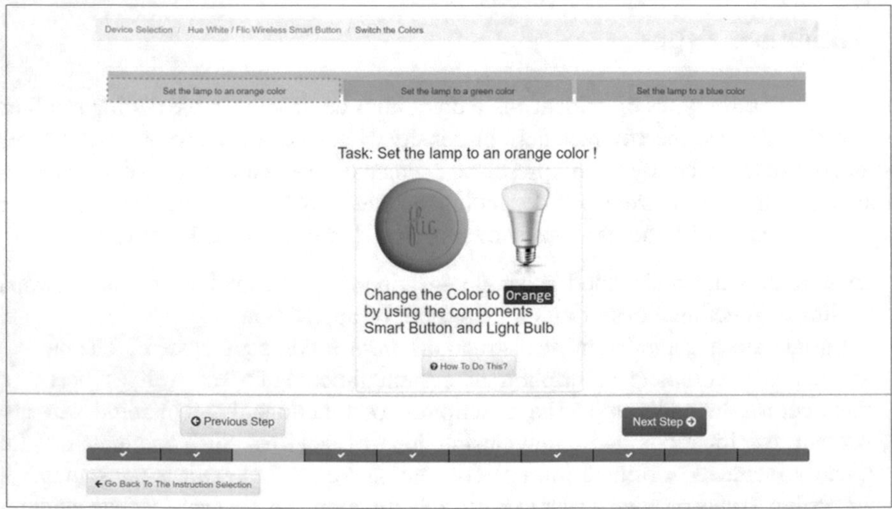

Fig. 4. Task to learn how to change the color of the lamp to orange using the Flic Smart Button.

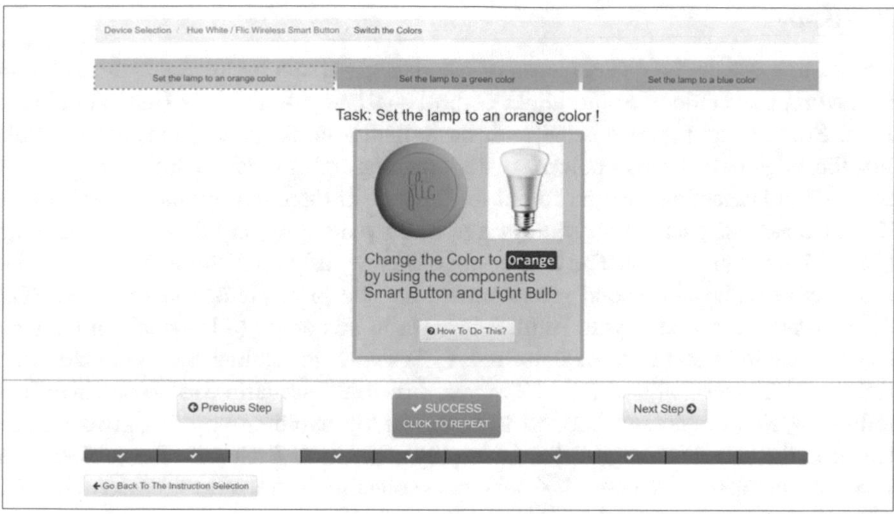

Fig. 5. Successful finished learning task.

6 Evaluation

In order to determine the impact on learning the interaction with smart objects, we have developed a first prototype in form of a tutorial (see Fig. 3) in a participatory design process, which presents user tasks based on their current knowledge of an ensemble. By using whole architecture of the Ambient Reflection framework, the tutorial is able to generate learning tasks from aggregated self-descriptions of smart objects at runtime and present them to the user (see Fig. 4). The tutorial adapts the tasks and shows more detailed or less detailed information as soon as a user has successfully completed or failed a task. Further, the tutorial responds to the changing system states and can give the user visual feedback on the successful execution of an action (see Fig. 5).

Table 1. Mapping of interactions to functionality.

Philips hue color ambience	Flic smart button
On	Hold
Off	Hold
Dim	Double click
Change color	Click
Toggle running light	(1) Double click (2) Click

6.1 Setup

In order to evaluate the effectiveness of tutorials generated by the Ambient Reflection framework, a scenario of a smart light control consisting of a Philips Hue White[1] and a Flic.io Smart Button[2] was created with basic (lamp on, lamp off), and more complex (dim the brightness, change colors, change running light) functionality. A mapping of functions to interactions can be found in Table 1. In total, two groups of test persons (10 and 6 persons each) at an average age of 27 years were asked to carry out simple tasks with the light control. The first group has previously gone through a tutorial (G1) on a display while the second group should use the principle of trial-and-error (G2). We've asked all test persons to fulfill four tasks in this order: (T1) switch on the lamp, (T2) dim the light and turn the color red, (T3) switch to running light with maximum brightness, (T4) turn off the lamp. However, this simple scenario was chosen in order to facilitate a simple entrance for the test persons by avoiding any cognitive exertion regarding the understanding of the devices as such. After each task, the subjects were asked to complete a NASA-TLX raw questionnaire in order to measure efforts in different areas ranged from 1 to 20 each [6].

Table 2. Evaluation results for the NASA-TLX questionnaire.

Groups		Mental demand		Physical demand		Temporal demand		Performance		Effort		Frustration	
		G1	G2	G1	G2	G1	G2	G1	G2	G1	G2	G1	G2
T1	$\bar{x}$	2.00	5.17	1.70	4.83	2.30	4.33	1.30	2.17	1.40	4.00	1.90	4.33
	μ	1.49	4.17	1.25	5.12	3.13	3.44	0.95	1.83	1.26	2.53	1.73	4.90
T2	$\bar{x}$	4.50	7.83	1.60	5.83	1.80	8.83	4.10	4.76	3.40	8.00	4.00	7.33
	μ	3.78	7.05	1.07	7.03	1.03	8.75	6.14	7.61	4.53	9.32	5.25	8.16
T3	$\bar{x}$	4.50	4.83	2.60	4.33	3.80	4.83	4.30	3.00	4.90	5.00	4.90	4.67
	μ	3.89	2.71	2.55	3.89	4.85	2.14	6.46	3.36	6.10	2.68	6.69	3.93
T4	$\bar{x}$	1.00	1.17	1.50	1.33	1.20	1.17	1.20	1.00	1.40	1.00	1.00	1.00
	μ	0.00	0.41	0.97	0.82	0.42	0.41	0.42	0.00	0.97	0.00	0.00	0.00

6.2 Results

As in Table 2 presented with mean values and standard deviations, all tasks were evaluated by the subjects according to their effort in different categories. Although the evaluation used a very first prototype of a tutorial generator using the Ambient Reflection framework, it is clearly recognizable, that previous training using an automatically generated guidance leads to a clear relief in all areas except temporal demand in comparison to trial and error usage.

It is striking that, especially in tasks 2 and 3, a higher perceived physical demand was encountered without the previous use of the tutorial, although all interactions involve only pressing a button at different intervals or lengths. Task 3 challenged both,

[1] http://www2.meethue.com.

[2] https://flic.io.

test persons with and without previous tutorial use, which is reflected in similar results in the evaluation.

It was observed that the difficulty of this was the composition of two single interactions for starting the running light. In both groups, the correct timing must be found between the double click and the single click.

In order to improve the learning effect of a tutorial, it would be possible to repeat the learning task multiple times for more complex interaction sequences, giving users a correct feeling for the timing. An exception is task 4, since turning the light on and off requires the same interaction to control binary functionality.

Independent of the group, it was possible for the test persons to fulfill this task with very low demand, although the initial hurdle was even higher for test persons without using the tutorial.

7 Conclusion

In this paper we have presented the Ambient Reflection framework, which enables smart objects to explain their functionality and interaction possibilities independently on the basis of generated guidance, based on combined self-descriptions depending on the current interconnection state.

Upon this, we also introduced a possibility for developers to implement generators for guidance in a simple way and integrate them into the framework. In addition, we presented a typology to classify different types of guidance and carried out an evaluation with a very first prototype of a tutorial generator using the framework within a smart light scenario.

Furthermore, we have analyzed the results regarding the learning effect and compared the results with a control group. Currently, we are developing a range of different types of guidance for different media, which could be evaluated within a large-scale study.

References

1. Babaoglu, O., Jelasity, M., Montresor, A., Fetzer, C., Leonardi, S., van Moorsel, A., van Steen, M.: The self-star vision. In: Self-star Properties in Complex Information Systems, pp. 397–397 (2005)
2. Burmeister, D., Altakrouri, B., Schrader, A.: Ambient reflection: towards self-explaining devices. In: Proceedings of the 1st Workshop on Large-scale and Model-based Interactive Systems: Approaches and Challenges, LMIS 2015, Co-located with 7th ACM SIGCHI Symposium on Engineering Interactive Computing Systems (EICS 2015), 23–26 June 2015, pp. 16–20. ACM, Duisburg (2015)
3. Burmeister, D., Burmann, F., Schrader, A.: The smart object description language: modeling interaction capabilities for self-reflection. In: 2017 IEEE International Conference on Pervasive Computing and Communications Workshops (PerCom Workshops), pp. 503–508, March 2017
4. Cormen, T.H., Stein, C., Rivest, R.L., Leiserson, C.E.: Introduction to Algorithms, 2nd edn. McGraw-Hill Higher Education, New York (2001)

5. Freeman, D., Benko, H., Morris, M.R., Wigdor, D.: ShadowGuides: visualizations for in-situ learning of multi-touch and whole-hand gestures. In: Proceedings of the ACM International Conference on Interactive Tabletops and Surfaces, ITS 2009, pp. 165–172. ACM, New York (2009)
6. Hart, S.G., Staveland, L.E.: Development of NASA-TLX (task load index): results of empirical and theoretical research. Adv. Psychol. **52**, 139–183 (1988)
7. Liu, C., Huot, S., Diehl, J., Mackay, W., Beaudouin-Lafon, M.: Evaluating the benefits of real-time feedback in mobile augmented reality with handheld devices. In: Proceedings of the 30th International Conference on Human Factors in Computing Systems, CHI 2012, pp. 2973–2976 (2012)
8. Nielsen, J.: A virtual protocol model for computer-human interaction. DAIMI Rep. Ser. **13** (178) (1984)
9. Norman, D.A.: The way I see it: signifiers, not affordances. Interactions **15**(6), 18–19 (2008)
10. Raymaekers, L., Vermeulen, J., Luyten, K., Coninx, K.: Game of tones: learning to play songs on a piano using projected instructions and games. In: CHI 2014 Extended Abstracts on Human Factors in Computing Systems, CHI EA 2014, pp. 411–414. ACM, New York (2014)
11. Sodhi, R., Benko, H., Wilson, A.: LightGuide: projected visualizations for hand movement guidance. In: Proceedings of the SIGCHI Conference on Human Factors in Computing Systems, pp. 179–188. ACM (2012)
12. Stanton, N.A.: Hierarchical task analysis: developments, applications, and extensions. Appl. Ergon. **37**(1), 55–79 (2006). Special Issue: Fundamental Reviews
13. Youngblood, G.M., Heierman, E.O., Holder, L.B., Cook, D.J.: Automation intelligence for the smart environment. In: Proceedings of the 19th International Joint Conference on Artificial Intelligence, IJCAI 2005, pp. 1513–1514. Morgan Kaufmann Publishers Inc., San Francisco (2005)

Holistic Scenarios by Using Platform Technologies for Small Batch-Sized Production

Yübo Wang[1(✉)], Michaela Dittmann[2], and Reiner Anderl[1]

[1] Department of Integrated Design, Technische Universität Darmstadt, Otto-Berndt-Straße 2, 64287 Darmstadt, Germany
{y.wang, anderl}@dik.tu-darmstadt.de

[2] Schuler Pressen GmbH, Schuler-Platz 1, 73033 Göppingen, Germany
Michaela.Dittmann@schulergroup.com

Abstract. Industrie 4.0 defines a new level of organization and management in the entire value chain for the product lifecycle. The basic technologies include cyber-physical systems, embedded systems and Internet technologies. The reflected benefits are more communicable, flexible, and networkable products and production environments. The purpose of the paper is to present holistic scenarios by using platform technologies. The combination of defining scenarios by using data flow diagram with business process modeling notation is used to establish Human-Machine-Collaboration for small batch-sized production. Cornerstones of this approach are boundary-crossing models in the distributed Industrie 4.0 infrastructure from product development into the production environment.

Keywords: Industrie 4.0 · Human Machine Collaboration
Platform technology · Small batch-sized production

1 Introduction

The Internet of Things is moving to the factories. This creates a new product and production paradigm: rigid value chains create highly flexible value creation networks that are optimized in real time. Products become intelligent by the principle component as an information carrier. Machines will be connected with each other and with the internet. The communication between man and machine will be like a social network. New business models are becoming possible and opening up new potential, especially for small and medium-sized enterprises (SMEs).

Industrie 4.0 represents a new level of organization and control of the entire value chain across the lifecycle of products. The basis is the use of cyber-physical systems (CPS), which includes embedded software systems, an internet address and capability of communication and flexibly network [1].

This change is a result of the increasing flexibility requirements of production facilities as well as the need of short product life cycles and product innovations, which cannot be supported by conventional production facilities [2].

Customized mass production up to batch size one are the potential results from these requirements. It is possible for companies in many industries to realize numerous product variants at low costs. With optimization of the use of resources, a higher added value of

H. Ayaz and L. Mazur (Eds.): AHFE 2018, AISC 775, pp. 297–306, 2019.
https://doi.org/10.1007/978-3-319-94866-9_30

processes can be created. With integrated digitization and the real-time evaluation of data, value characteristics can be retrieved. The embedded systems realize transparency and provide information about the state of all actuators. A flexible and short-term response to the changing product requirements through modified market developments and fluctuating raw material and energy prices is possible. Even events such as technical malfunctions or delays in delivery can be limited by the intelligent factory [3, 4].

However, according to surveys, only one of four companies has networked and digitized their production processes. Industrie 4.0 has not yet arrived in a large number of companies or its coverage has not yet been recognized [1, 5].

2 Challenges for Introducing Industrie 4.0 in SME

The first challenge for small and medium-sized enterprise is to achieve an Industrie 4.0 competence overview of enterprise areas. Herein, the "Generic Procedure Model to introduce Industrie 4.0 in Small and Medium-sized Enterprises" is one of the most successful methods to get a competence overview of enterprise areas [6]. The core element of this method is to use Industrie 4.0 Toolboxes. These Toolboxes are structured as a matrix. The vertical columns represents application level, which displays Industrie 4.0 themes. Every application level is broken down in five horizontal rows, which displays technological and sequential stages. The highest stage represents the vision Industrie 4.0. Figure 1 is showing the Toolbox Industrie 4.0 Production [7]. After a SME identifies his current development state for each application level, fields of action will be defined. The current development state is shown in Fig. 1 as full lined profile and the fields of action profile is dotted. This method is validated in several project formats: holistic project format over one year, workshop concept, competence-building or coaching event for trainers [6]. The results of these projects can be summarized in two proposals: The first proposal is that SME needs a "digital twin" for their product and has to establish for their production the principle "component as an information carrier". The second proposal is that SME needs a "platform strategy". After the SME is solving the first challenge, there are facing with the problem to solve the second challenge of defining scenarios, which includes the proposals.

In the following component as an information carrier and the digital twin will be described as well as a method for defining scenarios will presented to build up a basement for the second proposal.

2.1 Component as an Information Carrier

Components collect knowledge about themselves when they go through their lifecycle. The components that act as information carriers are physically and virtual linked to the product, which is in the manufacturing process. The data of the product either are recorded directly as a shopping list or is referenced to the data in a server [1, 8]. The information components could be QR code, RFID or other identifiers. When using identifiers, the actual data is stored in a data management system, which can reached via a Management Order System (MOS). Through their connection to product model and manufacturing processes, components should be able to control their own manufacturing process [9].

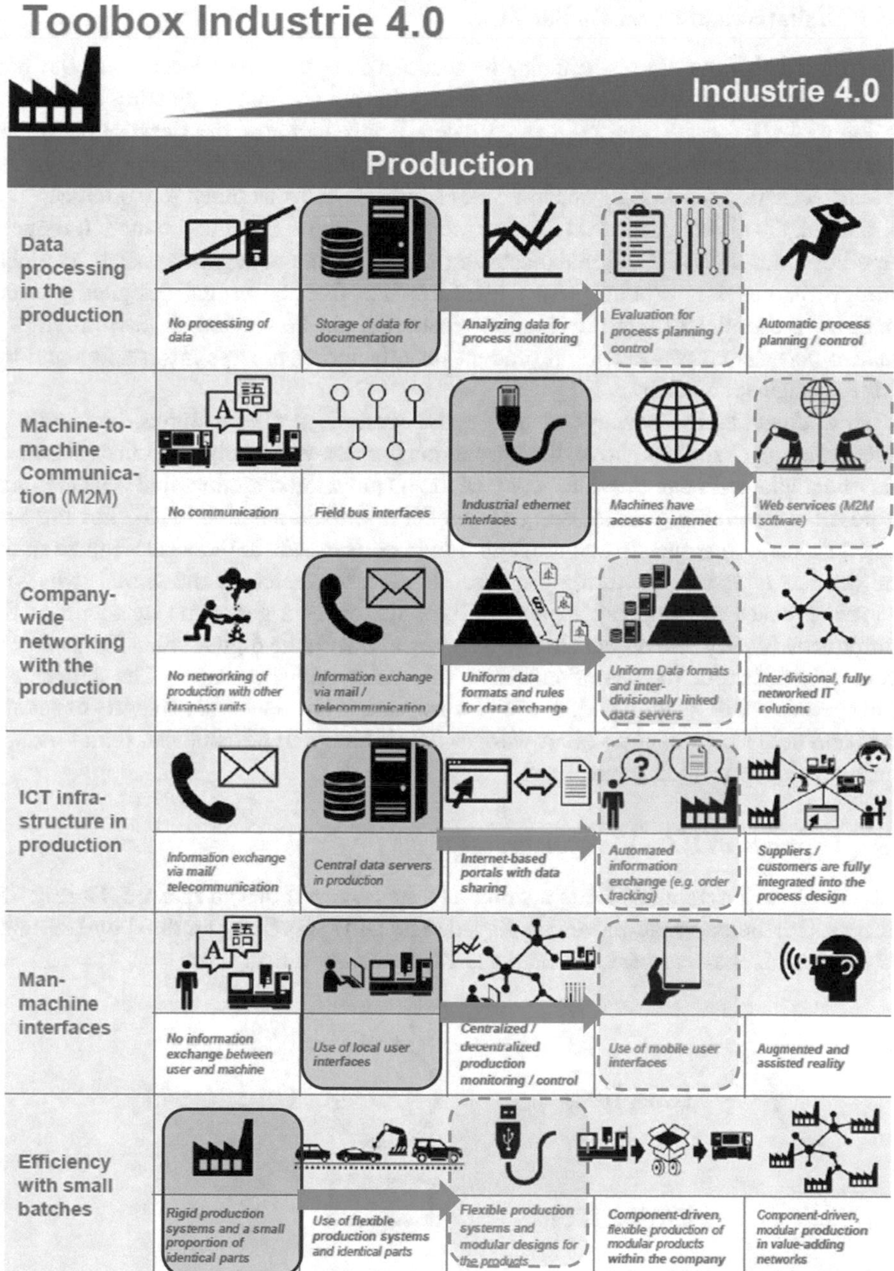

Fig. 1. Toolbox I4.0 full lined/dotted: current/advanced enterprise Industrie 4.0 competence of the production environment

2.2 Digital Shadow and Digital Twin

The "digital shadow" is the sufficiently accurate image of the processes in the production, the development and adjacent areas with the purpose of creating a real-time evaluable basis of all relevant data. In detail, this includes the description of the necessary data formats, the data selection and the data granularity level [10]. In the present scenario, the digital shadow refers only to the manufactured product. This means that the digital shadow is divided into the customer-specific production request, the component data and the manufacturing data. By connecting data that is available through the physical product with the information that the virtual designed product contains, a completely new level of application could be created. It also provides a concise perspective of the current product manufacturing process information and the virtual design specifications [11].

In addition to the concept of the digital shadow, the term digital twin will be introduced. As described above, the production process will be digitized first. Based on this, the digital twin can provide a copy of the manufactured product that is as identical as possible through a "virtual designed model and its simulation" [10]. For the presented scenario, this means that the digital twin corresponds to the virtual image of the real product after manufacturing process. The data for matching the digital twin with the real product will updated separately by a sensor, as a geometric detection of the component. Finally, the digital shadow is compared with the digital twin. The goal is to accurately determine the deviation within the framework of tolerance. The dimensions of all products will be digitized. A relevant use case is a product that consists of several parts and needs to be assembled. A waste minimization will be achieved, if all products can be assigned to their counterpart.

2.3 Elements of Data Flow Diagrams

The data flow diagram (DFD) is a structured analysis and design method. During this process, data transformations are displayed. The DFD was first described by DeMarco [12]. Figure 2 characterized the elements for modeling a data flow.

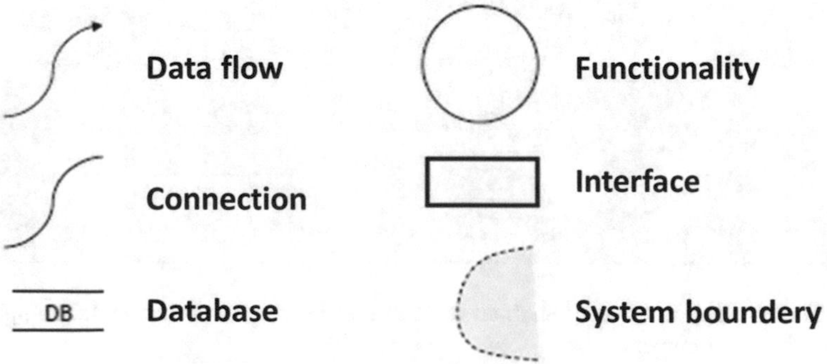

Fig. 2. Elements of data flow modeling

The essential elements of the DFD are data storage, function, data flow and the interface to the environment. Two parallel lines, wherein the name of the data store is shown between the lines, represent the data memory. An arrow represents the data flow. The function is represented by a circle, wherein the activity is described. The interface to the environment is represented by a rectangle. This contains the name of the interface [13]. Dashed lines are framed the systems, which distinguished systems from each other. The DFD serves in the context of this work for the representation of scenarios. A line without an arrow represents a relationship of elements to each other without data flow.

3 Combined Scenario Digital Shadow and Digital Twin and Component as Information Carrier

The scenario "digital shadow and digital twin" is shown together with the scenario "component as information carrier" as data flow diagram in Fig. 3. The circle in red and the rectangle in orange the distinction between the scenarios. Both scenarios are based on a production environment. The data flow diagram describes this with the sequence of a miniature production line. The data flow diagram represents four main areas that separated by system boundaries with the environment. These systems are the human being, the manufacturing, the control and the delivery system. Customers and employees represent the human. They have the possibility to make an input to an input device, which can be a smart phone, a tablet or a computer. Instead of a device, a platform scenario as a further development will presented in the next chapter.

The production includes the elements machine, transport system, sensor and the manufactured component. The production works fully automatically and without human intervention after triggering the production order. A data exchange takes place between the production control computer and the production. The machine receives the time of production start-end as well as the machining data. The transport system needs the destination component and the transport time. The sensor requires the sensor triggering time. The machine returns the processing status to the production control computer at the beginning or end of a process. The machine returns the current component location and the sensor provides the sensor data. Important for the generation of the digital twin is the geometrical measurement of the component after processing. This data will be compares to the digital shadow.

The manufactured component will be delivered to the customer with a delivery system. The customer data is available to the delivery system. The challenge of the scenario is the complexity of geometrically capturing and then reconciling of the data.

The customer defines his customer data and his customized production request as the input, for example a corresponding letter. The desired lettering will be sent after the order confirmation of the customer with the customer data to the production control computer. Simultaneously, the IBAN of the customer is also required for payment.

The employee of the producing company also stores the component data and processing data as an input to the device. He also passes the instruction to the production with the starting time of the production. This data will be sent to the production control computer. An encrypted transmission takes place with the TCP/IP protocol. The production control computer manages and executes the production order. In addition, a database is connected. The customer-specific production request is combined with the component data and machining data in order to create the numerical control program. Depending on the degree of automation, an employee can also carry out this step. Furthermore, the data created by the customer and the employee as well as the generated processing data will be assigned to an identifier.

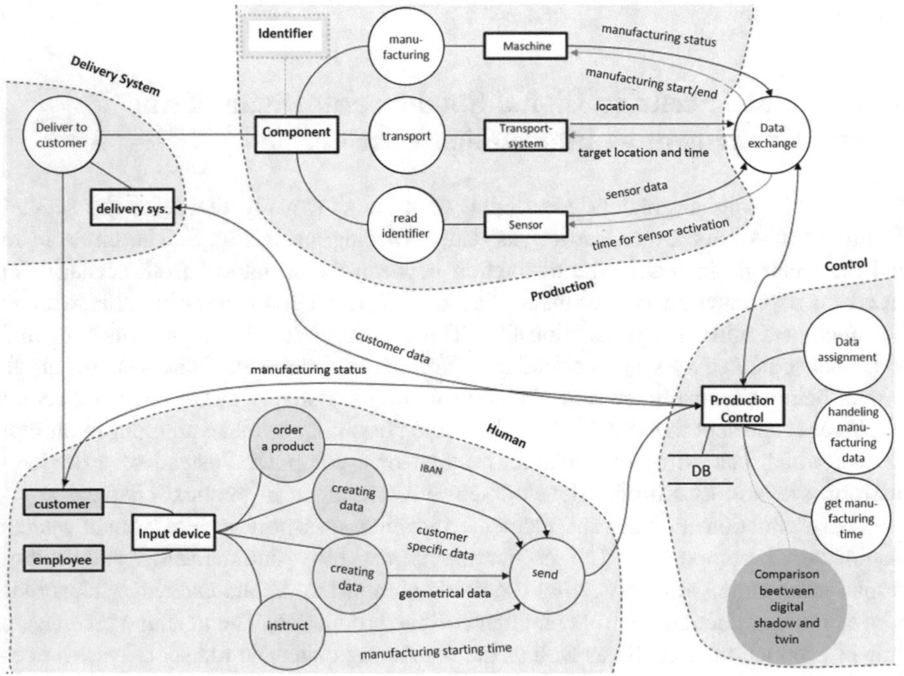

Fig. 3. Data flow diagram of combined scenarios

The second scenario, where the component serves as an information carrier, will presented in Fig. 3. For this purpose, the red circle will be omitted, and the orange box is considered. Herein, only the implementation of the component as an information carrier will discussed. The identifier can be occurred in various forms. These include barcode, QR code or RFID chips. According to the application, depending on the desired storage capacity and costs the identifier can chose individually. If a small, relatively uncomplicated product will be produced, e.g. a pen, a barcode or a QR code can be use. The identifier will be interlinked with the component; therefore, the component has now a digital product memory. It knows the past and further manufacturing and processing steps. At the manufacturing processing, the identifier can read by a sensor. Therefore, the transport system knows the transport route, and on which machine the component will manufactured next.

4 Technology Platform Scenario

4.1 Platform Data Flow Diagram

The third scenario describes the operations of a platform shown in Fig. 4. Platform participants are technology data producers and recipients, e.g. partners and customers. The platform operators provide the platform, which can be reached via a website. The platform has a dashboard. The partner accesses this dashboard to set up its products. For this process, he will be enabled by a rights assignment from the platform operator.

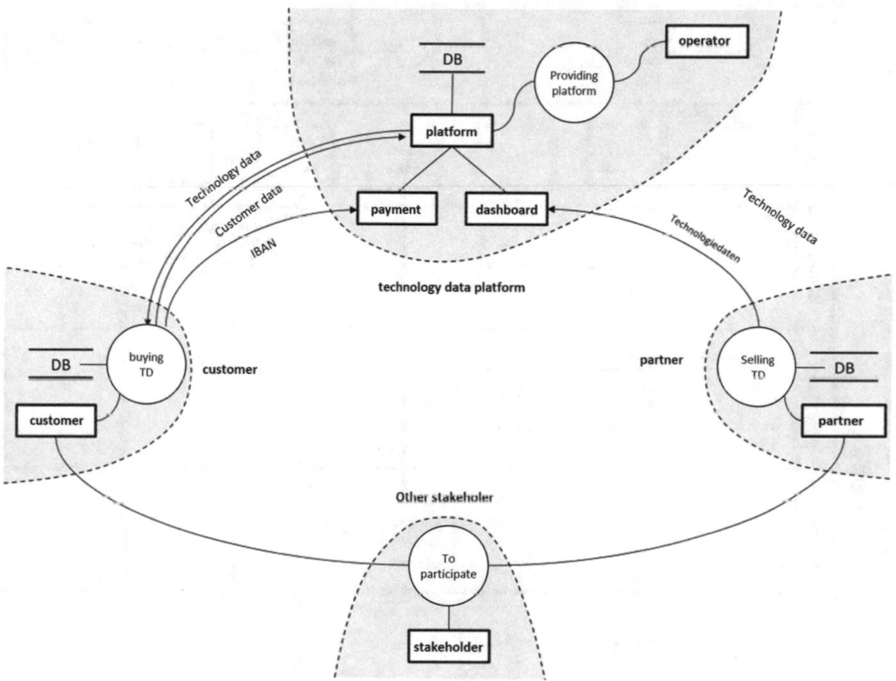

Fig. 4. Data flow diagram of platform scenario

The partner stores the technology data that he would like to offer to the database of the platform. The customer can now acquire technology data via the platform. For this purpose, customer data is initially stored for registration. With the payment system of the platform, payment information such as the online bank information will send securely to the platform. After these processes, the customer receives his desired product, the technology data. This data can be use directly in the production for manufacturing components.

4.2 Platform Business Process

After the description of the platform scenario the business process will presented in Fig. 5. The holistic platform scenario is divided in partner, platform, customer, dashboard, simulation, parser, payment system, database, server and laser cutter. Each of the lanes displays the activities of the elements in order to explain their interaction. As a partner of the platform, he will visit the platform and upload svg data as design parameters at the platform dashboard.

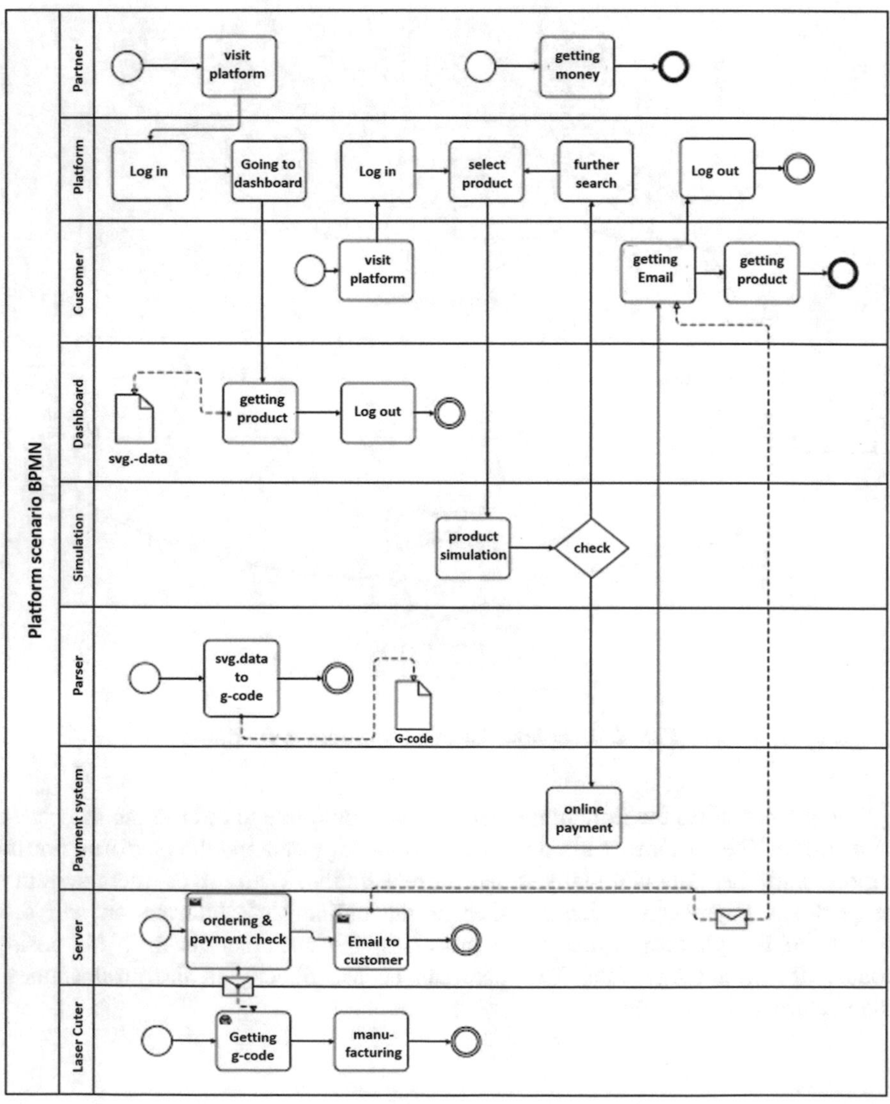

Fig. 5. BPMN of the platform scenario

The customer visits the platform and has to log in. He selects a product and places it in the shopping cart. Now he can look at a simulation to check the quality of the laser processing. If he decides against the purchase, he can continue to search for suitable products at the platform. If he chooses to buy, he will be directed to the payment system. In this case, the customer pays by online payment. After a successful payment process, the server sends the customer a confirmation e-mail.

Now, the svg.-file will converted to G-code by the parser. The parser accesses the database file and stores the target file. The server sends the G-code to the laser cutter, which begins editing. The partner receives his share after the payment process of the customer, the rest goes to the platform operator and the customer receives his product.

5 Further Works and Conclusion

Necessarily, the preparation and the hardware order to build up the developed scenarios is taking place. An ordered laser marker has been allocated to the Human-Machine collaboration. The programming process of the front- and backend of the platform is in the final development process. Therefore, the described scenarios will be built in a prototypical platform to validate and verify itself.

The basement of this work was the challenge for introducing Industrie 4.0 to SME. Over years after several project validations, the first challenge can be solved by the Toolbox method to identify the current Industrie 4.0 status of a SME and defines further development stages to get a fields of action profile. However, the results of the first challenge merge in two proposals. This work starts methodologically with the construction of Industrie 4.0 scenarios, which includes the two proposals. The Human-Machine-Collaboration is defined as the connectivity of interactions between the human engagement in development process and the machine application in production environment. This includes all levels from data exchange, to simulation and data conversion, and manufacturing of components.

References

1. Kagermann, H., Anderl, R., Wahlster, W., Gausemeier, J., Schuh, G.: Industrie 4.0 in a Global Context. Strategies for Cooperating with International Partners. Acatech, Berlin (2016)
2. ten Hompel, M., Büchter, H., Franzke, U.: Identifikationssysteme und Automatisierung. VDI-Buch. Springer, Berlin/Heidelberg (2008)
3. BMBF: Zukunftsbild "Industrie 4.0". Bundesministerium für Bildung und Forschung (2013). https://www.bmbf.de/pub/Zukunftsbild_Industrie_4.0.pdf. Accessed 02 Feb 2018
4. BSI: ICS-Security-Kompendium. Bundesamt für Sicherheit in der Informationstechnik (2013). https://www.bsi.bund.de/DE/Themen/Industrie_KRITIS/Empfehlungen/ICS/empfehlungen_node.html. Accessed 20 Feb 2018
5. Zentrum für Europäische Wirtschaftsforschung - ZEW: IKT-Report (2015). http://www.zew.de/fileadmin/FTP/div/IKTRep/IKT_Report_2015.pdf. Accessed 05 Jan 2018

6. Wang, Y., Wang, G., Anderl, R.: Generic procedure model to introduce Industrie 4.0 in small and medium-sized enterprises. In: Proceedings of the World Congress on Engineering and Computer Science, San Francisco (2016). http://www.iaeng.org/publication/WCECS2016/WCECS2016_pp971-976.pdf. Accessed 04 Feb 2018
7. Anderl, R., Picard, A., Wang, Y., Fleischer, J., Dosch, S., Klee, B., Bauer, J.: Guideline Industrie 4.0 - Guiding Principles for the Implementation of Industrie 4.0 in Small and Medium Sized Businesses. VDMA, Frankfurt (2015)
8. Anderl, R.: Industrie 4.0–technological approaches, use cases, and implementation. at-Automatisierungstechnik **63**(10), 753–765 (2015)
9. Anderl, R.: Industrie 4.0 - advanced engineering of smart products and smart production. In: 19th International Seminar on High Technology, Technological Innovations in the Product Development, Piracicaba, Brazil (2014)
10. Bauernhansl, T., Krüger, J., Reinhart, G., Schuh, G.: WGP-Standpunkt Industrie 4.0. Wissenschaftliche gesellschaft für produktionstechnik WGP e.V. (2016). http://www.wgp.de/uploads/media/WGP-Standpunkt_Industrie_4-0.pdf. Accessed 18 Feb 2018
11. Grieves, M.W.: Digital Twin: Manufacturing Excellence Through Virtual Factory Replication. A Whitepaper by Dr. Michael Grieves. Florida Institute of Technology (2014). http://innovate.fit.edu/plm/documents/doc_mgr/912/1411.0_Digital_Twin_White_Paper_Dr_Grieves.pdf. Accessed 18 Nov 2017
12. Ward, P.T.: The transformation schema. An extension of the data flow diagram to represent control and timing. IEEE Tran. Softw. Eng. **SE-12**(2), 198–210 (1986)
13. Chen, Y., Li, Q.: Modeling and Analysis of Enterprise and Information Systems. From Requirements to Realization, vol. XVIII, p. 397 ff. Higher Education Press und Springer, Beijing (2009)

Internet of Things - New Paradigm of Learning. Challenges for Business

Łukasz Sułkowski[1(✉)] and Dominika Kaczorowska-Spychalska[2]

[1] University of Social Sciences in Lodz, ul. Sienkiewicza 9, 90-113 Lodz, Poland
lsulkowski@san.edu.pl

[2] University of Lodz, ul. Matejki 22/26, 90-237 Lodz, Poland
dominika.spychalska@uni.lodz.pl

Abstract. The Internet of Things (IoT) is a new technology paradigm envisioned as a global network of machines and devices capable of interacting with each other. It can change everything - including ourselves because technology and accompanying digital transformation change a character and a way of the current learning process. On the one hand, it refers to a man that acquires more and more advanced digital skills and competencies supported by smart education systems. On the other hand, it refers to smart machines based on artificial intelligence (AI) which analyse people's opinions, decisions and behaviour, thanks to which they acquire knowledge and learn their emotions. As a result, they perform activities that used to be characteristic only for people. In such an approach, education becomes a natural activity at the same time being a natural sequence of remaining in the ecosystem of mutual connections. But it is also a source of dangers of informational rubbish bin and post-true messages that could spoil a real learning effort. Undoubtedly, it creates a great number of new challenges facing enterprises which while combining innovative technologies with their core resources will build their own ecosystem of Smart Business Intelligence. The aim of the paper is to identify and explore possibilities of IoT as a new paradigm for learning, with particular focus on its significance for business.

Keywords: Business · Management · Internet of Things
Artificial intelligence · Learning

1 Introduction

Technology and accompanying digital transformation change a character and a way of the current learning process. It becomes increasingly complex and related to innovative technologies that break a traditional approach to acquiring knowledge and education. This process evolves continuously towards the concept of constructivism according to which people acquire knowledge by means of active participation in a process of learning about the surrounding reality. Dualism of the world in which they remain – the real world and the digital world – generate an interdisciplinary approach to the role and consequence of using advanced technologies in the process of shaping information society and proactive attitudes towards technologies as such as well as ways of using them in practice. New technologies such as the Internet of Things enable to redefine

H. Ayaz and L. Mazur (Eds.): AHFE 2018, AISC 775, pp. 307–318, 2019.
https://doi.org/10.1007/978-3-319-94866-9_31

this process through observed relations and exchange of information between sensors, machines, smart devices and people. Artificial intelligence (AI) or computational intelligence (CI) which increasingly often become an indispensable element of the ecosystem of IoT, multiply consequences of the evolution of Homo Sapiens concept through Homo Oeconomicus toward Homo Cyberoeconomicus. As a result, the process of learning that is understood as an education cycle (schools, universities), a cycle of self-education and self-development (learning by acquiring knowledge resulting from Big Data and new skills conditioned by the process of interaction with machines and smart devices) is modified as well as it creates a new one – the third pillar of education process: machine learning, including deep learning. Continuity of interactions in which a man remains, their diversity and heterogeneous character have an impact not only on their profile and a way of behaviour of an individual (smart person) – a narrow approach but also on shaping a new system of thinking and behaviour in a social, cultural and behavioural (smart society) – a broad approach. A man will definitely design, manage and coordinate work of multitask smart systems including business ones while determining their further autopoiesis and at the same time, accepting and allowing for a possibility of autonomic activity of "smart" machines and devices - Human - to - Machine (H2 M) and Machine - to - Human (M2H) interactions. It constitutes an enormous challenge for enterprises that will have to live up to dynamically developing processes of digital transformation. New competencies and digital skills, digital workers or hybrid employment models that combine work of a man and its digital counterpart can determine a process of building a competitive advantage while influencing both, costs incurred by a company and increasing effectiveness and efficiency of activities undertaken by them. Therefore, it forces duality of an approach to various aspects of using IoT in education process on the one hand as a consequence of changes in contemporary business, as well as on the other hand, as a premise creating these changes. The first part of this paper will present the potential of the discussed concept, not only as an innovative technology but first of all, as the philosophy of thinking and acting in the process of broadly understood education. The second part will be based on qualitative expert studies that enable to diagnose key areas of changes in the analysed area. Obtained results will constitute a starting point for further critical exploration of the analysed issues, reflections, discussions and research studies.

2 Internet of Things - Business for Learning, Learning for Business

Nothing stands still in the world of the Internet. As a result of constant transformations of technologies directions of activities that not long ago were considered good, tomorrow they may prove inappropriate [1]. It means that both enterprises as well as their strategies of building a competitive advantage cannot remain static. Technology definitely determines changes at all levels and in all areas of enterprise activity It changes the nature of our reality. As a result we become participants of metamorphosis of the reality surrounding us from virtual toward ubiquitous, from observing toward creating, from smart toward adaptive [2]. Consequently, the new technologies like the

Internet of Things (IoT) or Artificial Intelligence (AI) create the new technology paradigm envisioned as a global network of machines and devices capable of interacting with each other. They are recognized as one of the most important areas of future technology and are gaining vast attention from a wide range of industries. As they will spread, the implications for business model innovation will be huge. Filling out well-known frameworks and streamlining established business models won't be enough. To take advantage of new opportunities, today's companies will need to fundamentally rethink their orthodoxies about value creation and value capture [3]. As the Internet of Things (IoT) techniques mature and become ubiquitous, emphasis is put upon approaches that allow things to become smarter, more reliable and more autonomous. In this paper we present challenges and enablers as technologies that will allow things to evolve and act in a more autonomous way, becoming more reliable and smarter [4]. For this reason, future of IoT is AI. Artificial Intelligence will be playing a starring role in IoT because of its ability to quickly wring insights from data. Machine learning, an AI technology, brings the ability to automatically identify patterns and detect anomalies in the data that smart sensors and devices generate. Companies are finding that machine learning can have significant advantages over traditional business intelligence tools for analyzing IoT data. The powerful combination of AI and IoT technology is helping companies avoid unplanned downtime, increase operating efficiency, enable new products and services, and enhance risk management.

Nowadays enterprises are able to determine shopping paths of individual consumers to a greater extent than before while identifying stimuli they react to, to indicate potential and actual consumer touch points. New technologies prove perfect also in production area (intelligent manufacturing) and logistics (intelligent warehouses, warehouse space management, management and fleet optimization).. There are a lot of arguments proving that a future work model will also change and will be closely related to increasing significance of information as well as new means of communication [5]. In the next few years artificial intelligence and robots can lead to disappearance of up to 30% of traditional workplaces. Digital workers will not replace all traditional specialists but using them on a wider scale will enable companies to achieve an advantage over competition, especially as far as efficiency, quality and financial effectiveness are concerned. At this point – as the Accenture research states – only 34% companies are ready to "employ" digital workers although digital transformation is already planned by a half of global enterprises and organizations. Gartner Institute predicts that till 2025 one third of traditional workforce will be replaced by digital workers [6]. As a result, it can lead to creating so-called hybrid employment models that will combine work of an employee – a man and their digital counterpart as well as to disappearance of some jobs and creating new, closely connected with skills and competencies of digital technologies. It creates an opportunity to develop the ecosystem of innovative technologies for an enterprises - Smart Business Intelligence (SBI), in which the main pillar is the web and devices and smart systems based on it [7]. A dogma of functionality of smart business based on the one hand, on its isomorphism and on the other, on its constant dynamics and diversification leads to increasingly bigger metamorphosis of business models in which individual elements interact. As the literature on the subject states there are numerous model approaches that diversify a degree of absorption of new technologies, including the Internet of Things, in particular. They refer both, to offering additional services related to a product, smart products creating additional benefits as well as so-called extended services, complex IoT infrastructure solutions and those concerning directly IoT platforms. Classifications also include hybrid models based on behavioral profiling. Models based on using data and models determined by the philosophy of "everything as a service" [8]. Acceptance of technological progress and commercialization of new technologies change their position in the process of management while linking them with core resources and leading to conceptualization of new ways of building a market position of enterprises. In such an approach innovation is often an imperative for business survival but also development and expansion. Digital Technologies, like Internet of Things (IoT), Artificial Intelligence (AI), Cloud Computing or Machine Learning open new possibilities for innovation and can make them better, cheaper and faster. However, innovativeness must be perceived in a broad sense, not only as new technologies available in the market but first of all, as new ways of using them in practice. Enterprises that want to adapt them in the process of building a competitive advantage must perceive them as an integral part of their strategies. Yet, it requires to understand the essence of digital transformation as well as shaping knowledge and competence that enable to create optimum architecture of smart business. As a result, the current education process must evolve. As a result, current education process evolves at every level of education, starting from primary, secondary through tertiary education to continuous learning leading to a process of constant refreshing, developing and improving general and professional qualifications for the whole life.

New technologies become a specific catalyst of knowledge between enterprises, entities of education world, especially secondary schools or universities and smart people (customers, employees, managerial staff, etc.). It mainly concerns applied tools in that area, methods and techniques of teaching as well as a broadly understood education process. The world in which we are an element of network of connected people, machines or tools, in which sensors constantly communicate requires education that is directed at active looking for knowledge, its independent creating as a result of multilateral and complex interactions (smart society). While dealing with a new technology a man acquires new knowledge and competence at the same time reflecting their own world of ideas and looking for new multiple experiences while putting them into a kind of coherent entirety. As a result, knowledge is not acquired only passively as an effect of numerous but single interactions often occurring at repeatable intervals between a learner and a teacher (a teacher passes knowledge). This is a dynamic process based on constant adaptation of a learner to reality in which they remain. However, it requires transformation in education process from a reactive to a proactive attitude. Instead of replicating current patterns and schemes of behaviour it is necessary to look for own solutions, to initiate mechanisms of their exploration and to interpret phenomena independently (then a teacher becomes a coach, a guide in the world of science). A crucial point in this approach is an ability to activate learners, to create conditions in which learning is a natural process, not a forced process directed at creativity and engagement of a learner.

Internet of Things and Artificial Intelligence can be used to create more significant learning spaces. Education, as any human activity nowadays, has not been immune to this phenomenon dating from the e-learning, m-learning up to the u-learning this finally as the leap to the pervasiveness of knowledge. The potential of ubiquitous learning is reflected in increasing access to learning content and collaborative learning environments supported by computers anytime and anywhere. It also allows their combination of virtual and physical spaces. The purpose of ubiquitous computing technology is basically improving learning processes, which are trying to adapt learning resources to different contexts of use of apprentices [9]. New technologies can also help schools streamline mundane operations such as attendance, fee alerts, and student reports which can be automated easily. It can also bring down energy costs. Used wisely, they can also become a platform to conduct exams [10]. New technologies enable to enhance a process of education by using intelligent classrooms and intelligent equipment which enable to intensify experiences collected by learners (gamification, multimedia, interactions in real time with simultaneous elimination of space barriers – participants of education process can be in different places). An interesting idea in this area is the initiative of the European Commission's Erasmus+ of supporting the creation of an online education module focused on the Internet of Things. Students from multiple European universities connect to a remote lab to learn about IoT hardware, infrastructure, and mobile applications. Open access learning materials can be integrated into courses across various disciplines. - The European Commission's Erasmus + initiative is supporting the creation of an online education module focused on the Internet of Things. Students from multiple European universities connect to a remote lab to learn about IoT hardware, infrastructure, and mobile applications. Open access learning materials can be integrated into courses across various disciplines [11]. As a result, this process becomes more authentic accelerating acquiring new - subsequent skills and competencies. The IoT removes the traditional barriers to teaching and learning, providing faculty with the same flexibility to provide better learning experiences for students and allowing them to connect with experts from around the world and create robust, hybrid learning environments [12]. Thanks to that people can share their ideas, discuss studies and the latest events in the field of their interest while creating increasingly connected communities of practice. Offering access to a possibility of education fulfilling their needs will make education more effective in motivating students to learn. A student could e.g. observe their own ranking in the real time while comparing themselves with other students from the same level. A model of measurement could be available at any time and could constantly transfer directed and individualized remarks on what a student must do in order to improve their results [13]. Applications of IoT or AI have also potential to enhance many aspects of campus life, including safety and efficiency. As IoT proliferates, institutions are partnering with industry to enable student innovation and developing new programming to equip learners with the latest skills. The interdisciplinary course of study can for example encompass electrical and computer engineering, wireless communications, and data analytics. Through exposure to cutting-edge technologies, the program prepares students to develop new products and ideas for fields such as healthcare, utilities, transportation, retail, resource management and strategy management [11]. Consequently, the IoT in education can help solve challenges across a wide array of topics, from logistics to administration to student life. Streamlining and optimizing the utilization of facilities can help achieve financial savings (e.g., responding to weather events, automating operations). Smart devices can alert staff and providers about when to service equipment before a problem even presents itself. Smart doors, locks, and cameras can be used to monitor and control movement in different facilities. As more devices become connected, campus leaders will be able to extract much more value from the continuous stream of data and information, helping them move from a transactional relationship with students, faculty, administrators, and providers to an iterative process in which micro-decisions can be made on an ongoing basis [12].

Changes in education resulting from implications of new technologies can, yet, stimulate a need for creating dynamic systems modifying functionality of current solutions while posing new challenges for business (e.g. software, hardware, network architecture) at every level (data layer, analytical layer, interaction layer). In the next years significance of analysed interdependencies and relations between the world of business and education will undoubtedly intensify the development of machine learning, in particular deep learning, becoming intelligent support for entities creating or implementing diverse projects based on the IoT and AI. Designing machines that systematically learn our – human behaviour and use it e.g. in recommendation systems, managerial processes or fulfilling routine, repeatable activities can prove an integral part of smart enterprises, enabling to think about new markets, new business models and new professions. However, it is necessary to be aware of potential threats or fears connected with new technologies, especially in the area of losing personal data or confidential information as well as in the ethical area concerning excessive control of our attitudes and behaviour. Fear may be also evoked in case of potential losing control over "thinking" devices or machines, both in terms of their growing autonomy as well as a possibility of taking control over them by the third party as a result of hacker attacks. On the one hand it requires intensive work on safety systems integrated with solutions based on the IoT and AI, on the other hand, promoting knowledge enabling development of smart society. Business for learning, learning for business seem a natural synergy process in that approach.

3 IoT in Smart Education in the Light of Own Research

Conducted research is a subsequent element of studies and analyses carried out by the authors since 2016 on the degree of the development of the Internet of Things in Poland with particular focus on its significance for management of enterprises of the future. This is another area of discussions directed at further exploration of determinants, dependencies and trends influencing dynamics of absorption of the IoT in business. The main aim of presented studies was to identify mutual connotations of the IoT and AI between smart business, smart enterprises and smart education. The points of crucial significance included: identification of areas in which higher education institutions could use the concept of the Internet of Things and artificial intelligence; determination of potential benefits that college and students could gain; determination of barriers and restrictions connected with potential application of solutions based on the concept of the Internet of Things in colleges; recognition of challenges facing enterprises that can require implementation of solutions based on the concept of the IoT and AI in higher education institutions.

Conducted studies were alongside available literature of the subject a basis for preparation of the main assumptions for further research in the discussed area. They were of interpretative and idiographic character, which means that obtained results do not allow to formulate general conclusions, but they refer only to an examined group of subjects and their local context. Therefore, they present informative value and require further depended analyses and studies.

The research process comprised a number of stages. Its initial stage, based on available literature and results of own studies conducted between 2016 and 2017, determined the range and research objectives, research techniques as well as a measurement tool was designed. The stage of realization of research started from studies conducted among students and research and development staff of Polish universities. It was acknowledged that these two groups of respondents will be the first beneficiaries of changes resulting from implementation of new technologies to the process of education. Both studies were conducted by means of the CAWI method based on prepared research questionnaires.

The group of students comprised 231 people, out of which 209 correctly filled questionnaires were accepted for further analysis. The authors deliberately chose students of the last year of the 1st cycle studies from departments connected with business, for whom the IoT and AI can prove significant in the next years: Logistics (almost 26%), Human Resources Management (16%), Business Analytics and Finance and Real Estate (respectively: 12.4%) Management (11.5%), Public Management (10.5%) as well as Entrepreneurship and Innovation Management (6.7%) and Accounting (almost 5%). What was of crucial importance for the authors was to obtain information about a degree of recognition of such issues as the IoT or AI among people finishing their next level of education, who may have started their professional career or will start it soon. It was important to assess possibilities of using analyzed technologies both, in the process of education as well as in various areas of enterprise functioning, related fears as well as reflections concerning mutual dependencies between the world of science and business in the area of designing and implementation of related innovations. Women were a prevailing group accounting for almost ¾ of respondents as well as people with permanent residence in towns of above 30 thousand of residents (56% of respondents). At the same time every fourth respondent was living in a small town – below 5 thousands of residents. All respondents were below 25, which allows to assume that they belong to a generation that widely use new technologies in diverse areas of their life, thanks to which they possess sufficient knowledge to talk about this topic. At first students were asked whether such issues as the IoT and AI are familiar to them. More than a half admitted that they encountered such terms (IoT - 53% of respondents, AI - 55%) and they understand their essence. However, next the authors decided to present the definitions of both terms to all respondents and explained their meaning. It enabled to unify a way of perception of analyzed concepts in the examined group and opinions obtained in that way were of objective character. At subsequent stages of studies and analyses of obtained results the awareness of these two notions in a group of respondents will become an independent variable.

In case of research and development staff sampling was of purposeful character. The authors concentrated on the people who have examples of digital technologies application and/or conduct related classes in their academic track record. 43 workers of

Polish universities were asked to participate in the research and the final stage included 39 respondents, mostly from Łódź, Małopolska, Wielkopolska and Dolny Śląsk provinces. Nearly 70% of respondents were employed in public universities and conducted classes at economic, social, science and technical departments. All respondents conducted classes at the 1st and 2nd cycle studies, and every third also at uniform M.A. studies. Men constituted a prevailing group (slightly above 69%) and people between 45–54 slightly above 38.5% of respondents. Almost ¼ of respondents were between 35 and 44, and people below 35 and at the age of 55–64 included respectively 15.4% of respondents. The smallest group were people above 65 (7.6%). The prevailing group were holders of PhD (slightly above 46%) as well as holders of postdoctoral degree (30.7%). Almost 15.5% of respondents were holders of the title of professor and almost 8% - were holders of M.A.

Subsequently, studies concentrated on enterprises which were considered, on the one hand, a catalyst of changes resulting from implementation of solutions based on the IoT and AI in education, and on the other hand, their initiator. Studies were based on qualitative expert interviews. That group, similar to year 2016, comprised enterprises connected with practical application of the IoT, namely: realization of projects of so-called smart home, producing white goods, dealing with provision of electricity, installing solutions in the area of alternative energy sources as well as media (water, sewage, air-conditioning). Their choice was purposeful, as it was agreed that enterprises that have knowledge and competence in the discussed field at least at a basic level can discuss the essence of the analyzed concept, conditions of development, barriers and influence on management process. Finally, the studies comprised 11 enterprises, but the authors deliberately decided to include independent variables at this stage such as enterprise size, time of market presence, origin of capital, addressees of conducted activities (B2C, B2B) as well as a number of employees. Respondents were allowed to express their opinions freely, treating them due to a relatively small sample, as complementation to research material. The objective was just to obtain the widest possible spectrum of opinions, experiences, observations and suggestions, which are connected with using the IoT and AI in education and business. Particular focus was put on a problem of initiating changes connected with new technologies. Business for learning or learning for business became a crucial point of interest of the authors in the context of changes in a way of management and market orientation of enterprises facing challenges of digital transformation.

Slightly above 92% of scientific-didactic employees claim that the Interned of Things and artificial intelligence are technologies of the future from the point of view of development strategies of universities and 84.6% are convinced that the fact that objects and devices can communicate without a man can be of crucial significance in the education process in the next years. Students also notice a potential for education in the analyzed technologies. 68% believe that these technologies will influence the future of universities and 67% are certain that it will be extremely important in further development of education processes. What seems interesting is that enterprises claim that *"IoT and AI technologies are considerably significant in technical colleges and at the same time they are neglected by universities and humanistic courses"* although *"both technologies can totally change the ecosystem of education"*. However, the horizon for these changes is estimated by practitioners to take the period of 5 to 15 years.

Almost 62% of lecturers assume that it is justified to introduce such solution at universities but only one in every three respondents would be happy to use these technologies in their institutions. Students are more enthusiastic about taking advantage of using solutions based on both technologies at colleges where they study, which was indicated by almost 63% of respondents and slightly above 65% think that this implementation is justified. Among the most important benefits university employees indicated financial aspects resulting from optimization of system management in a building (heating, ventilation, air-conditioning, lighting, smoke removal, audio-visual services, etc.) as well as their better adjustment to users' needs (14% of respondents) and optimization of education process thanks to integration of all devices offering access to educational contents within a common network of university and/or a department (14%). What proved also really important was improvement of educational experiences of students (almost 12%) and related growth in their involvement in managing their own learning process and achieved results (10.3%). Lecturers participating in the study indicated also an increase of significance of education methods enabling students' interactions, created virtually for specific education processes (nearly 10%), improvement of educational experiences of college employees (nearly 10%) as well as image benefits connected with perception of a given university as a modern one implementing innovative IT, communication technologies, etc. (8.5%). Aspects that were the least important concerned an increase in automation of learning and teaching process thanks to using smart devices monitoring a pace of acquiring knowledge and its absorption as well as automation of systems of monitoring and evaluating education process and its quality (4%). Students noticed the biggest benefits from implementation of new technologies mainly in possibilities of improving own educational experiences (15.6%), improvement of attractiveness of classes they can participate in (13.6%) as well as an increase in a degree of effectiveness of education process thanks to simplified adaptation of educational resources to various contexts of their usage (12.8%).

The greatest fear connected with the implementation of the IoT and AI at universities was evoked in both groups of respondents by a potential risk of: losing confidential students and workers' data, e.g. personal data, confidential information, etc., loss of privacy due to constant monitoring and loss of users' control of devices, especially in case of a failure. Both groups also agreed that the greatest barriers connected with implementation of analysed technologies in education can be found in a lack of sufficient users' knowledge on the possibilities of the IoT in education process and their fears concerning potential surveillance of their activities in college area. What was also mentioned by didactic-scientific workers was a lack of understanding of the significance of the concept of the Internet of Things or artificial intelligence for an education process at various managerial levels of universities.

Lecturers agreed that college could use analyzed technologies in the area of monitoring of system administration process such as lighting, heating or air-conditioning (15% of respondents), increase in using smart equipment that automatically recognizes a lecturer (slightly above 10% of respondents), including good equipment and devices in technical colleges as well as automation of a number of processes e.g. checking attendance register at individual classes (almost 10% of respondents) and individualization of an education process – a mode of education adjusted to a pace of learning of individual students (8.6% of respondents). In contrast, students were looking for the IoT and AI potential in possibilities of creating new courses and specializations by universities connected with using new technologies (15% of respondents) as well as solutions increasing their comfort such as solutions of wearables type replacing a student's card, a library card, etc. (14% of respondents). They also emphasized their interest in participation in classes conducted in smart lecture halls while using smart equipment that would enable them to participate much more actively in education process.

Almost 70% of workers of universities think that using solutions based on the concept of the Internet of Things and artificial intelligence in their institutions and related change of education process itself (methods, tools, etc.) will become significant for enterprises. Students expressed a similar opinion. However, only 7.6% of examined lecturers and 8.5% of students notice in the analysed technologies a possibility of potential intensification of current cooperation of college and enterprises in the area of e.g. creating new technologies and their implementation or intensification of research and educational needs important for enterprises and realized in the real time. It was emphasized that colleges, especially those educating students in technical and science courses *"can become a background ensuring highly qualified staff responsible for building, development and research area of solutions based on the IoT and AI at the scientific-technical level"*. Moreover, creating new competencies and skills, also in case of economic, social or humanistic faculties, allows to construct new business models, explicitly shape a role of digital workers, including hybrid work models and *"formulate new professions such as data scientist, scenario planner, digital influencer or market learner/teacher"*. Enterprises taking part in the study agreed that *"implementation of changes connected with analysed technologies will undoubtedly require significant changes in workers' competencies, including university workers"* because *"incapacity of using them can be considered a kind of illiteracy"*. Moreover, they paid attention to the fact that an increasing interest in the IoT and AI is definitely expressed by representatives of the world of business and observed changes unfortunately very often are not a result of a mutual synergy of science and practice. And although they are implemented simultaneously, they are not related in a deliberate manner. As a result, they claim that *"this is the business that will force changes at universities and education process"*, which can mean *"a need for reorientation and changes in paradigm of management both, in case of smart business as well as smart education toward actions and decisions undertaken in real time based on analytics of billions of interactions of individual events and operations"*.

4 Conclusions

Both learning and teaching have benefited from integrating new technologies, like Internet of Things and Artificial Intelligence, into the educational framework. Their application creates a wide spectrum of financial, social, behavioral possibilities, etc., influencing optimization of many aspects connected with education and a role that it can play in the next years. But integration by itself does not lead to a smart society, smart business or smart enterprises. That requires much more. Learning is no longer encapsulated by time, place, and age but has become a pervasive activity and attitude that continues throughout life and is supported by all segments of society. Teaching is no longer defined as the transfer of information, learning no longer as the retention of facts. Rather, teachers challenge students to achieve deeper levels of understanding and guide students in the collaborative construction and application of knowledge in the context of real world problems, situations, and tasks. Education is no longer the exclusive responsibility of teachers but benefits from the participation and collaboration of business people, scientists, and students across age groups [14]. But this definitely requires mutual openness of representatives of the world of science, education and business, understanding their own diversity, needs and expectations. New technologies provide only tools and create a new space for Human to Human, Human to Machine and Machine to Machine interaction, but it depends on us – the people how we use them and how effectively we create a smart ecosystem of business for learning and learning for business connections.

Conducted studies enabled to describe a preliminary picture of ways of perceiving the potential of IoT and AI in tertiary education as well as possibilities of its absorption with particular regard to potential benefits, barriers and threats. Special focus was put on consequences of this process for enterprises while trying to indicate related key challenges. Yet, they did not explain all aspects concerning the analyzed problems but they became an inspiration for the authors for further studies and analyses, including some of international dimension.

References

1. Afuah, A., Tucci, C.: Biznes internetowy. Strategie i modele. Oficyna Ekonomiczna, Kraków (2003)
2. Sułkowski, Ł., Kaczorowska-Spychalska, D.: Internet of Things – w poszukiwaniu przewagi konkurencyjnej. In: Sułkowski, Ł., Kaczorowska-Spychalska, D. (ed.) Internet of Things. Nowy paradygmat rynku, Warszawa, Difin (2018)
3. Hui, G.: How the Internet of Things Changes Business Models. Harvard Business Review, Watertown (2014)
4. Kyriazis, D., Varvarigou, T.: Smart, autonomous and reliable Internet of Thing. Procedia Comput. Sci. **21**, 442–448 (2013)
5. Galaske, N., Arndt A., Friedrich, H., Bettenhausen, K.D., Anderl R.: Workforce management 4.0 - assessment of human factors readiness towards digital manufacturing. In: Advances in Ergonomics of Manufacturing: Managing the Enterprise of the Future, Proceedings of the AHFE 2017 International Conference on Human Aspects of Advanced Manufacturing, Los Angeles, California, USA (2017)
6. https://biznes.newseria.pl/news/do-2025-roku-30-proc,p676710462. Accessed 12 Feb 2017
7. Sułkowski, Ł., Kaczorowska-Spychalska, D.: Management of enterprise of the future in the ecosystem of the internet of things. In: Schlick, C., Trzcieliński, S. (eds.) Advances in Ergonomics of Manufacturing: Managing the Enterprise of the Future. Advances in Intelligent Systems and Computing, vol. 490. Springer, Cham (2016)
8. Wielki, J.: Internet Rzeczy i jego wpływ na modele biznesowe współczesnych organizacji gospodarczych, no. 281. Zeszyty Naukowe Uniwersytetu Ekonomicznego w Katowicach, Studia Ekonomiczne (2016)

9. Gomeza, J., Hueteb, J., Hoyosa, O., Perezc, L., Grigorid, D.: Interaction system based on internet of things as support for education. Procedia Comput. Sci. **21**, 132–139 (2013)
10. Charmonman, S., Mongkhonvanit, P., Dieu, V.N., van der Linden, N.: Applications of Internet of Things in e-learning. Int. J. Comput. Internet Manag. **23**(3), 1–4 (2015)
11. NMC, Horizon Report: 2017 Higher Education Edition. The New Media Consortium, Austin (2017)
12. Asseo, I., Johnson, M., Nilsson, B., Chalapathy, N., Costello, T.J.: The Internet of Things: riding the wave in higher education. EDUCAUSE Reviewer, July/August 2016
13. Cisco Systems, Report - Internet przedmiotów wspomagających uczenie się (2014)
14. Kozma, R.B., Schank, P.: Connecting with the twenty-first century: technology in support of educational reform. In: Dede, C. (ed.) Technology and learning. American Society for Curriculum Development, Washington, D.C. (1998)

Inferring a User's Propensity for Elaborative Thinking Based on Natural Language

Veena Chattaraman[1(✉)], Wi-Suk Kwon[1], Alexandra Green[1], and Juan E. Gilbert[2]

[1] Auburn University, Auburn, AL 36849, USA
{vchattaraman, kwonwis}@auburn.edu,
aag0042@tigermail.auburn.edu
[2] University of Florida, Gainsville, FL 32611, USA
juan@ufl.edu

Abstract. Natural language-based aids (e.g., intelligent cognitive assistants) that assist humans with various tasks and decisions, often need to recognize the user's propensity (low-high) to elaborate on the task or decision, to ensure that the information provided matches the user's thinking level. We conducted two qualitative studies of natural language usage in customers' written product reviews (Study 1) and conversational transcripts of customer-store associate interactions (Study 2) to generate (Study 1) and validate (Study 2) four rules that can be employed to infer a user's propensity for elaborative thinking. These include: consideration of multiple (2+) attributes/alternatives; detailed description (word count) about a single attribute/alternative; demonstration of specific knowledge (use of specific terms) about an attribute/alternative; and consideration of pros and cons about an attribute/alternative. Implications for natural language-based, intelligent cognitive assistants emerge as a result of this work.

Keywords: Elaboration level · Natural language
Rules for elaboration inference · Intelligent cognitive assistants

1 Introduction

Humans differ in their propensity for elaborative thinking on a specific topic, task, or decision. When designing natural language-based aids (e.g., intelligent cognitive assistants) that assist humans with various tasks or decisions, it is important for the system to recognize the user's propensity (low to high) to elaborate on the task or decision, to ensure that the information provided matches the user's thinking level. Human-to-human interaction reveals that natural language cues can be employed to infer an individual's level of elaboration; however, no published work has examined this. To address this gap, the purpose of this research is to generate and validate rules that can be employed to infer a user's propensity for elaborative thinking in the specific context of consumer decision-making.

The Elaboration Likelihood Model (ELM), a theory of consumer psychology that has seen important applications in consumer and health research [1–3], maps humans on a quantitative elaboration continuum between effortful and heuristic processing

H. Ayaz and L. Mazur (Eds.): AHFE 2018, AISC 775, pp. 319–324, 2019.
https://doi.org/10.1007/978-3-319-94866-9_32

based on a person's ability and motivation to engage in an object-relevant elaboration [4]. The theory postulates that (1) amount and nature of object-relevant elaboration is contingent on individual and situational factors, and (2) people will rely on peripheral cues in forming their evaluations as motivation and/or ability to evaluate an object decreases, whereas people will employ extensive, effortful, and object-relevant information processing as motivation and/or ability increase [4]. The theory also makes a clear case that low and high elaboration along the continuum differs based on both quantitative and qualitative aspects of thinking [4].

Based on ELM [4], the quantitative aspect is defined as the amount of object-relevant information or the specific number of arguments considered during an evaluative task or decision; whereas the qualitative aspect refers to the type of object-relevant information and quality of arguments considered. For example, when deciding to buy a product, a high elaborator may consider multiple arguments for why the product may or may not satisfy his or her needs, whereas a low elaborator may consider only a few arguments (quantitative difference). In the same scenario, while a high elaborator may consider all the specific merits and demerits of a brand in addressing his or her needs, a low elaborator may simply consider the brand's holistic image and make a decision based on it (qualitative difference). Given these explanations for qualitative and quantitative differences between high and low elaborators, there may be many other context-specific scenarios that give rise to differing elaboration inferences. To account for these, this paper conducts and reports the results of two qualitative studies of natural language usage in (1) customers' written product reviews and (2) conversational transcripts of customer-store associate interactions to delineate rules that can be employed to infer a user's propensity for elaborative thinking in the context of consumer decision-making.

2 Study 1

2.1 Method

In Study 1, two coders (male and female) independently coded 400+ verbal transcripts of customer review data on air filters, drawn from a large online retailer, for elaboration level (low vs. high) and qualitatively described their reason for inferring 'high elaboration'. In reasoning the elaboration level, the coders were instructed to assume that a transcript was low elaboration unless they saw a clear rational to categorize it as 'not low'. They then wrote the rationale for this inference. Hence, all transcripts were initially categorized as 'low' or 'not low'. The 'not low' category was later recoded as a 'higher' elaborator.

2.2 Results

The initial inter-coder consistency between the two coders for elaboration level coding of the customer reviews was 89%. The discrepancies were negotiated by the coders to arrive at complete agreement. In addition to the elaboration level, the coders also provided the reason or rationale for coding a review as originating from a higher elaborator. It was possible that a single customer review was categorized as high elaboration due to multiple reasons. These reasons were then content-analyzed by another coder to generate more abstract rules, which encompass the specific reasons. Based on this thematic content analysis of the reasons, four rules were evident for inferring high elaboration: (1) consideration of multiple (2+) attributes/alternatives; (2) detailed description (word count) about a single attribute/alternative; (3) demonstration of specific knowledge (use of specific terms) about an attribute/alternative; and (4) consideration of pros and cons about an attribute/alternative. Examples of customer reviews for each of these rules are included in Table 1.

Table 1. Rules for inferring high elaboration from customer reviews

Rule	Description	Customer review example
1	Considering multiple (2+) attributes or brands/models together	Customer: Purchased these to help eliminate smoke odors [Odor - attribute]. So far, so good. Much more effective than MERV 7 [MERV rating - attribute]. I change filters every 30 days [Replacement time - attribute] because of the cigarette smoke [Tobacco smoke - attribute], so no comment on their lasting power
2	Detailed thoughts about a single attribute (word count – over 100)	Customer: One can immediately see that this product is higher quality than the cheap stuff found elsewhere [filter quality – attribute]. Thinking that it would stop more dust and particles (and it did), I installed it on the air intake of my HVAC. A few days later, I noticed that my furnace would sometimes shut itself down before reaching the desired temperature on the thermostat. Brought in a professional to have a look at the furnace thinking there was something wrong with it, the guy told me that these very high MERV filters reduce air flow considerably because they are denser. With less air incoming, the core of the furnace would heat up considerably until it shuts itself down to protect itself. So for us, he advised us to go back the cheap stuff that interferes less with air flow. Bottom line, it works very well but there is the possibility that it could not work for everyone. Make sure that your HVAC can handle these high-density filters
3	Showing specific product knowledge about an attribute or a brand/model	Customer: The filter looks like built in good quality [filter quality – attribute]. I have seen someone else complained that there is no marking on the actual filter, and yes, he is right. However, after doing some research under [brand name] web site, it appears that it is the norm nowadays for [brand name] filters to save cost of printing elaborate markings on the filter
4	Thinking about pros and cons about an attribute or a brand/model	Customer Review: They really catch the dust, but they also decrease the airflow through the unit

3 Study 2

3.1 Method

Study 2 involved observational studies in the form of in-store purchase decision-making simulations with a sample of 48 consumers representing both genders and varied age groups, with the purpose of eliciting qualitative interactional data between customers and a retail service associate. A simulation home improvement store environment was created for the purpose of this study. In each simulated session, a consumer shopped for an air filter under one of two conditions, low and high risk, randomly assigned. In the low risk condition, participants were asked to imagine that they were going to buy an air filter just to replace their current filter without special needs to meet. On the other hand, in the high-risk condition, participants were told to imagine that they were recently diagnosed with severe allergy and asthma and thus must buy the right kind of air filter to supply clean air, which is of utmost importance to their health and safety. These varied risk conditions were created to ensure that there were equal numbers of consumers who were either more or less motivated to elaborate on their decisions. Each consumer was paired with a retail service associate (trained research assistant), who interacted with the consumer as he or she reached a choice. The conversations between customers and store associates were audiotaped using nonintrusive voice recorders worn by both. Following the observational studies, the recordings were transcribed verbatim, and two separate coders (different from Study 1) independently coded the 48 conversational transcripts, for customer elaboration level (low vs. high) based on the four rules identified in Study 1. Prior to this, the coders were trained on identifying the high elaborators based on clear definitions (see Table 2) generated from the four rules developed in Study 1 (see Table 1). In addition to coding each transcript, the coders were also asked to indicate all the rules that they applied in inferring high elaboration within each transcript. Hence, each customer could be classified as a high elaborator due to multiple reasons.

Table 2. Definitions of high and low elaboration levels in consumers

Code	Name	Description
H	High elaboration consumer	The consumer who thinks extensively to make a purchase decision. This thinking is reflected in different ways, such as (1) considering *multiple product attributes or brands/models* together, (2) thinking *in detail about a single* attribute or brand/model such as considering supportive arguments about an attribute or a brand/model, (3) demonstrating *specific product knowledge* related to an attribute or a brand/model, and (4) comparing *pros and cons* about an attribute or a brand/model
L	Low elaboration consumer	The consumer who does not think extensively before making a purchase decision. This type of consumers make a quick decision based on a single product attribute or brand/model without much thought or involvement. This type of consumers does not show any sign of the four thinking patterns listed above for the high-elaboration consumer

3.2 Results

The inter-coder consistency in the coding of the elaboration level of the conversational transcripts was 89.6%. Further, all four rules were applied in coding the 48 conversational transcripts for elaboration (Rule 1: $f = 41$; Rule 2: $f = 27$; Rule 3: $f = 21$; Rule 4: $f = 13$). These results demonstrate that the elaboration inference-rule-based coding had equivalent reliability in identifying high and low elaborators as natural inference coding conducted in Study 1. The results also demonstrated that elaboration inference rules, although developed from non-interactional customer review data, was applicable to human-to-human interactional data as well. Hence, Study 2 extends the validity of the elaboration inference rules to interactional contexts in consumer decision making.

4 Conclusions

The elaboration inference rules generated and validated through two qualitative studies employing both narrative and interactional natural language data has important implications for intelligent cognitive assistants that may need to infer a user's elaboration level. Many scenarios may require intelligent cognitive assistants to infer the user's elaboration. For example, an intelligent consumer decision aid may retrieve more or less information for a consumer who seeks to make a decision on a product, depending on the consumer's propensity to elaborate. Presenting more than the requisite information to a low-elaboration consumer may lead to information overload and a desire to abandon the decision at the given moment. Similarly, presenting less than the requisite information to a high-elaboration consumer may lead the consumer to question the usefulness of the decision aid and seek other approaches to securing more information on the product. The critical point here is that the amount of information provided by an intelligent cognitive assistant should match the thinking propensity of the user. Hence, there is a need for the cognitive assistance provided by an intelligent aid to 'fit' with the cognitive ability of the user at the given time.

Prior works have emphasized the concept of 'fit' from multiple perspectives (1) a resource matching theory perspective such that decision performance is greater when the cognitive resources provided by the decision aid match those demanded for the task than when resources provided were higher or lower than the task demands [5]; (2) a cognitive fit theory perspective such that the problem representation of the decision aid should match the mental representation for decision-making [6–8]; (3) task-technology fit perspective such that the functionality of the decision aid should match the task and ability of the decision-maker [9]; and (4) similarity attraction perspective such that the decision process of the decision aid should align with the process by which users make decisions, especially for users with low domain knowledge [10]. The studies described in this paper help delineate the rules for identifying the thinking level of the user, so that 'fit' between the user and intelligent cognitive assistant can be achieved.

Acknowledgments. This material is based in part upon work supported by the National Science Foundation under Grant Numbers IIS-1527182 and IIS-1527302; the Alabama Agricultural Experiment Station; and the Hatch program of the National Institute of Food and Agriculture, U. S. Department of Agriculture. Any opinions, findings, and conclusions or recommendations expressed in this material are those of the authors and do not necessarily reflect the views of the funding agencies acknowledged above.

References

1. Angst, C.M., Agarwal, R.: Adoption of electronic health records in the presence of privacy concerns: the elaboration likelihood model and individual persuasion. MIS Q. **33**(2), 339–370 (2009)
2. Kreuter, M.W., Wray, R.J.: Tailored and targeted health communication: strategies for enhancing information relevance. Am. J. Health Behav. **27**(3), 227–232 (2003)
3. Hinyard, L.J., Kreuter, M.W.: Using narrative communication as a tool for health behavior change: a conceptual, theoretical, and empirical overview. Health Educ. Behav. **34**(5), 777–792 (2007)
4. Petty, R.E., Cacioppo, J.T.: Attitudes and Persuasion: Classic and Contemporary Approaches. Brown, Dubuque (1981)
5. Tan, C., Teo, H., Benbasat, I.: Assessing screening and evaluation decision support systems: a resource-matching approach. Inf. Syst. Res. **21**(2), 305–326 (2010)
6. Hong, W., Thong, J.Y.L., Tam, K.Y.: The effects of information format and shopping task on consumers' online shopping behavior: a cognitive fit perspective. J. Manag. Inf. Syst. **21**, 149–184 (2004)
7. Huang, Z., Chen, H., Guo, F., Xu, J.J., Wu, S., Chen, W.-H.: Expertise visualization: an implementation and study based on cognitive fit theory. Decis. Support Syst. **42**, 1539–1557 (2006)
8. Vessey, I.: The theory of cognitive fit. In: Zhang, P., Galletta, D. (eds.) Human-Computer Interaction and Management Information Systems: Foundations, pp. 141–183. M.E. Sharpe, Armonk (2006)
9. Goodhue, D.L.: Task-technology fit. In: Zhang, P., Galletta, D. (eds.) Human-Computer Interaction and Management Information Systems: Foundations, pp. 184–204. M.E. Sharpe, Armonk (2006)
10. Al-Natour, S., Benbasat, I., Cenfetelli, R.: The effects of process and outcome similarity on users' evaluations of decision aids. Decis. Sci. **39**(2), 175–211 (2008)

Building Consumer Trust to Improve Internet of Things (IoT) Technology Adoption

Areej AlHogail(✉) and Mona AlShahrani

Information Systems Department, College of Computer and Information Science, Al Imam Mohammad ibn Saud Islamic University (IMSIU), Riyadh, Saudi Arabia
{aalhogail,mjshahrani}@imamu.edu.sa

Abstract. The Internet of Things (IoT) has gained a huge importance in the recent years among the other latest Internet evolutions. Internet of Things aims to provide people with the innovative and intelligent technologies and services where all the physical objects around their world are connected to the Internet and communicate with each other. IoT products and services span several fields such as, healthcare, hospitality, transport, infrastructure, education and frontline social services. Trust plays a crucial role in the adoption of Internet of things technologies and services for ensuring a satisfactory and expected transaction result. It helps customers to overcome perception of risk and uncertainty related to IoT technology and enhances the customers' level of acceptance that have a positive impact in adoption intention. However, current literature of the IoT services still lacks studies on the behavioral aspect that clarify the perception of customers about the adoption and usage of the IoT technologies and services. In this paper, we develop a conceptual trust model that contains the major factors affecting trust towards IoT technology adoption. The model is presented in three domains and 11 dimensions and designed based on the theory of Technology Acceptance Model (TAM). The result of this paper is a conceptual framework that classified the factors into three main domains, namely: product related factors, Social influence related factors and Security Related Factors. This framework aims to serve as a foundation for future studies about trust in IoT technology adoption.

Keywords: Internet of Things IoT · Trust · Adoption · Information security TAM · Human behavior

1 Introduction

During the last few years, Internet of Things (IoT) approached our daily lives gradually through the wide spread of wireless communication and smart devices such as phones, home appliances, health care devices, hospitality, transport and so on. Smart devices that are often interconnected with cloud services promise easy usage and global access and make more consumers engage in such technology.

An IoT system can be described as a collection of interconnected smart devices that interact on a collaborative basis to fulfill a common goal. It embraces many different technologies, services, and standards [1]. IoT involves three major agents i.e. people,

H. Ayaz and L. Mazur (Eds.): AHFE 2018, AISC 775, pp. 325–334, 2019.
https://doi.org/10.1007/978-3-319-94866-9_33

objects and data. It is expected that tens of billions uniquely identifiable objects will be a part of this global computing network by the year 2020 [2]. Undeniably, this high level of heterogeneity, coupled to the wide scale of IoT systems and the increased intractability of humans, machines, and robots, in any combination is expected to expand security threats. Moreover, the traditional security countermeasures and privacy enforcement cannot be directly applied to IoT technologies due to their limited computing power, high number of interconnected devices and scalability issues. At the same time, to reach a full acceptance and adoption by users it is mandatory to define valid security, privacy and trust models suitable for the IoT application context [1, 3].

An IoT applications, such as smart home, use sensors to control the room smartly and remotely. These sensors collect information in very sensitive and personal domains, i.e. a bedroom. In this automated communication the owner takes no active role in communication and relies on services to act on person behalf [4]. In such situation, security, and in particular trust, remains major challenge for consumers and developers of IoT application and services.

Different security challenges could face the adoption of the IoT. Trust is a fundamental issue since the IoT environment is characterized by different devices which have to process and handle the data in compliance with user needs and rights. This paper is concerned with this challenge, namely, trust. As consumers believe in the potential IoT benefits to them, they are also concerned about the security and privacy of their data and the data breach of sensitive personal information can be exploited to harm them; the trust of these consumers on the IoT can be tampered. The success of IoT ultimately depends upon the consumer perceptions about security and privacy in IoT and the level of digital trust of its consumers [2]. In a global rush to promote IoT products and services, the industry is falling behind on building consumer trust. The wide adoption of IoT devices is highly reliant on establishing consumer's trust.

A deep understanding of the aspects and variables that lead to customers' trust in IoT enabled product and services could help developers in implementing an efficient and widely accepted and adopted IoT services [5]. This study contributes to the body of literature of trust for IoT adoption by presenting a comprehensive view about related factors that affect the consumers' trust of IoT services and products, and study how factors might be useful to implement marketing strategies focused on consumer's demands and requirements. Moreover, this paper proposes a conceptual model to instruct future research directions and developers in understanding the factors that could affect IoT adoption.

2 Literature Review

2.1 Internet of Things

The Internet of Things (IoT) has become a research focus for both industry and academia. It promises to create a world where all the objects around us are connected to the Internet and communicate with each other with minimum human intervention [6]. The IoT referred mainly to RFID tagged objects that used the internet to communicate. The IoT technology is used across many areas at present, for instance, healthcare, supply chain management, smart cities, digital logistics, efficient transportation, home automation, mobile payment, warehouse management, and the private domain. The prediction of the devices that connected to the internet will be 100 billion devices by 2030 [7].

IoT consists of four levels that starts with perception layer that consists of different sensors and data collectors followed by network layer that is responsible for transmitting the data. The third layer is the middleware layer that consists of information processing systems followed by fourth level that is the application and services level [8].

The IoT enabled products and services affect consumers' behavior and choices on numerous aspects of their daily life. However, the convenience provided by interconnected smart devices is accompanied with a lot of security and privacy issues. For instance, the threat that an intruder takes control of IoT devices, access to sensitive information or disrupt services. On the ignorance of these issues in IoT, undesired consequences will be challenged like lack of trust, non-acceptance and damage to reputation [2].

IoT manufacturers are working effectively to add more intelligence and connectivity into devices. However, the reality is that the security and privacy concerns are poorly considered in the development, for instance, poor encryption or unencrypted communications, weak passwords, and defective user interfaces exposing the consumers and their sensitive private information to be compromised by the hackers [2]. A trustworthy system or service should preserve its users' privacy, which is one of the ways to gain user trust. Trust, security and privacy are highly related crucial issues in IoT.

2.2 The Role of Trust in the Adoption of IoT Technology

Trust is a complicated concept that is influenced by several measurable and non-measurable properties [9]. It is highly related to security since ensuring system security and user safety is a necessity to gain trust. However, trust is more than security as it relates not only to security, but also to many other human related factors.

Trust means beliefs that a person or technology has the attributes necessary to perform as expected in a situation [10]. Mayer et al. [11] have defined trust as "the willingness of a party to be vulnerable to the actions of another party based on the expectations that the other party will perform a particular action important to the trustor, irrespective of the ability to monitor or control that other party" (p. 172). Latest technology adoption research has directed attention to trust in as a critical driver of value-added IT usage behaviors [12].

Trust is an essential and important aspect that encourage people to easily adopt modern technology with unpredictable circumstances. In the cyber context, trust has been identified as a key driver for adoption [13] due to its relevance to deal with two critical conditions of digital means: uncertainty and risk of vulnerability [5]. In uncertain situations trust reduces vulnerability and helps the person to understand the social surrounding of the exchange of services [11]. Therefore, trust is considered as a critical variable in studies concerning online services. Studies of human behavior online emphasize the significance of including trust in adoption models to better comprehend the user acceptance of such services [5].

Users' assessments reflect their beliefs about technology's ability to deliver and perform of its objective characteristics which may differ based on their experience or the context for its use. Consumers appreciate the possible benefits and advantages of using IoT products and services. However, consumers feel uncomfortable as these smart IoT devices collect and share personal information. They are concerned about the security of devices and privacy of their personal information. If these devices or their personal information is exploited for criminal acts, then trust upon adopting these IoT devices could be compromised. Therefore, it is considerable that the success and adoption of IoT products and services is greatly dependent on establishing consumer's trust [14].

The literature in the technical aspects of IoT are more numerous than the behavioral and attitudinal aspects. This indicates the need for more investigation in that field. Few studies have investigated the role of trust in IoT adoption. For instance, the study by Gao and Bai [14] concluded that trust has a significant factor that affects the behavioral intention to adopt the in the IoT products and services. Han et al. [15] proofed that trust is an essential factor for the adoption of third party apps. The following section describe a conceptual model that combine the major factors that influence consumer trust towards adoption of IoT products or services.

3 Trust Factors for Improving IoT Technology Adoption Model (TFITAM)

Lee and Turban [16] have described social-psychological theoretical perspectives in trust related studies as it characterize trust in terms of the expectations and willingness of the trusting party in a transaction, the risks associated with acting on such expectations, and the contextual factors that either enhance or inhibit the development and maintenance of that trust. The social-psychological perspective appears to be relevant for understanding factors affecting consumer's trust in IoT because it focuses on transactions, and the work on the uncertainty situations that is usually associated with IoT transactions. To examine the importance of the trust factors on adoption of IoT technology, a research model has been proposed that draws from the diverse research on trust and will be based theoretically on social-psychological perspective and the technology acceptance model TAM.

Technology acceptance model (TAM) is the most influential and common theory in Information Systems field that models the way the user accepts and uses a technology [17]. TAM is created from the theory of reasoned action that is mainly about the impact of attitude on behavior [17]. TAM proposes that there are number of factors influence the users' decision regarding how and when they will use a recent technology presented to the field. In particular, TAM suggests that two factors are significant determinants of behavioral intention to use a system/technology perceived ease of use and perceived usefulness [14]. Researchers have verified and validated TAM which shown to be suitable as a theoretical foundation for the adoption of technology [18].

Different factors that have been collected from the literature are combined to form it. The conceptual model guiding this research is depicted in Fig. 1 that has been called Trust factors for improving IoT technology adoption model (TFITAM).

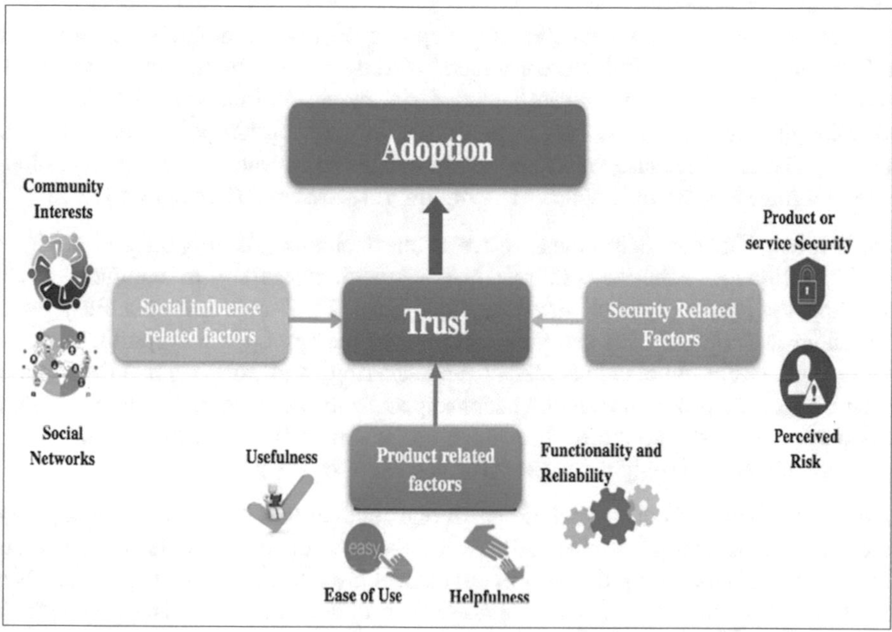

Fig. 1. Trust factors for improving IoT technology adoption model (TFITAM)

In this paper, we are targeting IoT technology adoption. Actually, there is a distinction between adoption and acceptance of a technology. Technology adoption is a process – starting with the user becoming aware of the technology, and ending with the user embracing the technology and making full use of it. Acceptance, on the other hand, is an attitude towards a technology, and it is influenced by various factors [19]. We are investigating the factors that influence the adoption.

3.1 Factors Affect Trust

Trust is impacted by numerous quantifiable and non-quantifiable factors. The conceptual model (TFITAM) has classified the factors into three main domains, namely: product related factors, Social influence related factors and Security Related Factors. Each domain consists of a number of factors.

3.1.1 Product Related Factors

A number of factors that are product-specific could influence the consumer decision to trust IoT product or a service. Different models and studies have suggested a number of factors that influence trust that affect the adoption decision. Based on literature we have collected a number of factors that are:

Functionality and Reliability: refers to whether a technology have the capacity or capability to complete a required task by providing required features and will consistently operate properly and predictably [10]. It must have the capacity detect data corruption and try to fix it. This feature is essential for IoT products and services to keep running efficiently.

Trust in a technology's functionality depends on that technologies' capability to perform properly. It is noted that consumers' trusting is based on that they perceive the product or service will perform its intended and requested functions [20]. Moreover, the reliability have a positive influence on trust in the adoption of IoT as reported by Lai et al. [12, 20]. Because errors are not acceptable to consumers on any technology, there is a huge impact of absence of errors on trust toward IoT adoption [21].

Helpfulness: Trust in helpfulness refers to the technology's support and ability to provide adequate, effective and responsive advice necessary to complete a task including tips, guidance, and help functions [10, 12]. People may not fully utilize technology as they may fear that they will not find the appropriate support if things go wrong. That might limits the benefits of the technology and usually affect the adoption of the technology [10, 12]. Providing such support to users can guide them by avoiding undesired surprises. Therefore, to achieve a better trust that lead to IoT adoption, a good investment in providing support for end users is crucial.

Ease of Use: refers to the degree to which one believes that using the technology will be effort free. According to Lai [20], the ease of use of the technology plays a significant role in building up the trust of consumers towards this technology. It has been found that IoT products or services convenience increase the satisfaction level of consumers and affects the intention of consumers to adopt them [14].

Different researches have emphasized the positive effect of ease of use towards improving the trust that lead to IoT adoption. For example, Gao and Bai [14] found that perceived ease of use has major effect on the IoT services adoption. Similar findings were Abu et al. [22] that shows that ease of use is one of the most important factors for the adoption. Consumers tend to trust commonly used IoT products and services and distrust cases that are perceived to be outside their control [23]. Thus, it is expected that the perceived ease of use has significant effect on trust toward IoT adoption.

Perceived Usefulness: is defined as the degree to which one believes that using the technology will enhance his/her performance [10]. Different researches have proofed the positive relationship between IoT products or services adoption rates and perception of consumers that it could facilitate their daily life [14]. The satisfaction level of consumers affects their trust that lead to intention of IoT technologies adoption. It is found that perceived usefulness is a significant predictor of the intention to adopt IoT products services in a study by Coughlan et al. [24]. The TAM model specifies that perceived usefulness is a significant determinant of behavioral intention to adopt IT [25, 18]. Moreover, the perceived usefulness of IoT services advocates the fact that individuals will perceive them useful as it enables them to enhance their overall performance in everyday situations [14]. Therefore, perceived usefulness of the IoT technology must be advocated to achieve a successful adoption.

3.1.2 Social Influence Related Factors

In case of IoT technology or service adoption, most users lack reliable information about product details to make a decision. Social influence is demonstrated as a person's perception of a product or a service that is highly influenced by the others' perceptions of whom who around him/her. The Unified Theory of Acceptance and Use of Technology (UTAUT) has considered the social influence as one of four factors that influence consumers' technology adoption. Gao and Bai [14] and Abu et al. [22] found a positive effect of social influence on the adoption of IoT services. Social influence has received considerable attention in the IS field [26]. We have divided the social influence into two precise factors, social network and community interest.

Social Network: the social network opinions for individual evaluation of the products will influence his/her decision. Therefore the model should incorporate social influence that is demonstrated by a users' perception of whether other important people in his/her community perceived they should embrace with this technology or service [14]. Social networks play a crucial role in Influencing the user adoption of IoT technology since users generally seek information from peers, family, and even famous social networks influencers reviews to reduce IoT product or services uncertainty prior to purchase [26]. Users usually trust relevant users' reviews and opinions because such content can be interpreted as a general evaluation of a product [26]. According to Gao and Bai [14], many users have considered mobile IoT devices as trustworthy because it was trending by the social networks [14]. However, customers tend to doubt or resist the review and evaluation persuasion by developed companies. Thus, social networks will have significant role on influencing consumer trust towards IoT adoption and must be taken into account for introducing IoT product or service to the market.

Community Interest: is an important factor that empowers trust and interaction between objects from the same community [27]. Community interest and culture could highly affect how individuals make their decisions. Although the world with globalization became closer, still culture differences can distinguish nation from nation. For instance, a reserved middle-eastern culture could react differently to video camera sensors devices than open cultures. Managing trust needs a lot of investigation of the domestic market, where culture might set great barriers to understanding and the legislation. National differences might have a positive or a negative effect on any new technology trust [28, 29]. It is also important to note that sometimes the lack of alternatives or necessity could influence that factor towards trust [23]. Studies on the influence of culture on trust has been underrepresented [29]. Consequently, it is evident that for any new IoT technology or services entering a new market must take into the account the local community interest and a deep awareness and understanding on the perceptions and expectations related to trust.

3.1.3 Security Related Factors

Security in this context indicates the extent to which a person believes that using an IoT service or product will be risk free. Security is a major concern of customer when adopting a new technology and it has significant influence on consumer's trust of a specific product or service and therefore the adoption of technology [16, 25]. In order to increase the adoption, users need to feel safe when interacting with these systems. Moreover, the lack of security were found to be a major issue that prevent customers from adopting the IoT product and services [25]. To achieve that several factors need to be taken care of. We have divided the factors into:

Product or Service (Trustee) Security this factor is concern with the ability of the trustee to achieve major security concerns such as confidentiality, integrity and availability. This factor could be affected by trustee reputation and earlier behaviors and performance [9]. Security is always a critical issue to which consumers are concerned when it comes to trust towards adoption [20]. The level of security and privacy are critical characteristics of IoT technology that affect the development of consumers' confidence in them as it gives consumers the assurance that they will be safe [20].

According to Koien [23], users tend to trust the IoT devices that use credible entity authentication and access control. Such devices that show ability and willingness to protect themselves are welling to be noted as trusted devices. Therefore, we can conclude that there is a positive relationship between trust and IoT product or service security level.

Perceived Risk Associated with a product or service. In the context of IoT usage, the perceived risk of IoT services usages is higher because of the unique characteristics of the IoT technologies and the high level of IT involvement. Users feel greater uncertainty and heightened risk in their adoption decision [14, 20, 30]. Trust is considered as one the most effective tools for reducing uncertainty and risks by generating a sense of safety [14], therefore plays a major role in adoption intention. Consumers tend to distrust IoT devices or services because they perceived to be outside of their control, as they consider these devices as a high risk [23]. There is a mismatch between the actual risk level and the human trust in the device that is usually built upon his/her perceived risk [23], therefore, we have separated the actual product or service security level and the perceived risk. However, there is an inverse relationship between the actual product security level and the perceived risk associated with that product or service.

4 Conclusions

To improve consumer trust towards his/her adoption of IoT technology or services, an appreciation of trust related factors is required. Trust has been identified as a key driver for adoption [13] due to its relevance to deal with two critical conditions of IoT: uncertainty and risk of vulnerability. Furthermore, it has been proofed that the success of IoT adoption positively depends upon the consumer level of trust [2]. The proposed conceptual model (TFITAM) provided a comprehensive view of the issues that influence human behavior toward IoT technology trust. TFITAM model has been developed with three main domains and 8 factors under these domains. The domains were related to social-psychological perspective and the main factors of the Technology Acceptance Model (TAM); and the refinement of the factors were related to previous work. The TFITAM model could be used by researchers to further investigate the trust issues and enrich the literature in the field of IoT trust. Moreover, IoT technology developers could use this model and its specified factors to create products and services that could be widely adopted through gaining consumers trust.

The next step is to evaluate the TFITAM model through further investigation by the authors. Data will be collected from consumers and service providers to validate the proposed model. Data then shall be statistically analyzed to confirm the results by using a useful variable analysis technique. The main goal of this step is to measure and understand the relationships among factors and variables. Moreover, future work can consider specific domain of IoT products or services and this may require additional elements to be added and may cause some changes in the subdomains. The model could be further investigated to recommend practical standards that can be enhanced with performance indicators and indexes for improving trust in IoT products and services.

References

1. Sicari, S., Rizzardi, A., Grieco, L.A., Coen-Porisini, A.: Security, privacy and trust in Internet of Things: the road ahead. Comput. Netw. **76**, 146–164 (2015)
2. Khan, W., Aalsalem, M., Quratulain, A., Khan, M.: Enabling consumer trust upon acceptance of IoT technologies through security and privacy model. In: Advanced Multimedia and Ubiquitous Engineering, vol. 354, pp. 479–485 (2016)
3. Weber, R.H.: Internet of Things - new security and privacy challenges. Comput. Law Secur. Rev. **26**(1), 23–30 (2010)
4. Daubert, J., Wiesmaier, A., Kikiras, P.: A view on privacy & trust in IoT. In: 2015 IEEE International Conference on Communication Workshop (ICCW), London, pp. 2665–2670 (2015)
5. Belanche, D., Casaló, L.V., Flavián, C.: Integrating trust and personal values into the technology acceptance model: the case of e-government services adoption. Cuad. Econ. y Dir. la Empres. **15**(4), 192–204 (2012)
6. Perera, C., Zaslavsky, A., Christen, P., Georgakopoulos, D.: Context aware computing for the Internet of Things: a survey. IEEE Commun. Surv. Tutor. **16**(1), 414–454 (2014)
7. Del Giudice, M.: Discovering the Internet of Things (IoT) within the business process management. Bus. Process Manag. J. **22**(2), 263–270 (2016)
8. Farooq, M.U., Waseem, M., Khairi, A., Mazhar, S.: A critical analysis on the security concerns of Internet of Things (IoT). Int. J. Comput. Appl. **111**(7), 1–6 (2015)
9. Yan, Z., Zhang, P., Vasilako, A.: A survey on trust management for Internet of Things. J. Netw. Comput. Appl. **42**, 120–134 (2014)
10. McKnight, D., Carter, M., Thatcher, J., Clay, P.: Trust in a specific technology: an investigation of its components and measures. ACM Trans. Manag. Inf. Syst. **2**(2), 12 (2011)
11. Mayer, R., Davis, J., Schoorman, F.: An integrative model of organizational trust. Acad. Manag. Rev. **20**, 709–734 (1995)
12. Tam, S., Thatcherb, J.B., Craigc, K.: How and why trust matters in post-adoptive usage: the mediating roles of internal and external self-efficacy. J. Strateg. Inf. Syst. (2017, in press)
13. Gefen, D., Karahanna, E., Straub, D.: Trust and TAM in online shopping: an integrated model. MIS Q. **27**, 51–90 (2003)
14. Gao, L., Bai, X.: A unified perspective on the factors influencing consumer acceptance of Internet of Things technology. Asia Pac. J. Mark. Logist. **26**(2), 211–231 (2014)

15. Han, B., Wu, Y.A., Windsor, J.: User's adoption of free third-party security apps. J. Comput. Inf. Syst. **54**(3), 77–86 (2014)
16. Lee, M., Turban, E.: A trust model for consumer internet shopping. Int. J. Electron. Commer. **6**(1), 75–91 (2001)
17. Chang, S.-H., Chou, C.-H., Yang, J.-M.: The literature review of technology acceptance model: a study of the bibliometric distributions. In: Pacific Asia Conference on Information Systems, pp. 1634–1640 (2010)
18. Cho, Y.C., Sagynov, E.: Exploring factors that affect usefulness, **19**(1) (2015)
19. Renaud, K., van Biljon, J.: Predicting technology acceptance and adoption by the elderly. In: Proceedings of the 2008 Annual Research Conference of the South African Institute of Computer Scientists and Information Technologists on IT research in Developing Countries Riding the Wave of Technology, SAICSIT 2008, pp. 210–219 (2008)
20. Lai, I.K.W., Tong, V.W.L., Lai, D.C.F.: Trust factors influencing the adoption of internet-based interorganizational systems. Electron. Commer. Res. Appl. **10**(1), 85–93 (2011)
21. Bart, Y., Shankar, V., Sultan, F., Urban, G.L.: Are the drivers and role of online trust the same for all web sites and consumers? A large-scale exploratory empirical study. J. Mark. **69** (4), 133–152 (2005)
22. Abu, F., Jabar, J., Yunus, A.R.: Modified of UTAUT theory in adoption of technology for Malaysia Small Medium Enterprises (SMEs) in food industry. Aust. J. Basic Appl. Sci. **9**(4), 104–109 (2015)
23. Koien, G.M.: Reflections on trust in devices: an informal survey of human trust in an Internet-of-Things context. Wirel. Pers. Commun. **61**(3), 495–510 (2011)
24. Coughlan, T., Brown, M., Mortier, R., Houghton, R., Goulden, M., Lawson, G.: Exploring acceptance and consequences of the Internet of Things in the home. In: 2012 IEEE International Conference on Green Computing and Communications, pp. 148–155 (2012)
25. Al-Momani, A., Mahmoud, M., Ahmad, S.: Modeling the adoption of Internet of Things services: a conceptual framework. Int. J. Appl. Res. **2**(5), 361–367 (2016)

Cognitive Design

EEG Technology for UX Evaluation: A Multisensory Perspective

Marieke Van Camp(✉), Muriel De Boeck, Stijn Verwulgen, and Guido De Bruyne

Product Development, Faculty of Design Sciences, University of Antwerp, Ambtmanstr. 1, 2000 Antwerp, Belgium
{Marieke.VanCamp, Muriel.DeBoeck, Stijn.Verwulgen, Guido.DeBruyne}@uantwerpen.be

Abstract. Along with a growing interest in experience-driven design, interest in measuring user experience has progressively increased. This study explores the use of EEG for empirical UX evaluation. A first experimental test was conducted to measure and understand the effect of sensory stimuli on the user experience. A first experimental test was carried out with eight participants. A series of videos, eliciting positive and negative emotional responses, were presented to the participants. Subsequently, auditory stimuli were introduced and the effect on the user experience was evaluated using EEG measurements techniques and analysis software. After the tests the participants were questioned to verify whether the subjective results matched the objective measurements.

Keywords: User experience · Empirical evaluation · Neuroscience
Cognitive and neuroergonomics

1 Introduction

Nowadays, the competitive landscape is transforming from a materialistic to a post-materialistic value system, where experiences are valued over ownership [1]. Consumers are no longer satisfied with technical features, but are seeking for multi-sensory experiences wherein the physical design performs as a carrier [2, 3]. As a result of the rise of the Experience Economy [2, 4], many industries and academic fields are showing a growing interest in experience-driven design [5–7].

Research on user experience started from the premise that existing usability research is solely task oriented, and that more encompassing notions of quality - such as aesthetics, meaning, pleasure and emotion [2, 8] - need to be taken into account [9, 10]. Yet, the 'fuzzy' and dynamic concepts of user experience are generally not easily understood [11]. To better understand and adopt the theories of user experience, empirical evidence and guidelines on how design can positively affect the user experience, are needed [1, 12]. Up to now, user experience is mostly measured using questionnaires and other self-report instruments. However, these type of subjective techniques for the assessment of the perceived user experience may be insufficient due to the inherent limitations of the subjects, and of the assessment tools themselves. Next to the development and validation of more objective assessment tools, the use of

H. Ayaz and L. Mazur (Eds.): AHFE 2018, AISC 775, pp. 337–343, 2019.
https://doi.org/10.1007/978-3-319-94866-9_34

physiological sensor technologies for empirical UX evaluation - such as eye tracking, electroencephalography (EEG), and galvanic skin response (GSR) - can be explored.

This study focuses on the use of EEG for empirical UX evaluation. The possibility to investigate brain potentials for quantifying user experience is initiated through the availability of portable EEG devices. Advances in EEG technology now provide wireless headsets with dry electrodes that seek to improve usability and portability [13, 14]. Dry electrodes have an increased ease-of-use compared to wet electrodes, since the electrodes can be directly placed on the head without the need to apply a conductive gel. In addition, dry electrodes are becoming popular for both lab-based and consumer-level research, as they allow moving from traditional lab-based research into the real world [15]. Collection via conventional lab-based, wet-electrode systems is too invasive and cumbersome, unpleasant and can negatively affect user experience, whereas the ease and efficiency of data collection with these dry electrode interfaces is well suited for user experience measurement.

Within this study it is assumed that incorporating multiple sensory modalities enhances the overall user experience [16]. More specifically, the provision and regulation of activity-dependent feedback from multiple sensory modalities, may reduce negative stimuli and increase positive stimuli in interaction with and perception of design interventions and consequently may have a positive effect on the user experience. With such efforts, designers aim to increase sensory pleasantness and the perceived quality of future designs [17].

A first experimental test was conducted to empirically measure the effect of sensory modalities on user experience. This study hypotheses that the higher the number of sensory modalities, the higher the effect on the emotional state of the user. To control the multitude of influencing variables and make the concept of user experience manageable, the number of measured sensory modalities was restricted to two, namely visual and auditory parameters. In total eight user tests were performed. A series of short videos were presented to the participants, which are expected to elicit negative and positive emotional responses. The videos were presented to the participants both with and without sound. It is expected that the addition of auditory stimuli will induce higher emotional arousal levels, for both positive and negative emotional states.

2 Methods and Materials

2.1 Acquisition Hardware and Software

The user tests were carried out using the QUASAR DSI-24 dry electrode EEG headset. The DSI-24 consists of 21 electrodes positioned according to the 10–20 International System. The DSI-24 can be easily placed on the head and is ready to record EEG in under five minutes. The headset should be comfortable for wearing continuously over eight hours and is adjustable to fit a wide range of head sizes (54–62 cm in circumference) [18].

EEG signals were recorded using the provided data acquisition software; DSI-streamer [19].

Furthermore, QUASAR's QStates was used to quantitatively analyze the collected, large-scale EEG datasets and translate these into easily interpretable data output. QStates is a software package used for real-time or off-line analysis of physiologic data collected during cognitive-specific tasks [20]. QStates has a trainable engine capable of creating individual models for any cognitive state measurable using EEG [21]. QStates produces two graphical outputs for data interpretation. A first output reflects the likelihood that a given epoch - 2s segments of EEG data - belongs to the high state, which is derived from the probabilities estimated using a multivariate normal probability density function (MVNPDF). Within the MVNPDF plot, any value greater than 0.5 would be classified as belonging to the high state. A second linearized output indicates the exact position of the epoch between a value of 0 (low state) and 1 (high state) [22]. In this study the linearized output is used, since the complete timespan of the videos is compared and the comparison of specific time frames and events are not evaluated.

2.2 Experimental User Study

To empirically measure the effect of sensory modalities on user experience, an experimental study was carried out. This study aims to objectively evaluate the effect of introducing auditory stimuli on video experience.

Participants. The user tests were conducted with eight participants - four women and four men - aged between 20 to 40 years old. To obtain valid data, individuals who suffer from specific disorders which could have an influence on the results such as brain disorders, epilepsy, head traumas or other cognitive complications, were excluded from the study.

Setting. The user tests were conducted in the usability lab of the University of Antwerp, isolated from any sources of distraction or interference.

Procedure. Each participant completed the test individually. The test started with placing the EEG headset on the participant's head according to the 10–20 reference system [23].

In total, five different videos were presented to the participants, each with a duration between one and two minutes. The first three videos were presented in a randomized order: one inducing a high positive emotional response (i.e. enjoyment), one inducing a high negative emotional response (disgust) and one inducing a neutral or low emotional response (serenity). The EEG signals recorded watching these videos, were later used to train the models. Next, two more videos were presented, one inducing a positive emotional response and one inducing a negative emotional response. These last two videos are presented first with sound and second without sound.

Next to the recording of EEG signals, the participants were questioned to determine their subjective perception. The participants were asked to rate the videos on a scale from one to ten, with one inducing no emotional response and ten inducing a high - positive or negative - emotional response. In addition, their findings and opinions to the presented videos, were discussed. These subjective results can be used to explain and gain insight into the interpreted data, and ultimately to verify whether the perceived user experience matches the objective EEG measurements.

Analysis. QStates was used to quantitatively analyze the collected, large-scale EEG datasets. First, the training models for both the positive and the negative emotional state, were created. For each of the training models the high and low state conditions were allocated. Useful classification can be obtained with models whose reported Cross-Validation accuracy is more than 80%. 50% classification accuracy indicates that the two training files (high and low) were not distinguishable [22]. After successfully training the models, the EEG datasets can be classified, this for the positive state as well as the negative state, both with and without sound. After data classification a graphical interpretation is displayed, where the classified EEG data is positioned between the predefined high and low state conditions/margins plotted over a timeline. Also an overall percentage is given between 0% (low state) and 100% (high state), as shown in Fig. 1.

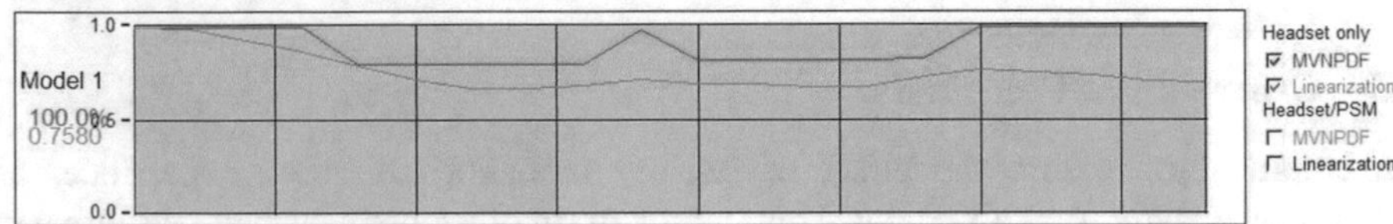

Fig. 1. Graphical output after data classification.

The independent variable in this study is the mode in which the video appears, namely with or without auditory stimuli, and the dependent variables are the calculated percentages. After data classification these variables are positioned on a scale (Fig. 2). The results are positioned between the low and the high state. It is expected that introducing auditory stimuli will induce higher emotional arousal levels, for both positive and negative emotional states. To confirm this hypothesis, each value of both α_p and α_n should be greater than 0.0.

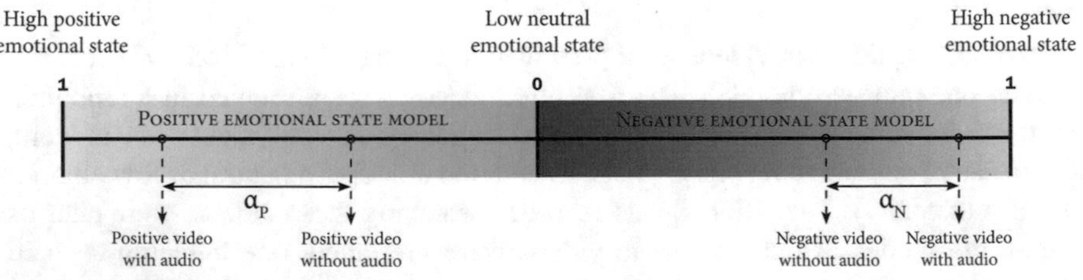

Fig. 2. Schematic representation of a positive and negative emotional state model.

3 Results and Discussion

3.1 Results

EEG Measurements. All of the eight participants managed to complete the test. None of the videos had to be interrupted or stopped. In all cases the EEG data was successfully recorded. Subsequently, the models were successfully trained, the lowest accuracy level measured is a value of 80.89%. This indicates that for each test the defined low and high states are sufficiently distinguishable and thus useful classification data can be obtained. Finally, the datasets for the positive state as well for the negative state, both with and without sound were successfully classified for each participant. α_p and α_n values are calculated for each participant. Six out of eight tests had a value for α_p higher than 0.0, and seven out of eight tests had a value for α_n higher than 0.0. What is notable is that for α_p generally higher values are measured than for α_n (Fig. 3). This can be explained by the effect of the auditory stimuli. For the positive video the humoristic effect completely disappeared without audio. For the negative video the auditory stimuli only had an intensifying effect on the emotional state. Ultimately, for 13 out of 16 tests the empirically measured results confirmed the hypothesis.

Positive emotional state					**Negative emotional state**				
with audio	*without audio*				*with audio*	*without audio*			
0.7687	*0.3052*	α_{p1}	*0.4635*	46.35 %	*0.9986*	*0.9944*	α_{n1}	*0.0042*	0.42 %
0.7580	*0.4638*	α_{p2}	*0.2942*	29.42 %	*0.1242*	*0.1197*	α_{n2}	*0.0045*	0.45 %
0.4304	*0.2810*	α_{p3}	*0.1494*	14.94 %	*0.8551*	*0.8182*	α_{n3}	*0.0369*	3.69 %
0.0616	*0.1228*	α_{p4}	*-0.0612*	-6.12 %	*0.7510*	*0.8755*	α_{n4}	*-0.1245*	-12.45 %
0.4065	*0.0824*	α_{p5}	*0.3241*	32.41 %	*0.7250*	*0.2647*	α_{n5}	*0.4603*	46.03 %
0.0674	*0.0766*	α_{p6}	*-0.0092*	-0.92 %	*0.4689*	*0.4508*	α_{n6}	*0.0181*	1.81 %
0.5788	*0.4211*	α_{p7}	*0.1577*	15.77 %	*0.8451*	*0.6561*	α_{n7}	*0.1890*	18.90 %
0.6725	*0.4490*	α_{p8}	*0.2235*	22.35 %	*0.3107*	*0.2038*	α_{n8}	*0.1069*	10.69 %

Fig. 3. Results EEG measurements and Qstates analysis.

Subjective Perception. The results from the interviews indicate that there are large individual differences among the participants. The positive videos were perceived as humoristic by most of the participants. Yet, one of the participants felt slightly uncomfortable watching the positive videos. Also, for participant number four the audio played in the positive video was experienced as irritating instead of enjoyable or humoristic. This can also explain why the value for α_{p4} is less than 0.0.

Therefore, without knowing the perceived user experience of the participant, it's difficult to understand the measured EEG signals. For example, high arousal both pleasant and unpleasant stimuli results in larger amplitudes and frequency's in the measured brainwaves, than less intensive pleasant and unpleasant stimuli. To know if the measured arousal is either induced by pleasant or unpleasant stimuli, subjective measurements need to be performed. Subjective measurements are required to gain insight and understand the empirically measured data. Therefore, to adequately evaluate the user experience, both objective and subjective measurements need to be combined.

4 Conclusion

To better understand the ‘fuzzy’ and dynamic concepts of user experience, empirical evidence and guidelines on how design can positively affect user experience are needed. Yet, objectively measuring user experience remains challenging. This study explores the use of EEG for empirical UX evaluation. Namely, advances in EEG technology have led to the development of headsets with an increased usability and portability, more suitable for user experience research.

This study hypotheses that increasing the number of sensory modalities, has an intensifying effect on user experience, this for both positive and negative emotional states. A first experimental test was conducted where a series of videos, inducing positive and negative emotional responses, were presented to eight participants. To measure the effect of introducing auditory stimuli on the user experience, the videos were played once with sound and once without sound.

For 13 out of 16 tests, the addition of auditory stimuli resulted in higher arousal levels and therefore confirmed the hypothesis. EEG technology thus demonstrated its usefulness within user experience research. Yet, to adequately evaluate user experience, both objective and subjective measurements need to be combined.

References

1. Desmet, P.M.A., Pohlmeyer, A.E.: Positive design: an introduction to design for subjective well-being. Int. J. Des. **7**(3), 5–19 (2013)
2. Dong, Y., Liu, W.: A research of multisensory user experience indicators in product usage scenarios under cognitive perspective. Int. J. Interact. Des. Manuf. **11**(4), 1–9 (2016)
3. Rousi, R.: Formidable bracelet, beautiful lantern. Studying multi-sensory user experience from a semiotic perspective. In: vom Brocke, J., Hekkala, R., Ram, S., Rossi, M. (eds.) Design Science at the Intersection of Physical and Virtual Design. DESRIST 2013. Lecture Notes in Computer Science, vol 7939, pp. 181–196, Springer, Heidelberg (2013)
4. Gardien, P., Djajadiningrat, T., Hummels, C., Brombacher, A.: Changing your hammer: the implications of paradigmatic innovation for design practice. Int. J. Des. **8**(2), 119–139 (2014)
5. Camargo, F.R., Henson, B.: Beyond usability: designing for consumers’ product experience using the Rasch model. J. Eng. Des. **26**(4–6), 121–139 (2015)
6. Park, J., Han, S.H., Kim, H.K., Oh, S., Moon, H.: Modeling user experience: a case study on a mobile device. Int. J. Ind. Ergon. **43**(2), 187–196 (2013)
7. Tonetto, L.M., Desmet, P.M.A.: Why we love or hate our cars: a qualitative approach to the development of a quantitative user experience survey. Appl. Ergon. **56**, 68–74 (2016)
8. Sun, L., Wu, J.: Total user experience design based on time dimension. Pack. Eng. **2**(35), 32–35 (2014)
9. Bargas-Avila, J.A., Hornbæk, K.: Old wine in new bottles or novel challenges: a critical analysis of empirical studies of user experience. In: CHI 2011 Proceedings of the SIGCHI Conference on Human Factors in Computing Systems, pp. 2689–2698 (2011)
10. Hassenzahl, M., Tractinsky, N.: User experience - a research agenda. Behav. Inf. Technol. **25**(2), 91–97 (2006)

11. Law, E.L.C., Roto, V., Hassenzahl, M., Vermeeren, A.P.O.S., Kort, J.: Understanding, scoping and defining user experience: a survey approach. In: CHI2009: Proceedings of the 27th Annual CHI Conference on Human Factors in Computing Systems, vol. 1–4, pp. 719–728 (2009)
12. Desmet, P.M.A., Vastenburg, M.H., Romero, N.: Mood measurement with Pick-A-Mood: review of current methods and design of a pictorial self-report scale. J. Des. Res. **14**(3), 241–279 (2016)
13. Chi, M., Jung, T., Cauwenberghs, G.: Dry-contact and noncontact biopotential electrodes: methodological review. IEEE Rev. Biomed. Eng. **3**, 106–119 (2010)
14. Ekandem, J.I., Davis, T.A., Alvarez, I., James, M.T., Gilbert, J.E.: Evaluating the ergonomics of BCI devices for research and experimentation. Ergonomics **55**(5), 592–598 (2012)
15. Mathewson, K.E., Harrison, T.J.L., Kizuk, S.A.D.: High and dry? Comparing active dry EEG electrodes to active and passive wet electrodes. Psychophysiology **54**(1), 74–82 (2017)
16. Cooper, N., Milella, F., Pinto, C., Cant, I., White, M., Meyer, G.: The effects of substitute multisensory feedback on task performance and the sense of presence in a virtual reality environment. PLoS ONE **13**(2), e0191846 (2018)
17. Özcan, E., Cupchik, G.C., Schifferstein, H.N.J.: Auditory and visual contributions to affective product quality. Int. J. Des. **11**(1), 35–50 (2017)
18. Wearable Sensing DSI-24 overview. http://www.wearablesensing.com/DSI24.php
19. Wearable Sensing DSI-Streamer overview. http://www.wearablesensing.com/DSIStreamer.php
20. McDonald, N.J., Soussou, W.: QUASAR's QStates cognitive gauge performance in the cognitive state assessment competition 2011. In: Proceedings of the Annual International Conference of the IEEE Engineering in Medicine and Biology Society, EMBS, pp. 6542–6546 (2011)
21. Wearable Sensing Dry Sensor Interface. http://www.physio-tech.co.jp/pdf/wearable-sensing/wearable-sensing_01.pdf
22. Wearable Sensing Qstates User Manual. http://wearablesensing.com/downloads/QStates%20User%20Manual_v1.6.2.pdf
23. Trans Cranial Technologies, 10/20 System Positioning. https://www.trans-cranial.com/local/manuals/10_20_pos_man_v1_0_pdf.pdf

Study on the Influence Mechanism Between Ordinal Factors and Cognitive Resource Consumption

Wenqing Xi[1], Lei Zhou[2(✉)], Xingyuan Ma[2], Yuqi Liu[2], and Huijuan Chen[1]

[1] Chinese Aeronautical Radio Electronics Research Institute (CARERI), Science and Technology on Avionics Integration Laboratory, 432 Gui Ping Road, Shanghai, China
[2] School of Mechanical Engineering, Southeast University, 2 Dong Nan Da Xue Road, Jiangning District, Nanjing, China
jjjjj0823@sina.com, zhoulei@seu.edu.cn

Abstract. *Destination* as the cognitive tasks have relatively fixed resource consumption, whether the ordinal factors influence the cognitive resource consumption or not is an important question, which means the threshold of dangerous performance may be anticipated accurately if the answer is found for true. *Methods* Two experiments are designed for cognitive analysis, which were based on the pilot-climbing status tasks. Each experiment procedure contains three steps of mode decision, warning and broadcasting. Orders were changed in two procedures. What's more, warning task had two sub-steps, and climbing step had three sub-steps, the inside group orders were not changed while the steps transformed. The reaction time of each task, step and sub-step was checked and the implicit mechanism of the ordinal effect was drawn. *Conclusions* the sequence of the task implementation has a significant impact on the reaction time, which is a reflection of cognitive resources consumption. Although the repetition of tasks may cause the promotion of familiarity, the reaction time of the subsequent task is still longer than the previous one. The ordinal effect should be taken into account in the total amount of cognitive resources consumption and anticipation.

Keywords: Ordinal factors · Cognitive resource · Task · Reaction time Pilot

1 Introduction

Cognition is the information processing activity of human beings to the external things, including a series of processes such as information identification, transformation, synthesis, encoding, conversion, reconstruction and storage by individuals. As the resource theorists argue that cognition means to be vigilant to the task, maintaining vigilance is cognitively demanding and is thus resource dependent [1–4]. Participants' ability to maintain focused attention on a vigilance task is a function of the amount of mental resources available [5–7]. Thus, as task time progresses mental resources are depleted

H. Ayaz and L. Mazur (Eds.): AHFE 2018, AISC 775, pp. 344–352, 2019.
https://doi.org/10.1007/978-3-319-94866-9_35

more quickly than they are replenished, which is behaviorally manifested as the increasing lapses of attention (performance impairment). More and more proves show that sustained attention failures are primarily due to sustained cognitive load not task monotony, and the result support a resource theory perspective with concern to errors being a result of limited mental resources and not simply mindlessness per se [8–10].

The process of most of the task completion is time-series. As the cognitive resource is limited, once beyond, it is difficult to ensure the reliability of the ordinary information processing. Well, how the limitation of cognitive resources in the human brain is arrived? Is it fixed by addition or changing based on the environment, time and other factors? On these issues, few references can be found.

According to our speculation, the relationship between the ordinal information processing and cognitive resources consumption may exist two kinds of possibilities as following: (1) the task implementation needs relatively fixed cognitive resource, human errors turn to be frequently when the consumption exceeds the limitation of the cognitive resource safety; (2) the cognitive resource consumption occurs persistently, with a decay effect because of the time progress. In this paper, the ordinal influence on the human brain cognitive processing was explored and two hypotheses above were checked by the cognitive experiments.

2 Experiment

2.1 Experiment Design

In order to avoid the environment variable influence on cognition, the same experiment materials were chosen for different tasks. The experiment background was about the pilot interaction activities with PFD interface during climbing task stage, to mimic the flying operation environment. The original PFD interface sample of the task was shown in Fig. 1. The examinee was acquainted with the name and function of the plane instruments firstly and then operates. There were three parts in these two experiments, which contained mode decision, warning and broadcasting step.

The examinee needed to be acquainted with the experiment well by practicing first, the steps of experiment A were as follows:

Step1: Read the introduction of the experiment and pressed any key to enter.
Step2: Read the airspeed display and chose the climbing way (mode decision). If the value was greater than or equal to the 100, pressed key ‘A’ that means constant airspeed climbing way. If the value was less than 100, pressed key ‘L’, which means constant rate climbing way.
Step3: Monitored and made sure the airspeed value and course value within ± 2, or pressed key ‘SPACE’ for warning and reported the display which exceeding the range verbally. Pressed key ‘ENTER’ to stop climbing when reaching the specified height.
Step4: Read the instrument value cyclically. Read the first picture and remembered the three values, then pressed key ‘SPACE’ to enter the empty screen and reported them verbally. After then, pressed key ‘SPACE’ again to enter the next picture. Recycled the procedure, and there were three pictures in total.

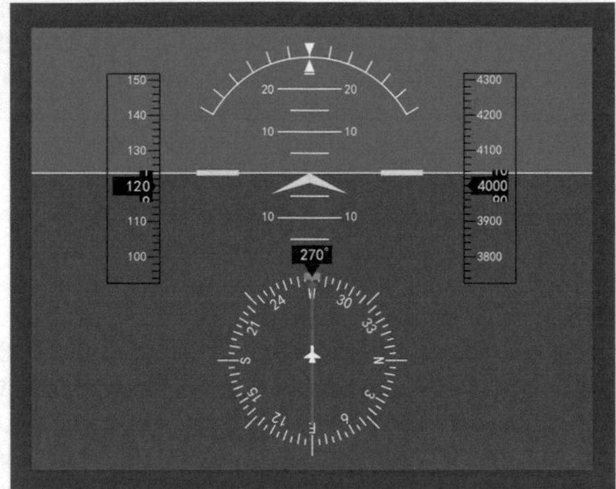

Fig. 1. PFD interface

Step5: Experiment was finish and ended automatically.

Experiment B adjusted the steps of experiment A, and the adjustments were as follows.

Step1: Read the introduction of the experiment and pressed any key to enter;
Step2: Read the instrument value cyclically. Read the first picture and remembered the three values, then pressed key 'SPACE' to enter the empty screen and reported them verbally. After then, pressed key 'SPACE' again to enter the next picture. Recycled the procedure, and there were three pictures in total. (Step4 in experiment A)
Step3: Read the instrument value cyclically. Read the first picture and remembered the three values, then pressed key 'SPACE' to enter the empty screen and reported them verbally. After then, pressed key 'SPACE' again to enter the next picture. Recycled the procedure, and there were three pictures in total. (Step 2 in Experiment A)
Step4: Monitored and made sure the airspeed value and course value within ±2, or pressed key 'SPACE' for warning and reported the display which exceeding the range verbally. Pressed key 'ENTER' to stop climbing when reaching the specified height. (Step 3 in Experiment A)
Step5: Experiment was finish and ended automatically.

2.2 Experiment Instrument and Program

The equipment, software and materials show in Tables 1 and 2. E-prime 2.0 was chosen for the experiment programming. The construct trail and a part of codes was shown in Fig. 2.

Table 1. Experiment instrument

Instrument	Application requirement
1 Computer (System WIN 7)	Available for E-prime 2.0
2 Display	Bigger than the Size Of PFD Interface
3 Keyboard	Basic use
4 Eye Tracker (Tobii X2-30)	Track the visual path of the subjects and collect the visual hot spots of the subjects

Table 2. Experiment software and material

Experiment software/material	Application requirement
1 E-prime 2.0	Loaded with video stimulation; data collection and integration
2 Video stimulation	(1) Duration: 17 s; frame rate: 30p; format: avi (2) Duration: 2 s; frame rate: 30p; format: avi
3 Picture stimulation	Format: bmp; size: 927 × 768 pixel; Resolution: 300 dpi

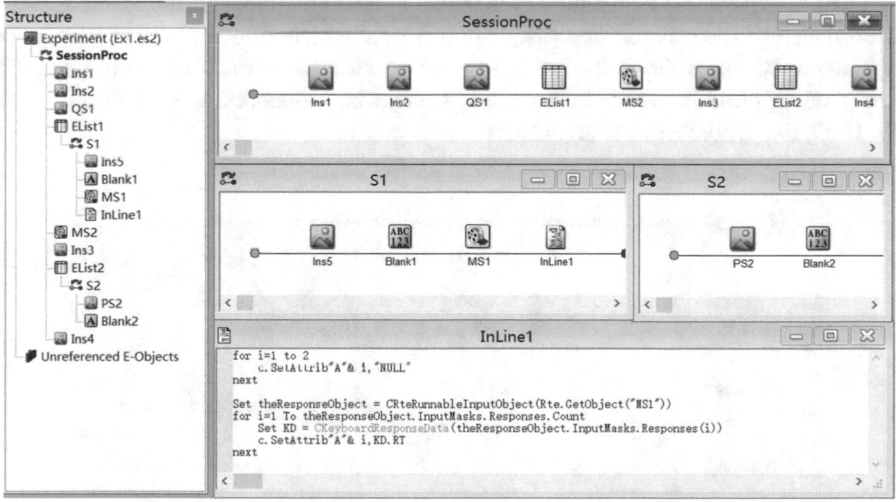

Fig. 2. E-prime 2.0 coding of experiment A

2.3 Examinee and Variable Control

20 undergraduate students and graduate students were invited between the ages of 23 and 25 as the examinees, 10 was male and 10 was female. All the examinees had normal vision or corrected vision without color blindness and color weakness, and

normal hearing. Experiments run in two days, the samples of each examinee in two days were random.

Experiment variables controls were as follows: white and constant brightness fluorescent lamp; constant color and brightness computer display; no noise disturbing. Examinees sat 500 mm–550 mm before the screen.

3 Experiment Result

The original data selected from the experiment were examinee's number, reaction time of the examinee's determination of climbing way, the operate time when examinee monitor different instruments, the operate reaction time during report cyclically.

Two stimulate time points of video stimulation were 9500 ms and 10500 ms, and the stimulations reaction time were A1' = A1-9500 and A2' = A2-10500, respectively. The time unit was millisecond. Processed the experiments' data, calculated the average react time and analyzed the data result.

1. Reaction time in mode decision step in experiment A and experiment B

The react time of the climbing way determination in experiment A showed in Fig. 3 with blue line, which belong to the regular simple determination task, and the reaction time fluctuated around 800 ms. The reaction time in experiment B (green) was longer than that in experiment A, which was 907 ms. In the experiment A, the climbing mode determination task was the first one, while it was switched to the second one in the experiment B. Since the task of report the value circularly was done before, the efficiency of the climbing mode determination might be influenced. $t = -7.856$, $df = 19$, $P < 0.05$, the difference was significant.

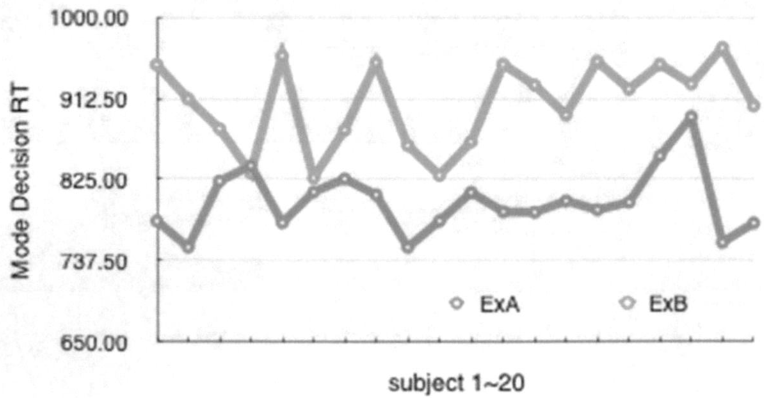

Fig. 3. Reaction time in mode decision step in experiment A and experiment B

2. Reaction time in warning steps in experiment A and experiment B

Figure 4 is drawn to show the reaction time tendency of the warning task when the airspeed indicator and acquired indicator appear unusual in experiment A and experiment B. The examinees had high cognitive resources consumption since they needed to complete two or more tasks at the same time. It was obvious that the react time of the second stimulation was longer than that of the first one in experiment A. The average react time of the second stimulation was 338.8 ms longer than that of the first one, which mean that the first warning task influence then second one. The stimulation time points were set at 9500 ms and 10500 ms after the video was played, and there was a 1500 ms difference between two time points. For subject 18, the react time of the first warning was much longer than 1500 ms, which mean that the examinee did not react to the first stimulation when the second one appeared. In addition, the average react time in experiment B was longer than that in experiment A. So the time sequence of the task highly influenced the efficiency during the tasks and between the tasks, but had little effect on the cross period.

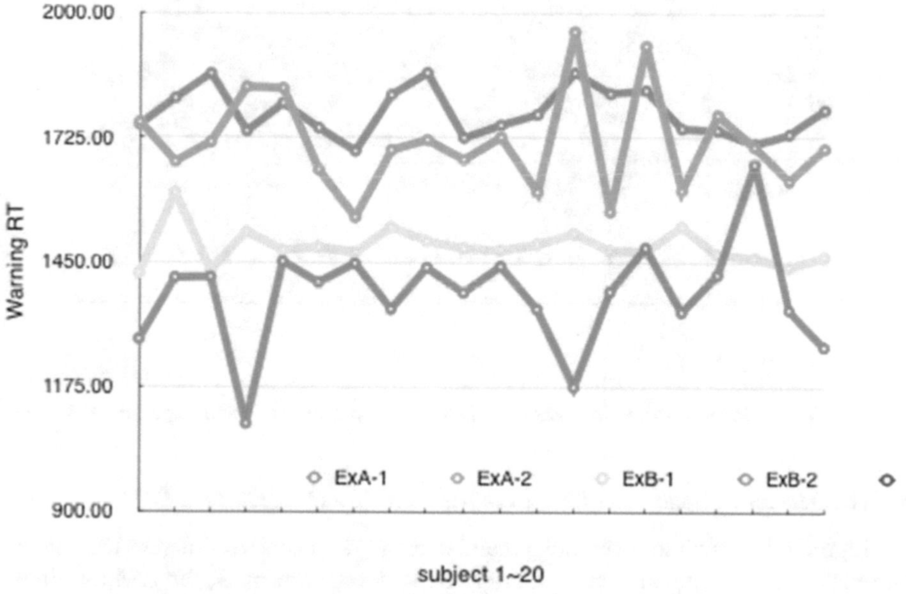

Fig. 4. Reaction time in warning step in experiment A and experiment B

It was noted that the adjacent tasks might be overlap in the real operation environment, which set higher requirement for the operators. What's more, the task flow should be improved to avoid this situation when task designing.

3. Reaction time in broadcasting steps in experiment A and experiment B

Figure 5 is drawn to show the react time of circularly report for three indicators in experiment A. It was obvious that the react time of report 3 was longer than that of report 2 which was longer than that of report 1 on the basis of the data of average react time: 2185.4 ms < 2290.7 ms < 2512.7 ms. The react time in experiment B: 2007.7 ms < 2154 ms < 2305.2.7 ms. The task belongs to the regular simple circular task, and the phenomenon in react efficiency indicated that the cognitive efficiency was not stable when operator completed the same task for a certain period. The familiarity of the task did not increase, while the cognitive load increased shown as the longer react time. The sequence of the circular task in the whole task influenced the cognitive consumption. But the data indicated that the sequence in the whole task had little effect on the circular task, F (1,18) = 0.481, P > 0.05, while the sequence inside the warning task and the cross influence were significant.

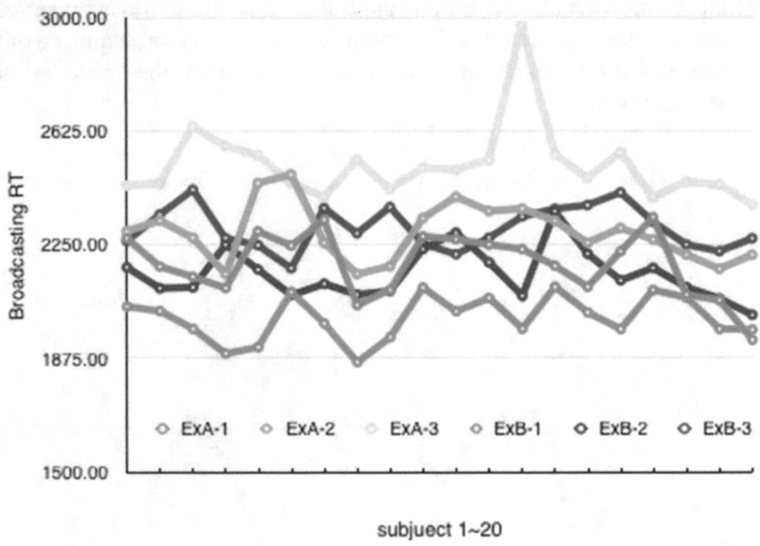

Fig. 5. Reaction time in broadcasting step in experiment A and experiment B

4. The general average react time in experiment A and experiment B

Figure 6 is drawn to show the general average react time of different task order in experiment A and experiment B. Compared with experiment A, the reaction time of climbing mode determination was 105.98 ms longer, and the react time of alarm 1 and alarm 2 were extended by 107.18 ms and 64.28 ms, respectively,while the react time of 3 circular reported shorter by 59.23 ms, 45.47 ms and 67.17 ms, respectively. It could be easily found that the sequence of the task influenced the single task operation clearly.

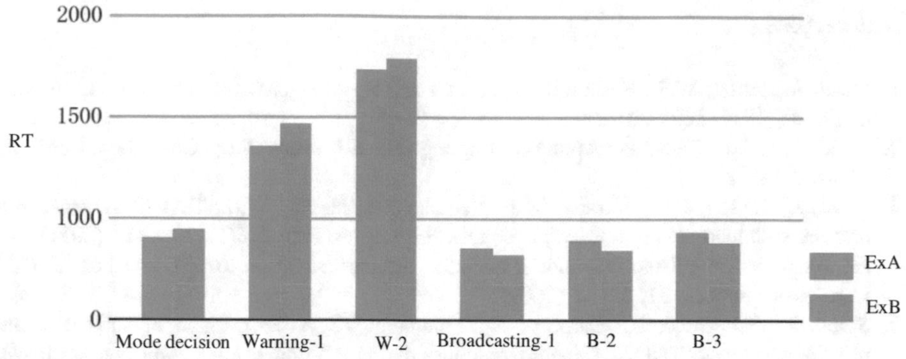

Fig. 6. The general average reaction time in experiment A and experiment B

4 Discussion

For a long time, the experimental order was considered as one of the external factors that affect the accuracy of the data, which can be eliminated by random, Latin square and other methods. However, this approach is equal to the first hypothesis that in the process of completing the task, the cognitive resource consumption is relatively fixed. Therefore, the so-called resource threshold should be a suddenly reached, limited value, which is determined by the sum of cognitive consumption of multiple tasks. In this paper, it can be found that the consumption of cognitive resources is mainly determined by the difficulty and familiarity of the task itself, but there is a decay effect. The performance of the same task may be different due to the different execution order.

In the experiments, the ordinal effect on the consumption of cognitive resources was studied by designing three consecutive tasks. Two tasks within the experiments also had repetitions, so as to learn whether the difference of cognitive resource consumption is caused by sequence within the task or not. The results show that the sequence of task implementation has a significant impact on the reaction time, which is a reflection of cognitive resources. Although the multiple executions of tasks may cause the promotion of familiarity, the final reaction time is still behind the previous executed one's, which is a very interesting discover. It verifies the second hypothesis presented at the beginning that in the process of cognitive load analysis and task flow design, the ordinal effect of the task implementation should be taken into account for the amendment of the cognitive resource consumption and anticipation, so as to reasonably analyze and predict human psychology status and behavior modes.

Acknowledgments. This paper is supported by Science and Technology on Avionics Integration Laboratory and Aeronautical Science Fund (No. 20165569019).

References

1. Head, J., Helton, W.S.: Natural scene stimuli and lapses of sustained attention. Conscious. Cogn. **21**, 1617–1625 (2012)
2. Head, J., Helton, W.S.: Perceptual decoupling or motor decoupling? Conscious. Cogn. **21**, 913–919 (2013)
3. Head, J., Russell, P.N., Dorahy, M.J., Neumann, E., Helton, W.S.: Text-speak processing and the sustained attention to response task. Exp. Brain Res. **216**(1), 103–111 (2011)
4. Helton, W.S.: Impulsive responding and the sustained attention to response task. J. Clin. Exp. Neuropsychol. **31**, 39–47 (2009)
5. Shaw, T., Satterfield, R., Ramirez, R., Finomore, V.: A comparison of subjective and physiological workload assessment techniques during a 3-dimensional audio vigilance task. In: Proceedings of the Human Factors and Ergonomics Society Annual Meeting, vol. 56, no. 1, pp. 1451–1455 (2012)
6. Helton, W.S., Warm, J.S.: Signal salience and the mindlessness theory of vigilance. Acta Physiol. **129**, 18–25 (2008)
7. Jonker, T.R., Seli, P., Cheyne, J.A., Smilek, D.: Performance reactivity in a continuous-performance task: implications for understanding post-error behavior. Conscious. Cogn. **22** (4), 1468–1476 (2013)
8. James, H., William, S.H.: Sustained attention failures are primarily due to sustained cognitive load not task monotony. Acta Physiol. **153**, 87–94 (2014)
9. Marissa, K., Daniel, E.F., Maria, M.L., Amishi, P.J.: The influence of time on task on mind wandering and visual working memory. Cognition **169**, 84–90 (2017)
10. Alessandra, V., Leonor, J.R.L., Rolando, B., Laura, Z., Laura, D., Manuela, B.: What is difficult for you can be easy for me. Effects of increasing individual task demand on prefrontal lateralization: a tDCS study. Neuropsychologia **109**(31), 283–294 (2018)

Wellness in Cognitive Workload - A Conceptual Framework

Eduarda Pereira[1(✉)], Susana Costa[2], Nelson Costa[2], and Pedro Arezes[2]

[1] DPS, School of Engineering, University of Minho, Guimarães, Portugal
pereira.eduarda@gmail.com
[2] ALGORITMI Centre, School of Engineering, University of Minho, Guimarães, Portugal
{susana.costa,ncosta,parezes}@dps.uminho.pt

Abstract. Driving is a highly complex task, comprising over 1600 separate tasks over five behavioural levels. 94% of road accidents occur by human faults. Drivers simultaneously control the vehicle, adjust speed and trajectory, deal with hazards, evaluate progress towards their goal, and make strategic decisions such as navigation. Novel technologies such as active cruise control and active steering are intended as comfort systems for the driver because they are designed to relieve the driver of workload. One could go further and imagine whether cars could be created so as to get to know their user and use that knowledge to recognise the user and be more safe. A cooperative driving leads to automotive systems with impacts on mental workload. Of particular concern in this study are the twofold areas. On the one hand the personalization of the car to the user and on the other, its customization. Using the principles of cognitive ergonomics, the main purpose is to allow more features to people, whilst providing comfort and trust in the human machine interface which (1) diagnoses the driver's state and uses additional monitoring devices (2) to provide feedback about driver´s behaviour as well as awareness and wellness.

Keywords: Wellness · Cognitive workload · ADAS · Driving
Human factors · HMI · Autonomous vehicles

1 Introduction

Road accidents have been numerous, accounting for 1.25 million deaths worldwide each year, and around 50 million injured people, with causes that are mainly attributed to victims themselves [1]. Indeed, the Executive Summary of the Federal policy entitled "Accelerating the Next Revolution in Roadway Safety", stresses out that 35,092 people died on U.S. roadways in 2015, and points that 94% of crashes can be tied to a human choice or error. The reasons underlying may be that driving is a highly complex task, comprehending over 1600 different tasks over five behavioural levels ([2] citing [3]). Unless preventive measures are taken, road death is likely to become the third leading cause of death in 2020 [4].

H. Ayaz and L. Mazur (Eds.): AHFE 2018, AISC 775, pp. 353–364, 2019.
https://doi.org/10.1007/978-3-319-94866-9_36

Driving support systems are increasingly involved in autonomous driving systems. It is not certain when cars with level 5 autonomous driving will be available to move legally on the roads. For now, the available autonomy levels require and/or allow driver intervention through the human-machine-interfaces (HMI). This dynamic environment with permanent changes, raises new issues from the human factors perspective, matters related to the limit and amount of mental effort (mental workload). The measurement of mental workload involves several processes, including cognitive, neurophysiologic and perceptual processes.

There are many reasons for assessing mental workload, but in this context it is imperative to do so mainly due to changes in driving conditions over the last decades: (1) the nature of the driving task has undergone changes; (2) accidents have been numerous; (3) human errors in road traffic related to mental workload (in the form of inaccurate perceptions, insufficient attention and inefficient information processing) which account for the majority of causes of most road accidents. So, the driving task and the driving environment place the driver under a wide array of physical and cognitive demands [5] in which brain continuously generates predictions of its own sensory input and uses the actual online sensory signal mainly to check and correct the predictions, i.e., as a statistical prediction through sensorimotor interaction [6]. This suggests that cognition and behaviour can be fundamentally understood in terms of the single principle of prediction error minimization [7]. Fridman used Hidden Markov Model to show that macro-glances (blink frequency and eye fixation measures) can be part of a multi-sensor system for predicting the state of the driver and the state of the driving environment, but on its own is only sufficient to predict a limited (70.9% to 88.3% accuracy) but important set of variables related to driver behaviour and state [5].

Drivers simultaneously control the vehicle, adjust speed and trajectory, deal with hazards, evaluate progress towards their goal, and make strategic decisions such as navigation.

In spite of all the advances that have occurred in the scope of the application of workload in HMI, it turns out that the term has become too subjective to describe the state of the driver. Thus, the concept of driver wellness arises, closely related to the ability to achieve optimal multiple performance goal. Integrating ambient intelligent concepts with advances in the vehicle systems presents an opportunity to develop a vehicle that is more than ‘simply’ a transportation mode, but to transform the car into a ‘wellness’ platform that supports optimal driver performance and improves overall traffic safety [8].

The focus of this article is to approach the wellness conceptual framework through cognitive workload by proposing a strategy for the development of an HMI, which adapts to the driver through driver recognition and monitoring.

2 Methods

This work was carried through scoping review of the workload concept integrating the incipient concept of wellness in the context of autonomous driving. According to Colquhoun [9], this methodology is “a form of knowledge synthesis that addresses an exploratory research question aimed at mapping key concepts, types of evidence, and gaps in research related to a defined area or field by systematically searching, selecting, and synthesizing existing knowledge". It is a very popular methodology for systematizing scientific knowledge, with a steep growing expression among the scientific community the last 4 to 5 years.

3 Results and Discussion

According to Hart [10] the workload of an individual can be defined as his/her cost for accomplishing a certain mission requirement, or, the portion of human resources he/she expends when performing a specific task.

Young et al. [11] stated that "Mental workload is a peculiar concept that has intuitive appeal" remaining difficult to define. Many definitions have been made but it is clear that a consensual definition does not exist. Processing human information is, generally, considered limited [12–14].

Nevertheless mental workload designates the amount of resources committed for the processing of a certain task [15], it is connected to the task characteristics, the situation and the person.

As it is very important to measure mental workload, in order to understand the problematic regarding the interaction between drivers and cars, it is important that the measurement methods are valid [12]. O'Donnell and Eggemeier [16] indicate that there are three main approaches for the measurement of mental workload:

1. subjective measures;
2. performance-based measures;
3. physiological measures.

As noted by Hancock and Desmond [17], while the terms stress and workload arise out of somewhat different traditions, there is a great deal of conceptual overlap in describing demands on the individual arising from both internal and external factors.

Indeed, the concept of workload is not only closely related to stress, but also to fatigue, burnout and inattention [8]. Now, these are concepts, which, in turn, are related to other major ergonomic indicators such as accidents.

The automotive industry is focused in developing advanced autonomous driving systems (ADAS) and its supporting technologies. A main condition for achieving this goal is to ensure drivers' safety and comfort during the ride. The driving task is often described as complex and dynamic and can be considered as the single most risky task that the individual has to perform on a daily basis [18]. According to SAE's International definitions of levels of automation, in the first four levels (level 0 to level 3), the human driver is responsible for taking back the control of the car, by request of the automated system, and in level 4, the automated system can only operate in certain environments and under certain conditions [19]. Therefore, the human driver attention to the driving task and environment must be assured and controlled, or managed, for all these first five levels of vehicle automation.

It is generally assumed that the deployment of automated driving will encompass six effects: eliminate or decrease congestion through management of the traffic flow, reduce traffic accidents by avoiding human errors, diminish environmental load through optimization of the traffic flow and reduction of the consumption of fuel, enhance the pleasure of driving by reducing the pressure associated with driving in stressful scenarios, strengthen international competitiveness and, adapt to demographic changes by supporting unconfident drivers and enhancing the mobility of elderly people [20, 21].

Even though autonomous cars have the potential to contribute with solutions to a number of transportation challenges, including improving road safety, optimizing traffic flow, allowing for transportation that is more efficient, new mobility models, and providing additional comfort for drivers and passengers [22, 23], they will only succeed if the prospective drivers trust the technology. Well, this trust will depend on the way the automation is designed, for providing enough information to create driver situation awareness and acceptance. In medium level automation vehicles, it can enhance safety by reducing workload or, if poorly designed, aggravate it [24].

The design attributes of new generation driving assistance systems should be influenced by human factors in driving. According to a recent news article, development of 'human-like' self-driving technologies is attracting investor capital [23]. Also, because the design of the driving assistance, if guided by human factors, is likely to enhance driver acceptance, safety benefits will be achieved.

There are already safety warning systems such as Lane Change Decision Aid Systems (LCDAS), Stop & Go, and Forward Vehicle Collision Warning Systems (FVCWS) which monitor the driving situation and provide the traffic situation information for drivers, whereby they may warn the driver proactively about a possible hazardous situation on the basis of the vehicle's current position, orientation, and speed, and the road situation [25]. According to Merat and Lee [26], these ADAS are projected to help users during the driving task. Many different parameters are being constantly verified and whenever thresholds are surpassed, the system informs the driver.

ADAS systems, mainly present visual, auditory and haptic warnings [27], nevertheless these systems, which are designed to help, if badly designed, may distract the driver and even increase workload, enhancing the possibility of driving distraction and for that making driving less safe [28].

Three functional abilities are required from the driver to perform the driving task: cognitive, visual and psychomotor [29].

Even though the driving task is considered one of the most risked that an individual has to perform daily, it is interesting to verify that daily use and associated risks seem to contradict each other. As driving is a complex task, performed by a large part of the population, which indicates global generalization, it should, therefore, be a simple and risk-free task. Yet this idea fades whenever the annual number of car accidents is verified [30].

Simultaneously, more entertaining applications are being made available for the driver to use inside the car. All things considered, one can assume that these infotainment applications are competing with the attention and awareness of the human driver to the driving task. Ryu et al. [31] refer that driver information systems require drivers' visual attention, for example, for selecting the desired functions, which in turn can increase the periods in which the driver eye-gaze is set off the road, consequently increasing the probability of accident occurrence.

SAE International [19] suggests a broader understanding of driving and for that purpose presents the concept of Dynamic Driving Task (DDT) as all of the real-time operational and tactical functions required to operate a vehicle in on-road traffic, excluding the strategic functions such as trip scheduling and selection of destinations and waypoints, and including without limitation: Lateral vehicle motion control via steering (operational); Longitudinal vehicle motion control via acceleration and deceleration (operational); Monitoring the driving environment via object and event detection, recognition, classification, and response preparation (operational and tactical); Object and event response execution (operational and tactical); Maneuver planning (tactical); and Enhancing conspicuity via lighting, signaling and gesturing, etc. (tactical). Taking level 3 [19] as an example, while the ADAS is engaged, the user must be receptive to a request to intervene and respond by performing DDT fallback in a timely manner. And/or, if DDT performance-relevant system failures in vehicle systems and, upon occurrence, performs DDT fallback in a timely manner. And/or, must determine whether and how to achieve a minimal risk condition. And/or becomes the driver upon requesting disengagement of the ADAS. All the components of the vehicle which provide feedback to the driver are sources of mental and manual workload.

In short, all external and internal stimuli (visual, auditory, haptic and cognitive) that add to the driving function affects the driver mental workload. The factors that compound for cognitive and mental workload are: traffic, poor visibility and lighting, bad weather, surface poor conditions, external noise, in-vehicle noise, poorly or partially functioning systems (e.g.,, windshield wiper), poorly responding vehicle dynamics (e.g., braking response), lost directions, sources of anxiety (e.g., urgency of trip) and the driver's health and mood [32].

Workload measures can be classified by nature as physiological, subjective or performance. Physiological measures include cardiac measures (heart rate, heart rate variability and blood pressure), respiratory measures (respiratory rate, volume and concentration of carbon-dioxide in air flow), eye activity measures (eye blink rate and interval of closure, horizontal eye activity, pupil diameter, eye fixations), speech measures (pitch, rate, loudness) and brain activity (measured through electroencephalogram and electrooculogram) [33].

Studies show that an increase of the workload to the driver will cause him to focus on the driving itself, relegating to a less demanding occasion the appreciation of his surroundings environment [34–38].

Despite the major social and economic benefits that ADAS technology enables, it is not without risk.

According to Son et al. [39] the increase of the use of in-vehicle technologies can increase driver distraction, and a significant proportion of the distraction may arise not from its manual manipulation but from cognitive consequences in their use.

At the same time, ADAS can fail from either random (e.g., communications interference and unexpected component failures) or systematic faults (for instance, software failures or overall failures in the design of the system). Adding that to the fact that the very own systems designed to alleviate driving strain might, themselves, increase stress by demanding the driver to execute other tasks (aggravated by the lack of user knowledge), the inefficacy in the use of such systems can, in fact, constitute a danger itself [40].

Although intending on minimizing the effects of in-vehicle systems overload on the driver, input sensory stimuli and sources of additional mental workload inevitably emerge during normal driving and may be compounded by the ADAS. This adding workload brought by the ADAS should not exceed the driver's visual, auditory, and haptic and cognitive capacities, but often arise from complex designs with poorly designed feedback mechanisms, which end up diminish the safety value of ADAS and the acceptability of the system [32].

As if it were not enough, evidence shows that excessive reliance on automated can damage the driving task performance. ADAS may encourage a driver to continue driving in situations he could have stopped otherwise, as is the case of collision avoidance systems in fog [41]. Claims are that autonomous performance will only relegate abnormal tasks to manual handling, which means that manual driving will be dealing with only sporadic situations for which the drivers will be highly estranged with, potentially showing reactions that will fall short of optimal [40].

The concept of vigilance refers to the degree of attention the driver is allocating to the primary task of driving.

Fatigue may be manifested through fixating on objects, frequent blinking, head nodding and unnatural facial features. The measurement of vigilance or statement of appropriate levels of vigilance is a challenge, because it varies among drivers.

Nevertheless, many development have been achieved through eye tracking and computer vision technology, that have enabled the measurement of some facial and eye features accurately and consequent correlation with vigilance.

The Japanese National Institute of Advanced Industrial Science and Technology developed a driver monitoring system integrated with driver assistance, the "ITS View-Aid System," in which driver warnings and displays are optimized according to measures such as level of alertness, gaze direction, age, road condition and inter-vehicle distance [42].

On high autonomy levels, for the majority of the time, drivers will be performing a monitoring task rather than a driving task, which has proven to induce high levels of workload. This shift of the attention gives origin to complacency, which negatively influences alertness and reaction time [40].

The visual space capacity of the driver allows him to deviate his gaze from the road ahead for certain periods, which may configure a visual diversion or a distraction, being that the latter emerges when this capacity is superseded by a stimulus (prolonged fixation on an object). Indeed, cognitive overload may arise from something as simple as driving and wondering about things other than driving and the driver experiences manual overload when he engages in secondary manipulation tasks during driving such as adjusting vehicle controls [32].

The auditory capacity of drivers with normal hearing exceeds their visual capacity (even though the first is the most difficult capacity to quantify), as they can receive more simultaneous distinct auditory inputs before achieving a level of (mental or cognitive) workload that may already pose a risk [32]. The HMI system, can therefore resort to alternative channels to communicate with the driver in order to minimize the workload he/she is under, without having to fail communication, but rather opting for a more suitable modality.

The design communicates necessary information to the user, regardless of ambient conditions or the user's sensory abilities. According to Dix et al. [43], redundancy of information is important, which means that information needs to be "represented in different forms or modes (e.g. graphic, verbal, text, touch) [44]. "Presentation should support the range of devices and techniques used to access information by people with different sensory abilities" and "designs should provide for multiple modes of output" [45].

One of the main concerns of automated vehicles are the trust issues. Many researchers have focused on this matter and it seems that one major key to the success of the autonomous driving paradigm is the resolution of trust issues among prospective users, a concept strictly related to the coherence of the system and also annoyance. Indeed, the design has to work in the expected way [45]. It seems reasonable to assume that, if the vehicle starts a frenzy of sounds, lights, colours and vibrations without nothing in its surroundings justifies this type of behaviour, it will leave the user with doubts about the efficiency of the system in detecting possible threats. Even without going to extremes, if one wishes to decrease the volume of the radio to request an indication on the street and, although pressing the correct interface, the sound instead of decreasing increases, it will cause stress, embarrassment to the user and the level of confidence in the whole system will suffer.

Similarly, the whole judgment about vehicle system will suffer if a little beep constantly alerts the user to things he/she considers insignificant, causing the user to fail to recognize value in the valences that the system supposedly offers. This lack of balance, which is often not anticipated by the designers/manufacturers, is responsible for the deactivation of these same features, which are classified as annoying and, as such, distracting rather than useful or beneficial to the driver. Likewise, one's confidence in the system will also be affected (in a more serious way) by the lack of a warning that should have been made by the system and that it failed to display. Clearly, if an event occurs without the driver having been advised in advance, confidence will be shaken because the design did not work as expected. In the advent of the new autonomous vehicles, some situations occurred (some fatal and wide-spread in the media), which put in question the future of the autonomous driving paradigm. Needless to say, since then, systems have been upgraded and technology has been refined so that, in fact, the design works as expected.

As Dix [43] posed it, "the design allows for a range of ability and preference, through choice of methods of use and adaptivity to the user's pace, precision and custom" [44]. Increasingly, car manufacturers are aware of the variability among their users, in fact, not everyone likes to listen to the same radio station, the same type of music, not everyone likes to answer mobile calls while driving (even with recourse to legal channels such as bluetooth) and manufacturers have already realized that. From there, the concepts of personalization and customization in the automotive context were born. Customization of the system's HMI is the process whereby the user may, for instance, predefine the HMI, so as to not display infotainment (warnings, information) during highly demanding, complex driving tasks. Another thing, however, has to do with when a regular user of a vehicle makes his/hers daily choices by giving priority to a certain radio station, music genre, temperature adjustment and information in a given situation, which the vehicle will apprehend. Here, machine learning plays a key role because it allows the vehicle to learn the user's coping mechanisms to deal with different levels of workload (from underload to overload). This modulation of the vehicle's response according to the user behaviour, is called personalization.

Given the diversity of the population of drivers, an effective cognitive workload management system is a flexible system, which adapts to the characteristics of its user. Well, a flexible system is a system that is able to read its user (driver) and adapt, by enabling the presentation of adequate features, i.e., the HMI will reflect the user's needs/expectations/likes. Older drivers, for instance, are less capable of engaging in secondary tasks. Indeed, older adults' lower levels of energetic arousal cause them to suffer vigilance loss, an aspect of monitoring found in highly reliable automated systems, making them more vulnerable than younger adults to automation-related reductions in monitoring efficiency [46].

Even though older drivers present decremented attentional capacity, their experience and age usually compensate, being the exception situations with high momentary mental workload, in which they fail with severe consequences. The majority of older drivers' accidents are caused by inattentive driving (e.g., failing to look forward, judgmental error and delayed discovery) [39]. This example mirrors the need for a flexible, adaptive HMI, which would present a different interface when being driven by a senior or a young person.

Costa et al. [4] stressed the need for a flexible cooperative vehicle-driver communication strategy, in response to the echoes that have been felt in the scientific community regarding the need for efficient responses, taking into account not only human factors related to the driving of semi-autonomous vehicles, but also the complexity of the AD systems themselves. A nexus diagram was developed that, in a comprehensive way, provides a variety of conditions in this driving context, rendering the necessary flexibility of the warning strategy (able to be customized in some specific traits) which can later be resorted to by programmers for expeditiously implementing this much needed strategy in the real context of semi-autonomous driving. Like future cooperative systems, the wellness feature shall, too, be cooperative, multimodal and adaptive. There is not a fixed formula for a wellness-provider HMI, the adequate HMI is one that adapts to a driver in a given situation.

Considering the aforementioned and the foreseen mobility modalities available: car sharing (growing commercial model for car manufacturers with fleets available for renting) and personally owned vehicles, it is proposed that the strategies to adopt will be customization (car sharing – by which the user will be able to configure the vehicle according to his/her needs/likes/dislikes/capabilities which will, then, promote optimized workload levels and, therefore enhancing wellness) but for personally owned vehicles, the strategy will be a combination between customization (mainly in the beginning of ownership), but it is expected that, with time, the vehicle will learn from the interaction with the user and will adapt its actions accordingly through personalization, thus rendering enhanced workload and therefore, wellness, through an optimized management of the inputs/outputs from and to the driver [8] and personalization.

The result will be the readjustment of the presentation of the HMI, emulating the driver successful workload-coping behaviour (in modulating his/hers workload to optimal levels in homologous situations) in a growing independency of the driver's intervention (gradually refraining from the need for the driver's action), by, for example, automatically lowering the radio volume, switching the music genre, adjusting temperature). Driving in highways promotes fatigue, given the monotony of the task, and therefore, the HMI workload manager shall reconfigure the vehicle so as to bring the driver to an optimal state of workload from the current underload (fatigue) situation, by alerting him/her, for instance, automatically lowering the temperature of the cockpit, increasing the volume of the radio and, even, switching to a different, more pungent music genre. In contrast, if the amount of information that is being conveyed to that specific driver is causing him/her an elevated level of workload (overload), the HMI response would be to adequate its presentation in order to calm the driver (by, for instance, lowering the music level, and cutting all infotainment), which would, then, bring the driver to an optimal state of workload – optimal wellness.

From an efficiency point of view, it is predictable that the personalization will render better results in wellness management, given that the HMI will be constantly receiving data from the workload measures from the driver and generating outputs which will, in turn, cause new workload levels in the driver, in the search for the optimal, wellness workload levels, all the while integrating this bulk of data and assessing whether these adjustments were successful. This feedback system is the basis for the modulation of the wellness system itself, and the construction of the driver behaviour emulation by the HMI. So, it is easy to understand why customizing will not be such an agile process: on the one hand, the driver needs to be sensitized to his/her safety in order to, willingly, make some concessions by virtue of his/her own safety but, on the other hand and more importantly, the driver also needs to know which workload-coping mechanisms really work for him/her, which is not an easy to know in depth, nor to be aware of in all kinds of situations. The personalization allows for the adjustment to different situations even when it concerns to the same driver.

4 Conclusions

Driving is a contemporarily almost unavoidable, necessary task. It is also one of the most dangerous activities a person can engage. As such, it is paramount that, when performing it, the person is in an optimal state of well-being, which has been referred to as wellness. This concept of wellness is intimately linked to cognitive workload of the driver, which, in turn, is closely related to human error, the major cause of road accidents. It seems, thus, reasonable to induce that, taking into account the foreseen globalized cooperative driving context, the most desired vehicles (the ones with commercial advantage) will be the ones that will provide the users a system that will place them in this desired wellness state. Given the upcoming driving paradigm, by which users will alternate between car sharing and car ownership, the strategies to grant the efficiency of these systems are bound to the users' trust in the autonomous vehicles. This, in turn, will depend on the ability of the vehicles' HMI to balance the information content and quantity to be conveyed to a diversity of users, by integrating inputs regarding their workload levels at a given moment and presenting information on a basis of as much as necessary, as little as possible. Therefore, the only approach feasible is a flexible approach, which means that the vehicles need to know its drivers' needs, capabilities, likes and expectations. When riding a rented or a new car, the only way the car can get to know its driver is having the driver communicate his/hers needs, capabilities, likes and expectation beforehand through customization. Personalization requires some continuous cooperative interaction through time. Nevertheless, it is the preferable strategy for having a driver-optimized HMI, where the vehicle recognizes and automatically adapts its state to the driver, i.e., such vehicle's HMI will reflect the cooperative interaction receiving constant inputs from the driver (different, relevant workload measures) and presenting diverse adaptive and adapted features, modulated by the driver's previous behaviour and coping mechanisms.

Acknowledgments. This work has been supported by COMPETE: POCI-01-0145-FEDER-007043 and FCT – Fundação para a Ciência e Tecnologia within the Project Scope: UID/CEC/00319/2013 and by the European Structural and Investment Funds in the FEDER component, through the Operational Competitiveness and Internationalization Programme (COMPETE 2020) Project nº 002797; Funding Reference: POCI-01-0247-FEDER-002797.

References

1. WHO: Work Health Organization (2013). http://www.who.int/en/
2. Young, M.S., Birrell, S.A., Stanton, N.A.: Safe driving in a green world: a review of driver performance benchmarks and technologies to support 'smart' driving. Appl. Ergon. **42**(4), 533–539 (2011)
3. Walker, G.H., Stanton, N.A., Young, M.S.: Hierarchical task analysis of driving: a new research tool. In: Hanson, M.A. (ed.) Contemporary Ergonomics 2001, Proceedings of the Annual Conference of the Ergonomics Society, Cirencester, April 2001, pp. 435–440. Taylor and Francis, London (2001)
4. Costa, S., Simões, P., Costa, N., Arezes, P.: A cooperative human-machine interaction warning strategy for the semi-autonomous driving context, pp. 1–7, November 2001
5. Fridman, L., Toyoda, H., Seaman, S., Seppelt, B., Angell, L., Lee, J., Reimer, B.: What can be predicted from six seconds of driver glances? pp. 2805–2813 (2016)
6. Colombo, M.: Andy Clark, surfing uncertainty: prediction, action, and the embodied mind. Mind. Mach. **27**(2), 381–385 (2017)
7. Engström, J., Bärgman, J., Nilsson, D., Seppelt, B., Markkula, G., Piccinini, G.B., Victor, T.: Great expectations: a predictive processing account of automobile driving. Theoret. Issues Ergon. Sci. **19**(2), 156–194 (2018)
8. Coughlin, J.F., Reimer, B., Mehler, B.: Driver wellness, safety & the development of an awarecar. Mass Inst. Technol. 1–15 (2009)
9. Colquhoun, H.L., Levac, D., O'Brien, K.K., Straus, S., Tricco, A.C., Perrier, L., Kastner, M., Moher, D.: Scoping reviews: time for clarity in definition, methods, and reporting. J. Clin. Epidemiol. **67**(12), 1291–1294 (2014)
10. Hart, S.G.: NASA-task load index (NASA-TLX); 20 years later. In: Proceedings of the Human Factors and Ergonomics Society 50th Annual Meeting, pp. 904–908 (2006)
11. Young, M.S., Brookhuis, K.A., Wickens, C.D., Hancock, P.A.: State of science: mental workload in ergonomics. Ergonomics **58**(1), 1–17 (2015). Taylor & Francis
12. Heine, T., Lenis, G., Reichensperger, P., Beran, T., Doessel, O., Deml, B.: Electrocardiographic features for the measurement of drivers' mental workload. Appl. Ergon. **61**, 31–43 (2017). https://doi.org/10.1016/j.apergo.2016.12.015
13. Broadbent, D.E.: Perception and Communication. Pergamon Press, Oxford (1958)

14. Kahneman, D.: Attention and Effort. Prentice-Hall, Englewood Cliffs (1973)
15. Eggemeier, F.T., Wilson, G.F., Kramer, A.F., Damos, D.L.: Workload assessment in multi-task environments. In: Damos, D.L. (ed.) Multiple Task Performance, pp. 207–216. Taylor & Francis, London (1991)
16. O'Donnell, R.D., Eggemeier, F.T.: Workload assessment methodology (1986)
17. Hancock, P.A., Desmond, P.A.: Preface. In: Hancock, P.A., Desmond, P.A. (eds.) Stress, Workload, and Fatigue, pp. 13–15. Lawrence Erlbaum Associates, Mahwah (2001)
18. Bellet, T., Tattegrain-Veste, H., Chapon, A., Bruyas, M.P., Pachiaudi, G., Deleurence, P., Guilhon, V.: Ingénierie cognitive dans le contexte de l'assistance à la conduite automobile. In: Boy, G. (ed.) Ingénierie cognitive. Lavoisier, Paris (2003)
19. SAE: Surface Vehicle Recommended Practice – J3016, SAE International, USA (2016)
20. Meyer, G., Deix, S.: Research and innovation for automated driving in Germany and Europe. In: Road Vehicle Automation, pp. 71–81. Springer International Publishing, Cham (2014)
21. Okumura, Y.: Activities, findings and perspectives in the field of road vehicle automation in Japan. In: Road Vehicle Automation, pp. 37–46. Springer International Publishing, Cham (2014)
22. Shaheen, S., Cohen, A.: Innovative Mobility Carsharing Outlook. University of California, Berkeley (2013)
23. Kyriakidis, M., de Winter, J.C.F., Stanton, N., Bellet, T., van Arem, B., Brookhuis, K., Martens, M.H., Bengler, K., Andersson, J., Merat, N., Reed, N., Flament, M., Hagenzieker, M., Happee, R.: A human factors perspective on automated driving. Theor. Issues Ergon. Sci. (2017). https://doi.org/10.1080/1463922X.2017.1293187
24. Jamson, A.H., Merat, N., Carsten, O.M.J., Lai, F.C.H.: Behavioural changes in drivers experiencing highly-automated vehicle control in varying traffic conditions. Transp. Res. Part C Emerg. Technol. **30**, 116–125 (2013)
25. Ho, D.C., Spence, P.C.: The Multisensory Driver: Implications for Ergonomic Car Interface Design. Ashgate Publishing Ltd., Farnham (2012)
26. Merat, N., Lee, J.D.: Preface to the special section on suman factors and automation in vehicles: designing highly automated vehicles with the driver in mind. Hum. Factors J. Hum. Factors Ergon. Soc. **54**(5), 681–686 (2012). https://doi.org/10.1177/0018720812461374
27. Meng, F., Gray, R., Ho, C., Ahtamad, M., Spence, C.: Dynamic vibrotactile signals for forward collision avoidance warning systems. Hum. Factors J. Hum. Factors Ergon. Soc. (2014). https://doi.org/10.1177/0018720814542651
28. Biondi, F., Rossi, R., Gastaldi, M., Mulatti, C.: Beeping ADAS: reflexive effect on drivers' behavior. Transp. Res. Part F Traffic Psychol. Behav. **25**, 27–33 (2014). https://doi.org/10.1016/j.trf.2014.04.020
29. Eby, D.W., Molnar, L.J., Zhang, L., St Louis, R.M., Zanier, N., Kostyniuk, L.P.: Keeping older adults driving safely: a research synthesis of advanced in-vehicle technologies, A LongROAD Study (2015)
30. Pereira, M.S.O.: In-vehicle information systems – related multiple task performance and driver behavior: comparison between different age groups. Doctoral Thesis, Universidade Técnica de Lisboa, Faculdade de Motricidade Humana, Lisboa (2009)
31. Ryu, J., Chun, J., Park, G., Han, S.H.: Vibro-tactile feedback for information delivery in the vehicle. IEEE Trans. Haptics **3**(2), 138–149 (2010). https://doi.org/10.1109/toh.2010.1
32. Eskandarian, A.: Fundamentals of driver assistance. In: Handbook of Intelligent Vehicles, pp. 491–535). Springer, London (2012)
33. De Waard, D.: The Measurement of Drivers' Mental Workload. Groningen University, Traffic Research Center, Netherlands (1996)

34. Tivesten, E., Dozza, M.: Driving context and visual-manual phone tasks influence glance behavior in naturalistic driving. Transp. Res. Part F: Traffic Psychol. Behav. **26**, 258–272 (2014)
35. He, J., Becic, E., Lee, Y.C., McCarley, J.S.: Mind wandering behind the wheel performance and oculomotor correlates. Hum. Factors: J. Hum. Factors Ergon. Soc. **53**(1), 13–21 (2011)
36. Ratwani, R.M., McCurry, J.M., Trafton, J.G.: Single operator, multiple robots: an eye movement based theoretic model of operator situation awareness. In: Proceedings of the 5th ACM/IEEE International Conference on Human-Robot Interaction, pp. 235–242. IEEE Press (2010)
37. Wang, Y., Reimer, B., Dobres, J., Mehler, B.: The sensitivity of different methodologies for characterizing drivers' gaze concentration under increased cognitive demand. Transp. Res. Part F: Traffic Psychol. Behav. **26**, 227–237 (2014)
38. Di Nocera, F., Camilli, M., Terenzi, M.: Using the distribution of eye fixations to assess pilots' mental workload. In: Proceedings of the Human Factors and Ergonomics Society Annual Meeting, vol. 50, no. 1, pp. 63–65. Sage Publications, Los Angeles, October 2006
39. Son, J., Lee, Y., Kim, M.H.: Impact of traffic environment and cognitive workload on older drivers' behavior in simulated driving. Int. J. Precis. Eng. Manuf. **12**(1), 135–141 (2011)
40. Planing, P.: Innovation Acceptance: The Case of Advanced Driver-Assistance Systems. Springer Science & Business Media, Heidelberg (2014)
41. Kompass, K., Huber, W., Helmer, T.: Safety and comfort systems: introduction and overview. In: Handbook of Intelligent Vehicles, pp. 605–612. Springer, London (2012)
42. Bishop, R.: Intelligent vehicle technology and trends (2005)
43. Dix, A., Finlay, J., Abowd, G., Beale, R.: Universal design. Human-computer interaction, pp. 365–394. Springer, US (2004)
44. Milakis, D., Snelder, M., van Arem, B., van Wee, B., de Almeida Correia, G.H.: Development and transport implications of automated vehicles in the Netherlands: scenarios for 2030 and 2050. EJTIR **17**(1), 63–85 (2017)
45. Gold, C., Körber, M., Hohenberger, C., Lechner, D., Bengler, K.: Trust in automation – before and after the experience of take-over scenarios in a highly automated vehicle. Procedia Manuf. **3**, 3025–3032 (2015)
46. Mouloua, M., Al-Awar Smither, J., Vincenzi, D.A., Smith, L.: Automation and aging: issues and considerations. In: Advances in Human Performance and Cognitive Engineering Research, pp. 213–237. Emerald Group Publishing Limited, Bingley (2002)

Supervising SSSEP Experiments with a Bluetooth Android Remote Control Application

José Rouillard(✉), François Cabestaing, Jean-Marc Vannobel, and Marie-Hélène Bekaert

University Lille, CNRS, Centrale Lille, CRIStAL, UMR 9189, 59655 Villeneuve d'Ascq Cedex, France
{Jose.Rouillard,Francois.Cabestaing, Jean-Marc.Vannobel,Marie-Helene.Bekaert}@univ-lille1.fr

Abstract. In this paper, we are presenting how we are controlling the vibration frequency sent from an Android smartphone to an Arduino board connected to a vibrator, in order to supervise more easily SSSEP (Steady State Somatosensory Evoked Potentials) experiments. Our researches are conducted in the context of hybrid Brain-Computer Interfaces for motor severely impaired patients, and our aim is to detect a physiological gating phenomenon on SSSEP responses when patients are trying to perform some small fingers moves while vibrations are emitted under their fingers.

Keywords: Hybrid BCI · SSSEP · Gating · Vibration · Smartphone Arduino

1 Introduction

The concept of Brain-Computer Interface (BCI), introduced by Vidal [1] in 1973, is a system that allows users to communicate between human's brain and external devices. Researchers are investigating different kinds of paradigm related to BCI (with or without stimuli, etc.) and several manners to measure the human activity during invasive or non-invasive BCI experiments (ElectroEncephaloGraphy – EEG, ElectroMyoGraphy - EMG, MagnetoEncephaloGraphy - MEG, etc.). They are also developing various kinds of BCI solutions according to different users (healthy or disable people, etc.). To improve the speed and robustness of communication, Leeb et al. have introduced the so-called "hybrid BCI" notion, in which brain activity and one or more other signals are analyzed jointly [2].

Since BCI is considered as an effective tool for rehabilitation and/or assistance of severely impaired patients, we are particularly working in our research team on hybrid BCI for Duchenne Muscular Dystrophy (DMD). DMD is a severe pathology of the skeletal musculature. This genetic disorder causes an absence of dystrophin, a protein that supports muscle strength and muscle fibers cohesion, which leads to progressive muscle degeneration and weakness. The residual motor ability of the most advanced DMD patients is characterized by very low amplitude movements associated with loss

H. Ayaz and L. Mazur (Eds.): AHFE 2018, AISC 775, pp. 365–375, 2019.
https://doi.org/10.1007/978-3-319-94866-9_37

of degrees of freedom and severe muscle weakness in the fingers. Specific brain activities can be detected in these patients. For instance, an event-related potential (ERP) is a measured brain electrophysiological modification, response to either an external stimulation (sound, image, vibration, etc.) or an internal event such as a cognitive activity (attention, motor preparation, etc.) [3].

Since the first recordings of electrical activity from the human brain, many studies have reported ERP as changes in brain signal expressing responses to sensory stimuli. Instead of observing sudden and time-lock responses to a transient event (for instance a P300 signal occurs 300 ms after a stimulus), it is also possible to detect more steady-state evoked potentials (SSEP). "*An increasing number of studies have used SS-EP to explore the neural activity involved in the cortical processing of visual and auditory sensory modalities and, to a lesser extent, the somatosensory modality*" [4].

Part 2 of the paper presents the background of this research within the BCI field and a state of the art concerning Steady State Evoked Potentials. Part 3 describes our contribution and methods to explore this field of study. The results, conclusion and perspectives are presented in part 4.

2 Steady State Evoked Potentials and BCIs: A Brief Overview

Steady state visually evoked potentials (SSVEP) are signals that are natural responses to visual stimulation at specific frequencies. Indeed, when the retina is excited by a visual stimulus (3.5 Hz to 75 Hz) the brain generates electrical activity at the same (or harmonic) frequency of the visual stimulus. Electrodes are positioned mainly on the occipital part of the human scalp in order to detect SSVEP. For example, BCIs used SSVEP to control a computer cursor [5, 6], an avatar [7], a robot [8, 9], a wheelchair [10] or a spelling system [11, 12].

Steady state auditory evoked potential (SSAEP) can be used to trace the signal generated by a sound (often at a frequency between 5 and 50 Hz) through the ascending auditory pathway when an evoked potential is generated from the cochlea to the cortex. Electrodes are positioned mainly on the temporal part of the human scalp in order to detect SSAEP. Steady state auditory evoked potential can be used to replace steady state visual evoked potential in brain–computer interface systems during visual fatigue periods [13]. Auditory ERP can also be used in auditory speller BCI or multi-choice based BCI [14, 15].

Steady state somatosensory evoked potentials (SSSEP) are detected as cerebral responses to vibratory stimulation applied on the body of the user (palm of the hand, wrist, finger, toe) with frequencies in the range of 5–250 Hz. Electrodes are positioned accordingly, for instance in C3, C4, or Cz location (see 10/20 international system) to detect a brain signal response to a vibration applied on right finger, left finger or toe, respectively. SSSEP based BCIs may reduce the fatigue usually induced by visual attention required in SSVEP based BCIs. Steady state somatosensory evokes potential are used for communication tools dedicated in complete locked-in syndrome (CLIS) patients for which SSVEP are inoperative [16]. The feasibility of SSSEP based BCIs for wheelchair control [17, 18], or task discrimination BCIs is also studied [19].

The part of the somatosensory system that transmits pain and temperature signals is monitored using laser evoked potentials (LEP) [20, 21]. Colon et al. explored the possibility to use SSEP in response to the thermal activation of cutaneous nociceptors in humans: "*Recently, we showed that it is possible to record SSEP in response to the rapid periodic thermal activation of cutaneous nociceptors in humans, as well as to the rapid periodic electrical stimulation of nociceptive intraepidermal free nerve endings*" [4].

Steady-state somatosensory evoked potential (SSSEP) can be produced by vibratory stimulation. Transducers provide different tactile stimulations in the resonance-like frequencies of the sensorymotor areas [22, 23]. Stimuli were applied on different parts of the body (e.g. palm of the hand, wrist, finger, toe) so that the user had to focus his attention either to one or other part. The increase of one of the elicited SSSEPs amplitude was detected in EEG signals [24].

In 2016, Ahn et al. [25] draw up a complete state of the art on SSSEP for BCI. After Müller-Putz et al. [24] who first defined the basic SSSEP-based BCI paradigm with index fingers stimulations, several studies on SSSEP based BCI have been published. Some stimulate the thumbs to discriminate between the two hands [26]. Breitwieser et al. [27] stimulate all the fingers of the right hand. In addition to finger stimulation, Yao et al. [28] applied stimulation on the skin of both wrists. Ahn et al. [29] performed a hybrid BCI based on motor imaging and SSSEP. This study demonstrated better classification performance in motor imaging using selective touch attention. These studies show the feasibility of BCIs based on the SSSEP but the performances achieved so far are quite low.

Otherwise, appears a phenomenon called "gating" or tactile suppression describing an attenuation of somatosensory input to the cerebral cortex during movement execution [30, 31]. This gating seems to be detected not only when users are moving while feeling a vibration under their fingers, but also when users are watching video showing hands moving or being touched by a virtual object: "*Observation of passive touch of the hand also gated the response (17% decrease). In conclusion, the results show that viewing a hand performing an action or being touched interferes with the processing of somatosensory information arising from the hand*" [32]. It seems that the "gating" phenomenon has not yet been used in BCI studies.

3 Methods

We try to exploit this physiological phenomenon within a BCI based on the SSSEP paradigm. Korean researchers have shown that a wheelchair can be driven by users asked to focus on certain vibrations emitted on left wrist, right wrist or toe, through SSSEP (with 30 electrodes) [18].

Canadian researchers have demonstrated that visualization by a subject of a video showing a movement of the hand, while some vibrations are sent to this patient's finger, interferes with the somatosensory process of information recovery for, in average, an amplitude reduction of 22% of SSSR (somatosensory steady-state response) [32];

We assume that it would be possible to use this physiological gating in a hybrid BCI as an interaction command.

We want to check if the following hypothesis is correct or not: by placing two vibrators under the left and right fingers of a DMD subject and by asking him to perform small finger movements, should we observe a decrease in the SSSR between 10 to 20% at the determined frequency (e.g. 7 or 13 Hz) at C3 and C4 (and maybe on a single Cz-electrode), and therefore, infer that the user moved (or wanted to move) to the right or left direction?

This preliminary study on healthy user (see Figs. 1 and 3) will be the first step to achieve a solution where the DMD patient will only think about the finger movement without really performing it.

3.1 Overall Architecture

As presented in Fig. 1, our overall architecture is organized in order to facilitate BCI experiments based on the acquisition of cerebral signals during SSVEP studies.

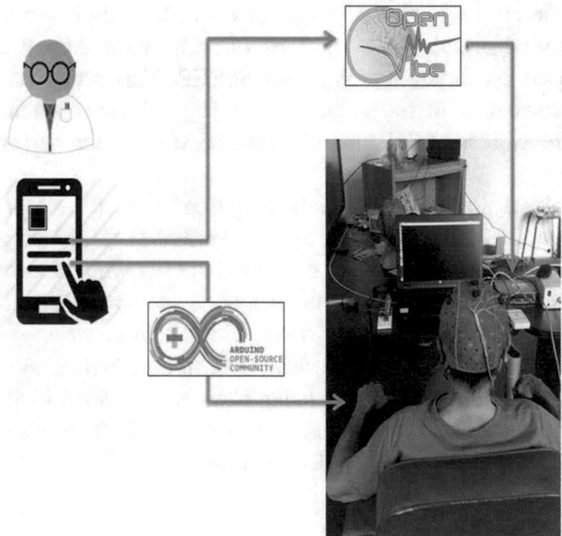

Fig. 1. Our overall architecture for the study of SSSEP.

The experimenter chooses a frequency (for instance 12 Hz) thanks to the mobile Android smartphone application that we have developed in our team. This frequency is sent by BLE (Bluetooth Low Energy) to an Arduino board connected to a vibrator device. At the same time, this frequency is also sent, via UDP, to the software in charge of the cerebral signals detection (OpenVibe), which can be found in an EEG when sending vibrations under the fingers or wrists of patients.

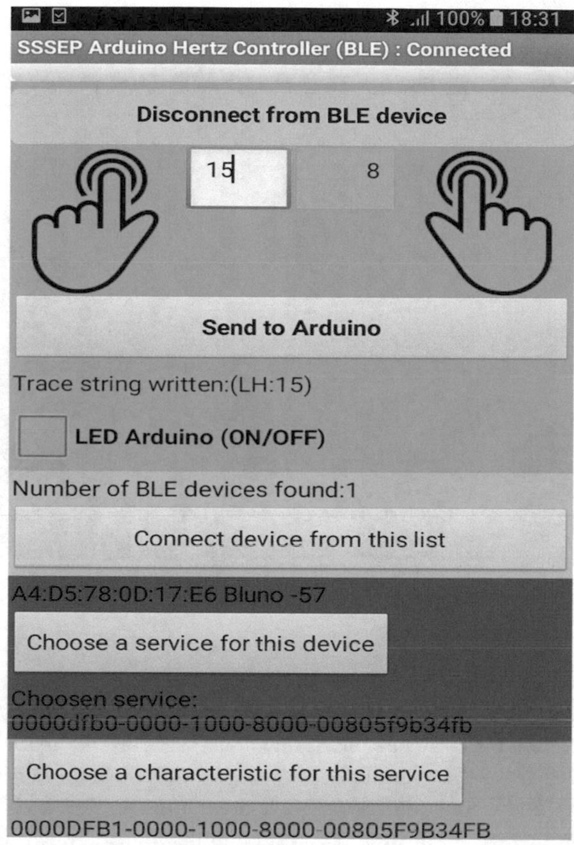

Fig. 2. Prototype of mobile application used in our SSSEP studies.

3.2 Technical Aspects

Mobile Application. Our mobile Android prototype application was developed with App inventor 2 which is a tool proposed jointly by Google and the MIT. We injected in this application both "BluetoothLE" extension [33] and "ClientUDP" extension [34] in order to communicate respectively through BlueTooth Low Energy protocol and User Datagram Protocol.

As we can see in Fig. 2, this mobile application prototype can be easily used by experimenters to enter the desired frequency and to send it wirelessly to the Arduino board and the BCI software.

3.3 Arduino Based Vibratory Stimulator

Linear Resonant Actuators (LRA) as shown in Fig. 3 are nowadays commonly used instead of eccentric rotating mass (ERM) motors for cell phones to vibrate. That makes them easy to find and affordable. A LRA can be compared to a miniaturized loud speaker making a round metallic pad moving from the front to the back instead of a cone. LRAs are then really convenient to use in haptic applications and best used within a narrow frequency vibration range centered around a resonant frequency close to 200 Hz (precisely 205 Hz for the 10 mm C10-000 LRA presented in Fig. 3).

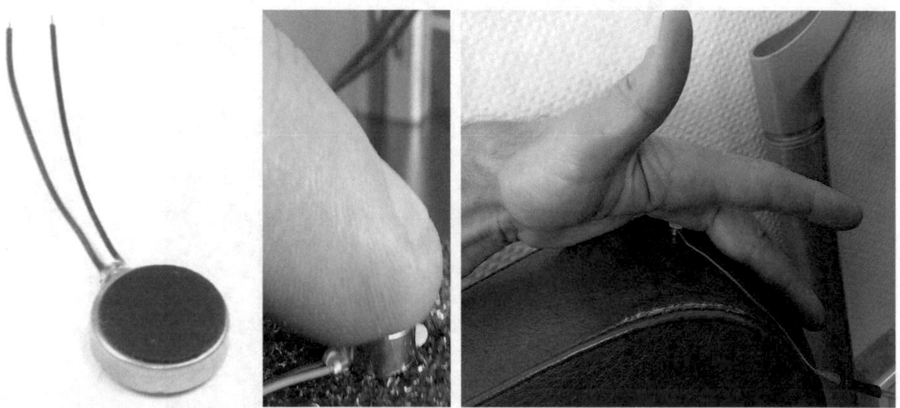

Fig. 3. LRA vibrator (*left*) applied under the finger (*center*) or under the hand palm (*right*).

An Adafruit DRV2605 board as shown in Fig. 4 (right) has been used to power up a 10 mm C10-000 LRA. This board is based on a TI DRV2605 motor driver [35]. As shown in Fig. 4 (left), an I2C (Inter Integrated Circuit) communication can be used to manage this driver from the vibration frequency control board. However, one has to note that the DRV2605 does not have I2C address pins neither a chip select one. To control more than one driver through I2C communication one have then to implement as much I2C communication buses as DRV2605 drivers used or to multiplex the existing I2C communication bus.

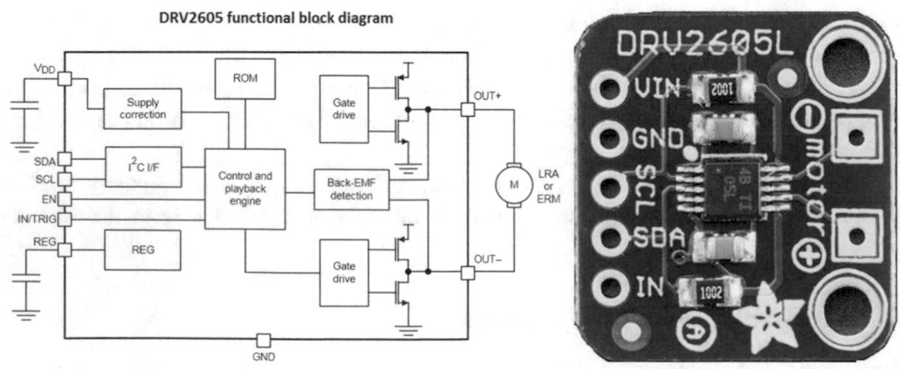

Fig. 4. TI DRV2605 driver functional block diagram (from [10], left) and Adadruit DRV2605 board (right).

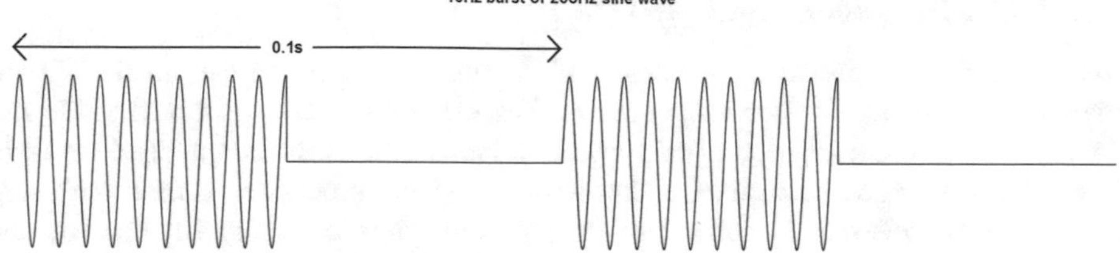

Fig. 5. TI DRV2605 10 Hz burst of 205 Hz sine wave output.

As previously said, an Arduino based board was used to control the low frequency burst of 205 Hz sine wave, as illustrated in Fig. 5, applied to the C10-000 LRA. Both BLE and I2C communication were needed. We could then use either an Arduino UNO and a BLE shield as the RedbearLab one for example (Fig. 6, left) or better a DFRobot Bluno Beetle board (Fig. 6, center) which offers an Arduino with Bluetooth Low Energy communication interface. In short, the Bluno Beetle is fully compatible with the Arduino Uno board except the number of inputs and outputs found in reduced quantity, uses the original Arduino IDE without the need of any external library for Bluetooth communication and can be used with any Bluetooth 4.0 compatible device.

At last, we present Fig. 6 (right) the whole development unit used in this haptic application. As one can see, this is really simple to implement since needing only a Bluno Beetle microcontroller board, a DRV2605 driver to drive a LRA and the LRA itself.

In haptic applications needing information coming from both hands, two LRA drivers and LRAs are of course needed. As previously said, one has then to take into account that a second I2C communication interface is needed. A software one can easily be computed in that way.

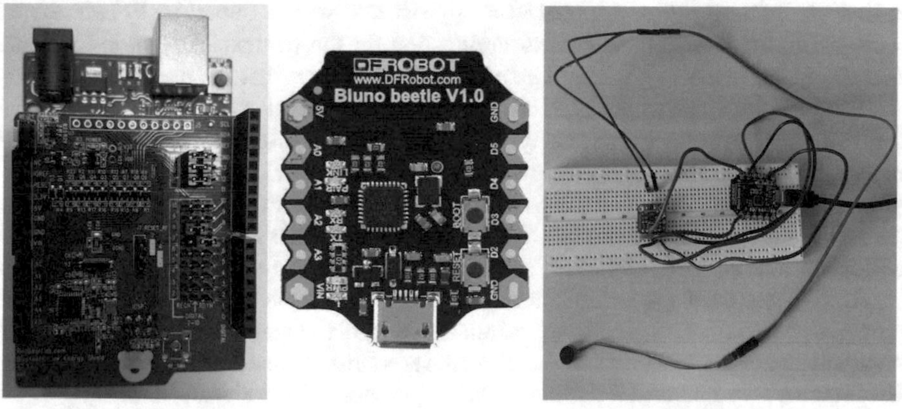

Fig. 6. Arduino UNO plus RedbearLab BLE shield (*left*), DFRobot Bluno Beetle BLE board (*center*), fully functional DIY SSSEP development board (*right*).

3.4 Signal Processing with OpenVibe

The electrodes are placed on the scalp of the user according to the classical 10/20 international system. For instance, in our studies, electrodes were positioned on C3 and C4 location in order to detect brain signal responses to vibrations applied on right finger and left finger, respectively. Currently, the signal processing carried out using the OpenVibe software is dependent on the parameters manually supplied by the experimenter. For instance, as we can see on Fig. 7, a Butterworth Band Pass filter between a minimum and a maximum value can be easily implemented in OpenVibe.

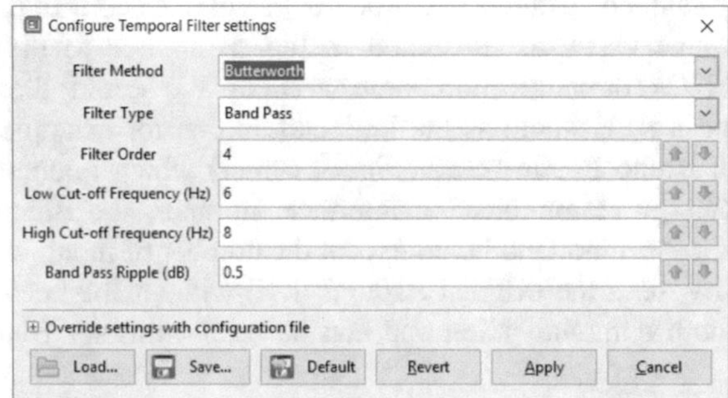

Fig. 7. Example of Butterworth Band Pass filter between 6 and 8 Hz usable in OpenVibe.

Being able to automatically filter the correct frequency to listen, according to the parameters indicated by the mobile application driven by the experimenter, should greatly accelerate the experiments and reduce the risk of error during BCI sessions.

4 Conclusion and Perspectives

In this preliminary study, our goal was to propose a way to supervise SSSEP experiments with a mobile remote-control application, in the context of a Brain-Computer Interaction, allowing to change dynamically the vibration that the user feels.

The technical part works correctly: the mobile application developed in our laboratory allows to quickly and easily choose a frequency (ex: 11 Hz) and send this information, via Bluetooth, to an Arduino board (Bluno in our study), which drives a vibration motor, oscillating at the chosen frequency, under the finger or the wrist of the user. This allows a smooth manner to try various frequencies in a same session, without to stop, recalibrate and send again another chosen frequency during BCI experiments.

At the same time, this information (ex: 11 Hz) is sent via UDP to the computer that manages the BCI scenario of the current experiment (OpenVibe in our study). We managed to recover this data in the OpenVibe scenario, thanks to a Python script, and it still remains to implement a dynamic filtering process of the signals, according to this parameter, by applying a bandpass filter (between 10 and 12 Hz, in our example).

In the short term, we will try to improve the prototype in order to drive two vibration motors simultaneously. Thus, sending a simple command (ex: "LH:7RH:9") will allow to send at the same time two different orders, one for each vibrator motors; example: Left Hand at 7 Hz and Right Hand at 9 Hz. We are also planning to let the experimenter choose and generate some rest time between two vibrations, in order to find more easily the researched patterns in the cerebral waves.

So, this preliminary study for the use of physiological gating in a hybrid BCI shows that the technical part is correctly implemented. In the next step, we are planning to use more powerful vibrator, like C2-Tactors, for instance [36], for human experimentation and validation of the signal processing chain.

Obviously, this laboratory work is considered preparatory, and will not be deployed in a medical field where the electronic waves used to communicate between smartphones and PCs (Bluetooth in particular) could disrupt the proper functioning of the biological signal recorders.

However, we believe that the use of such supervisory tools for BCI should greatly facilitate the exploratory work of the experimenters, since seeking the right frequency to use for each patient requires a lot of time when done manually. The risk of error should also be reduced, ultimately, thanks to the semi-automatic generation of scripts used to detect the reflex responses produced by the human body for such somatosensory experiments.

References

1. Vidal, J.J.: Toward direct brain-computer communication. Ann. Rev. Biophys. Bioeng. **2**(1), 157–180 (1973)
2. Leeb, R., Sagha, H., Chavarriaga, R., del R Millan, J.: Multimodal fusion of muscle and brain signals for a hybrid-BCI. In: 2010 Annual International Conference of the IEEE Engineering in Medicine and Biology Society (EMBC), pp. 4343–4346 (2010)
3. Luck, S.J.: An Introduction to the Event-Related Potential Technique. The MIT Press, Cambridge (2005)
4. Colon, E., Legraina, V., Mouraux, A.: Steady-state evoked potentials to study the processing of tactile and nociceptive somatosensory input in the human brain. Neurophysiol. Clin./Clin. Neurophysiol. **42**, 315–323 (2012)
5. Luzheng, B., Jinling, L., Ke, J., Ru, L., Yili, L.: A speed and direction-based cursor control system with P300 and SSVEP. Biomed. Sig. Process. Control **14**, 26–133 (2014)
6. Trejo, L.J., Rosipal, R., Matthews, B.: Brain-computer interfaces for 1-D and 2-D cursor control: designs using volitional control of the EEG spectrum or steady-state visual evoked potentials. IEEE Trans. Neural Syst. Rehabil. Eng. **14**(2), 225–229 (2006)
7. Faller, J., Muller-Putz, G., Schmalstieg, D., Pfurtscheller, G.: An application framework for controlling an avatar in a desktop-based virtual environment via a software SSVEP brain–computer interface. Presence: Teleoperators Virtual Environ. **19**, 25–34 (2010)
8. Deng, Z., Li, X., Zheng, K., Yao, W.: A humanoid robot control system with SSVEP-based asynchronous brain-computer interface. Robot **33**(2), 129–135 (2011)
9. Prueckl, R., Guger, C.: A brain-computer interface based on steady state visual evoked potentials for controlling a robot. In: Cabestany, J., Sandoval, F., Prieto, A., Corchado, J.M. (eds.) Bio-Inspired Systems: Computational and Ambient Intelligence. IWANN 2009. LNCS, vol. 5517. Springer, Heidelberg (2009)

10. Achic, F., Montero, J., Penaloza, C., Cuellar, F.: Hybrid BCI system to operate an electric wheelchair and a robotic arm for navigation and manipulation tasks. In: IEEE International Workshop on Advanced Robotics and its Social Impacts (ARSO), Shanghai (2016)
11. Vilic, A., Kjaer, T.W., Thomsen, C.E., Puthusserypady, S., Sorensen, H.B.: DTU BCI speller: an SSVEP-based spelling system with dictionary support. In: 35th Annual International Conference of the IEEE EMBS, Osaka, pp. 2212–2215 (2013)
12. Combaz, A., Chatelle, C., Robben, A., Vanhoof, G., Goeleven, A., Thijs, V., Van Hulle, M. M., Laureys, S.: A comparison of two spelling brain-computer interfaces based on visual P3 and SSVEP in locked-in syndrome. Research article (2013)
13. Punsawad, Y., Wongsawat, Y.: Multi-command SSAEP-based BCI system with training sessions for SSVEP during an eye fatigue state. IEEJ Trans. Electr. Electron. Eng. **12**(S1), S72–S78 (2017)
14. Höhne, J., Schreuder, M., Blankertz, B., Tangermann, M.: A novel 9-class auditory ERP paradigm driving a predictive text entry system. Front. Neurosci. **5**, 99 (2011)
15. Kim, D.W., Hwang, H.J., Lim, J.H., Lee, Y.H., Jung, K.Y., Im, C.H.: Classification of selective attention to auditory stimuli: toward vision-free brain–computer interfacing. J. Neurosci. Methods **197**(1), 180–185 (2011)
16. Guger, C., Spataro, R., Allison, B.Z., Heilinger, A., Ortner, R., Cho, W., La Bella, V.: Complete locked-in and locked-in patients: command following assessment and communication with vibro-tactile P300 and motor imagery brain-computer interface tools. Front. Neurosci. **11**, 251 (2017)
17. Kaufmann, T., Herweg, A., Kübler, A.: Toward brain-computer interface based wheelchair control utilizing tactually-evoked event-related potentials. J. Neuroeng. Rehabil. **11**, 7 (2014)
18. Kim, K.T., Suk, H.I., Lee, S.W.: Commanding a brain-controlled wheelchair using steady-state somatosensory evoked potentials. IEEE Trans. Neural Syst. Rehabil. Eng. **26**(3), 654–665 (2016)
19. Ahn, S., Jun, S.C.: Feasibility of hybrid BCI using ERD- and SSSEP- BCI. In: 12th International Conference on Control, Automation and Systems, Jeju Island, pp. 2053–2056 (2012)
20. Treede, R.D., Lorenz, J., Baumgärtner, U.: Clinical usefulness of laser-evoked potentials. Clin. Neurophysiol. **33**, 303–314 (2003)
21. Hsueh, J.J., Jason Chen, J.J., Shaw, F.Z.: Distinct somatic discrimination reflected by laser-evoked potentials using scalp EEG leads. J. Med. Biol. Eng. **36**(4), 460–469 (2016)
22. Regan, D.: Human Brain Electrophysiology: Evoked Potentials and Evoked Magentic Fields in Science and Medicine. Elsevier (1989)
23. Müller, G.R., Neuper, C., Pfurtscheller, G.: Resonance-like frequencies of sensorimotor areas evoked by repetitive tactile stimulation. Biomed. Tech. **46**, 186–190 (2001)
24. Müller-Putz, G.R., Scherer, R., Pfurtscheller, G.: Steady-state somatosensory evoked potentials: Suitable brain signals for brain computer interfaces? IEEE Trans. Neural Syst. Rehabil. Eng. **14**(1), 30–37 (2006)
25. Ahn, S., Kim, K., Jun, S.C.: Steady-state somatosensory evoked potential for brain-computer interface: present and future. Front. Hum. Neurosci. **9**, 716 (2015)
26. Haegens, S., Händel, B.F., Jensen, O.: Top-down controlled alpha band activity in somatosensory areas determines behavioral performance in a discrimination task. J. Neurosci. **31**, 5197–5204 (2011)
27. Breitwieser, C., Kaiser, V., Neuper, C., Müller-Putz, G.R.: Stability and distribution of steady-state somatosensory evoked potentials elicited by vibro-tactile stimulation. Med. Biol. Eng. Comput. **50**, 347–357 (2012)
28. Yao, L., Meng, J., Zhang, D., Sheng, X., Zhu, X.: Combining motor imagery with selective sensation toward a hybrid-modality BCI. IEEE Trans. Biomed. Eng. **61**, 2304–2312 (2014)

29. Ahn, S., Ahn, M., Cho, H., Jun, S.C.: Achieving a hybrid brain- computer interface with tactile selective attention and motor imagery. J. Neural Eng. **11**, 6 (2014)
30. Chapman, C.E.: Active versus passive touch: factors influencing the transmission of somatosensory signals to primary somatosensory cortex. Can. J. Physiol. Pharmacol. **72**(5), 558–570 (1994)
31. Cheron, G., Borenstein, S.: Specific gating of the early somatosensory evoked potentials during active movement. Electroencephalogr. Clin. Neurophysiol. **67**(6), 537–548 (1987)
32. Voisin, J., Rodrigues, E., Hétu, S., Jackson, P., Vargas, C., Malouin, F., Chapman, E., Mercier, C.: Modulation of the response to a somatosensory stimulation of the hand during the observation of manual actions. Exp. Brain Res. **208**, 11–19 (2010)
33. MIT App Inventor Extensions, BluetoothLE (edu.mit.appinventor.ble.aix). http://appinventor.mit.edu/extensions/. Accessed 14 Feb 2018
34. Andres Cotes, UDP Client Extension (co.com.dendritas.ClientUDP.aix). https://community.thunkable.com/t/udp-client-extension/5831. Accessed 14 Feb 2018
35. Texas Instruments, TI DRV2605 motor driver. http://www.ti.com/lit/ds/symlink/drv2605.pdf. Accessed 14 Feb 2018
36. C2 Tactors, provided by EAI (Engineering Acoustics, Inc.). www.eaiinfo.com, http://bdml.stanford.edu/twiki/pub/Haptics/VibrationImplementation/C2_tactor.pdf. Accessed 14 Feb 2018

Machine Usability Effects on Preferences for Hot Drinks

Hongjun Ye(✉), Jan Watson, Amanda Sargent, Hasan Ayaz, and Rajneesh Suri

Drexel University, 3141 Chestnut Street, Philadelphia, PA 19104, USA
{hy368, jlw437, as3625, ha45, surir}@drexel.edu

Abstract. Research suggests that usability of a machine affects consumers' preference for the machine. However, there is no research available to explain if evaluation of a product prepared using that machine, is likely to be impacted by the usability of the machine. In a controlled study using three trials, participants prepared hot beverages using two different beverage machines that also differed in their usability. In all three trials, machine usability, participants' drink selections and evaluations of hot drinks prepared on the machines were measured. Our results indicate that machine usability influences consumers' preference for the products prepared on these machines.

Keywords: Machine usability · Consumer behavior · Product preferences End products

1 Introduction

Prior research suggests consumers' cognitive and emotional states influence their evaluation of products and services [1–3]. Further, consumer satisfaction with electronic purchases and online websites is affected by usability of these products and websites [4–6]. With heightened automation, today, more and more consumers are using small machines and appliances at home and office to prepare products just like those that would otherwise be purchased from a store (e.g., beverages, or ice creams etc.). However, current research does not explain how small machine usability (tea or coffee machines, ice cream makers, microwaves etc.) will influence consumer assessment of the prepared products (e.g., beverages, ice cream, meals etc.).

The current research examines the effects of machine usability on consumers' evaluation of a prepared product. A lack of understanding of this relationship between machine usability and consumers' evaluation of prepared products could lead to misleading conclusions about consumer preferences for prepared products. A case in point is consumers preferring one hot beverage drink over another not because of its superior taste but simply because the machine used to prepare that drink was easier to use. Hence, final choices on drink selections might mislead a manufacturer to suggest a need to improve the product, which might not be a productive direction to go. This research investigates such a relationship between machine usability and consumers' evaluation of products they prepared themselves.

H. Ayaz and L. Mazur (Eds.): AHFE 2018, AISC 775, pp. 376–382, 2019.
https://doi.org/10.1007/978-3-319-94866-9_38

The present study used three trials involving drink preparation and consumption. Machine usability, drink selection, and drink evaluation were measured in each trial. Participants first prepared hot drinks using two machines that differed in their ease of use, and then selected one of these drinks for consumption. Our results suggest that machine usability influences consumers' preference for beverage prepared on the machines and an otherwise superior product prepared using a machine with inferior usability was negatively impacted by the user's experience.

2 Method

2.1 Participants and Product Stimuli

Ten hot beverage drinkers (6 females and 4 males; Age range: 26–53 years old) were recruited to participate in this study conducted at an East Coast U.S. university. All participants consumed at least one cup of hot beverage each day and had no smoking habit or declared neurological deficit. Tea and hot chocolate were used as beverages during this study. Prior to coming to the lab, participants completed a survey where they provided details about their daily drink consumption, and flavor preferences.

2.2 Environment Design

A flexible studio in the lab was created to simulate a real-world office environment. The office environment included a cubicle, a beverage bar area and a break room. In the cubicle, participants completed a battery of timed experimental tasks. On completion of these tasks, participants moved to the beverage bar area where they prepared and consumed one of the prepared hot beverages. Instructions to operate the two machines used in this study and the dependent variables were presented on computers. To prevent confounding effects of brand strength, brand names of the two beverage machines and the hot drink packets used to prepare the beverages were concealed.

2.3 Procedures and Experimental Design

Upon arrival in the lab, each participant was provided a walkthrough of the mock office space (cubicle, break room and beverage bar area) and an overview of study procedures. The tasks completed in the cubicle emulated cognitive, visual and verbal tasks performed daily in normal workplace office environments. The three tasks were presented in a random order using a computer. There were two levels of difficulty (easy vs. hard) for each task. Task performance was measured using accuracy and reaction times and each task was allocated 3–4 min to perform. The first trial prior to beverage consumption provided a baseline for participant's performance on the three tasks.

After completing each trial, participants entered the beverage bar area to prepare hot drinks on two brands of commercially available beverage machines popular in the offices in the US. Machine 1 provided more customized options but its functionality was more complicated and challenging for users than that for Machine 2.

The drinks included different tea flavors and a hot chocolate drink. Based on preferences for flavors indicated during the pre-screening, hot drink packets for both machines were preplaced on the beverage bar prior to participant entering that area. Drink packets were specific for each machine and the flavors of the drinks were matched for each trial.

In order to observe the effects of usage instructions, detailed preparation instructions were offered for both machines during the first and second trials. After drink preparation and prior to beverage consumption, participants responded to measures about machine usability (Appendix A) and then completed a filler task (Appendix B). The filler tasks provided time for both prepared drinks to cool down. The experimenter ensured that both drinks achieved a similar drinkable temperature in each trial. In first and third trials, participants chose a drink from the two machines. In the second trial, participants were constrained to consume the drink that was not selected during the first trial. The forced choice in the second trial ensured that all participants had tasted drinks prepared from both machines before making their choice decision for the third trial.

In each trial, participants were provided four minutes to consume their drink with no restriction on the consumption amount. The amount of drink consumed was recorded. Participants then rated the drink on three items measuring how strong, invigorating, and tasty the chosen drink was (Appendix C). At the end of the study, participants also completed an open-ended question that sought their comments on the study and the three trials.

3 Results

All analyses were run with the SAS software package (university edition). The default level of significance was set at $p < 0.05$.

3.1 Task Performance and Engagement

We expected participants to engage in all three tasks with difficult tasks being perceived more difficult to perform. Reaction times (RT) and accuracy were used to assess differences in performance (Table 1). For math tasks, there were significant differences in response times between the easy and hard levels of the math and the accuracy rates were lower for harder math tasks. For the visual processing tasks, there were significant differences in reaction times between the easy and hard levels but not on accuracy. Finally, the Stroop tasks did not exhibit any differences in participants' baseline performances. Lack of differences in Stroop tasks are found because these tasks are

Table 1. Performance in the tasks (baseline).

	Math (Easy)	Math (Hard)	RVP (Easy)	RVP (Hard)	Stroop (Easy)	Stroop (Hard)
Accuracy	0.97 ± 0.04	0.76 ± 0.12***	0.99 ± 0.02	0.95 ± 0.07	0.43 ± 0.27	0.42 ± 0.13
RT (ms)	2218.8 ± 487.5	4172.5 ± 736.9***	267.1 ± 54.2	333.5 ± 46.5**	822.4 ± 186.9	799.5 ± 178.7

*P < 0.05; **P < 0.01; ***P < 0.001 (two sample t-test)

inherently difficult. Overall, these results indicate that participants could engage in all three work-related tasks akin to a real work place and the difficult tasks imposed more needs for cognitive effort.

3.2 First Trial

There was a significant difference in usability for the two machines, with Machine 1 being rated lower than Machine 2 ($M_{\text{Machine1}} = 7.2$, $SD = 1.55$, $M_{\text{Machine2}} = 8.4$, $SD = 0.7$, $t(18) = 2.23$, $p = 0.03$).

Before participants made their drink selections in this first trial, the only information they had to provide a context for the drink's attributes was their impression formed due to their experiences with the machine. The lower usability rating of Machine 1 could place the machine in an inferior position in a user's mind vis-à-vis Machine 2. However, 80% of the participants chose the hot beverage produced by Machine 1 in the first trial. This could be curiosity-motivated [7], as many of them indicated in the exit interview that they were "familiar with Machine 2" and/or "wanted to try something new (drinks from Machine 1)" at the beginning of the study.

3.3 Third Trial

Since the second trial included a forced choice with participants being asked to consume the beverage from the machine that they did not choose in the first trial, we reviewed their decisions in the third trial. In this trial, participants chose a drink after they had used both machines and tasted drinks from both machines. Note that changes in task performance might have effects on subjects' affect states, which could in turn influence their decision making [8, 9]. Thus, task performance changes before the drink selection were compared and no significant differences were observed ($p > .10$).

Both coded categorical indicators and raw scores were used in data analyses. Three attributes (strength, activation, and flavor) to evaluate the hot beverage were aggregated at an individual level, and participants' preference for taste ("Taste") were coded categorically based on the comparison of the summed numbers (1 = prefer taste of drinks from Machine 1; 2 = prefer taste of drinks from Machine 2; 3 = no preference). Usability evaluations ("Usability") were coded in the same way (1 = considered Machine 1 easier to use; 2 = considered Machine 2 easier to use; 3 = see no difference in usability of both machines). Participants' final drink selections were also coded categorically ("Choice(F)"; 1 = selected the drink from Machine 1; 2 = selected the drink from Machine 2). None of the participants considered Machine 1 as easier to use than Machine 2. Further, all participants indicated a clear preference for the drink from one of the two machines with respect to taste. This suggests that participants could distinguish between the drinks from the two machines and exhibited a clear preference. Thus, both "Usability" and "Taste" only have two levels (i.e., "Usability" = 2 or 3; "Taste = 1 or 2).

Compared to trial 1, participants who chose drinks from Machine 1 decreased by 37.5% in trial 3 (50% chose drinks from Machine 1; 50% chose drinks from Machine 2). For participants who rated Machine 2 as being easier to use, their drink selections in trial

3 were also from Machine 2. On the other hand, for those who considered the usability of both machines as identical, all chose drinks from Machine 1 in trial 3.

Though due to a limited sample size, logistic regression was not considered suitable for this analysis [10], an exact match of values between "Usability" and "Choice(F)" suggest that machine usability had a strong influence on participants' final choices. To further examine the effect of taste, a two-tailed Fisher's exact test using "Taste" as a grouping variable and "Choice(F)" as an outcome variable showed a non-significant result (p = 1; exact equality). This result suggests that usability was a stronger predictor of product choice than its taste.

PROC GLM was ran for further examination on the dataset. Ratings of usability and taste (raw scores) were used as independent variables and participants' drink selection in trial 3 was used as the response variable with the CLASS option. The analysis showed similar results (R^2 = 0.74, F(4,5) = 3.58, p = 0.09; Table 2). It again showed that usability was a better predictor than taste when predicting participants' product choices. Similar to the results of the categorical data analysis, the result of the regression analysis also suggests the impact of machine usability on product preferences.

Table 2. PRCO GLM results.

Source	*B*	*SE B*	Β	*t*	*p*
Usability (Machine 1)	−0.37480	0.18905	−0.73061	−1.98	0.1042
Usability (Machine 2)	0.57101	0.16609	1.13630	3.44	0.0185
Taste (drinks from Machine 1)	−0.00565	0.02651	−0.06749	−0.21	0.8395
Taste (drinks from Machine 2)	−0.01439	0.02580	−0.15459	−0.56	0.6010

4 Discussion

When asked to choose a beverage to drink, people are likely to choose the one that tastes better, because taste is directly related to the process of consumption. Since drink preparation and consumption processes were separated in this human ergonomic motivated study design, our results suggest that participants' drink selection is influenced by their experiences with the machines on which it is prepared.

Small beverage machines are usually categorized as physical goods and it has been suggested that benefits from physical products should be from consuming outcomes of the preparation process rather than the process itself [11]. However, our results demonstrate that the ultimate benefits from beverage machines could come from the consumption of the process (i.e., an individual's interaction with the machine).

Blind taste tests and manipulation of involvement in preparation could be additional approaches for further examinations. Consumer research in do-it-yourself (DIY) and co-production behavior [12–14] suggests that invested effort plays a role in affecting the evaluation of the final outcomes. Bendapudi and Leone [12] reported a self-serving biased satisfaction evaluation with a firm and pointed out that higher rating is associated with higher level of customer participation in the production process. Similarly,

Troye and Supphellens' experiment [13] showed that consumers' active engagement in the product creation process positively biases their evaluation of the outcome products. In addition, age and gender differences might also influence preferences, as people could have dissimilar needs for empowerment [14] and identity formation [15], which could shape the DIY behaviors. These results suggest that additional research is needed to understand how people separate machine usability from product consumption.

Appendix A: Machine Usability Survey

Item No.	Description
1	Was Machine 1 easy to use?
2	Was Machine 2 easy to use?

Note: items were rated on a nine-point scale (1 = not at all; 9 = extremely)

Appendix B: Filler Task

Choice No.	Description
1	Color a book
2	Solve a crossword puzzle
3	Play an I Spy game

Appendix C: Post-consumption Survey

Item No.	Description
1	The hot beverage activated me
2	The hot beverage was strong
3	The hot beverage was good

Note: items were rated on a nine-point scale (1 = not at all; 9 = extremely)

References

1. Creusen, M.E., Schoormans, J.P.: The different roles of product appearance in consumer choice. J. Prod. Innov. Manag. **22**(1), 63–81 (2005)
2. Zajonc, R.B., Markus, H.: Affective and cognitive factors in preferences. J. Consum. Res. **9**(2), 123–131 (1982)
3. Keinonen, T.: Expected usability and product preference. In: Proceedings of the 2nd Conference on Designing Interactive Systems: Processes, Practices, Methods, and Techniques, pp. 197–204. ACM, New York, August 1997

4. Akrimi, Y., Khemakhem, P.R.: An analysis of perceived usability, perceived interactivity and website personality and their effects on consumer satisfaction. Int. J. Manag. Excell. **2**(3), 227–236 (2014)
5. Koufaris, M.: Applying the technology acceptance model and flow theory to online consumer behavior. Inf. Syst. Res. **13**(2), 205–223 (2002)
6. Bai, B., Law, R., Wen, I.: The impact of website quality on customer satisfaction and purchase intentions: evidence from Chinese online visitors. Int. J. Hosp. Manag. **27**(3), 391–402 (2008)
7. Steenkamp, J.B.E., Baumgartner, H.: The role of optimum stimulation level in exploratory consumer behavior. J. Consum. Res. **19**(3), 434–448 (1992)
8. Gardner, M.P.: Mood states and consumer behavior: a critical review. J. Consum. Res. **12**(3), 281–300 (1985)
9. Cohen, J., Pham, M., Andrade, E.: The nature and role of affect in consumer behavior (2008)
10. Albert, A., Anderson, J.: On the existence of maximum likelihood estimates in logistic regression models. Biometrika **71**(1), 1–10 (1984)
11. Grönroos, C.: Marketing services: the case of a missing product. J. Bus. Ind. Mark. **13**(4/5), 322–338 (1998)
12. Bendapudi, N., Leone, R.P.: Psychological implications of customer participation in co-production. J. Mark. **67**(1), 14–28 (2003)
13. Troye, S.V., Supphellen, M.: Consumer participation in coproduction: "I made it myself" effects on consumers' sensory perceptions and evaluations of outcome and input product. J. Mark. **76**(2), 33–46 (2012)
14. Wolf, M., McQuitty, S.: Understanding the do-it-yourself consumer: DIY motivations and outcomes. AMS Rev. **1**(3–4), 154–170 (2011)
15. Moisio, R., Arnould, E.J., Gentry, J.W.: Productive consumption in the class-mediated construction of domestic masculinity: Do-It-Yourself (DIY) home improvement in men's identity work. J. Consum. Res. **40**(2), 298–316 (2013)

Cognitive Architecture Based Simulation of Perception and Behavior Controls for Robot

Yanfei Liu[1(✉)], Zhiqiang Tian[2], Yuzhou Liu[3], Jusong Li[1], and Feng Fu[1]

[1] Department of Computer Science and Technology, Zhejiang Sci-Tech University, No. 928, 2nd Avenue, Xiasha, Hangzhou 310018, China
yliu@zju.edu.cn, ljscg@sina.com, fufeng@zstu.edu.cn

[2] National Key Laboratory of Human Factors Engineering, Astronaut Research and Training Centre of China, No. 26, Beijing Road, Haidian District, Beijing 100094, China
tianzhiqiang2000@163.com

[3] Department of Computer Science and Software Engineering, Auburn University, Auburn, AL 36849, USA
yzl0217@auburn.edu

Abstract. This paper firstly proposes a basic framework to form a cognitive robot – the primary functional units and cognitive abilities. Then it focuses on the module's composition and the basic functions for a cognitive architecture by merit attention of neurocognitive research achievements and the research results of existing cognitive architecture. Robot's motion/control simulation is implemented by applying the robot simulator v-rep, the cognitive architecture is conceive and developed by using python program language, and cognitive robot perception and behavior control based on hybrid cognitive architecture is realized through an interface call to v-rep API. As a case, a robot to avoid obstacles and its behaviors controlled by the cognitive process is conducted.

Keywords: Cognitive robotic · Cognitive architecture · Barrier avoidance
Cognitive abilities · Cognitive processes

1 Introduction

Artificial Intelligence has increasingly become a topic of conversation recently, and people were convinced how robotics can improve work, education, and life. More ideas will be transformed into actions and real advances for robotics, artificial intelligence will be one of the biggest trends and robots will become more intelligent [1]. Robots will to be capable of understanding their environment, adapting their behavior, generalizing ideas, predicating faults to accomplish task like human being, even learning from their experiments and mistakes [2]. Cognitive robotics is an approach to creating artificial intelligence in robots by enabling them to learn from and respond to real-world situations, as opposed to pre-programming the robot with specific responses to every conceivable stimulus [3].

H. Ayaz and L. Mazur (Eds.): AHFE 2018, AISC 775, pp. 383–393, 2019.
https://doi.org/10.1007/978-3-319-94866-9_39

A cognitive robot is a brain-inspired robot system being capable of inference, perception, and learning mimicking the cognitive mechanisms of humans. The fundamental theories and methodologies underpinning cognitive robotics are the cognitive learning engines and cognitive knowledge bases supported by a collection of contemporary mathematics known as denotational mathematics [4]. The field of cognitive robotics has made significant progress in equipping robots and software agents with high-level cognitive functions. Merrick conducts studies on value system, which permits a biological brain to increase the likelihood of neural responses to selected external phenomena for developing cognitive robotics by examining existing value systems [5]. Ratanaswasd implements an idea in a humanoid robot cognitive architecture ISAC (Intelligent Soft Arm Control) by using the Central Executive Agent and the Working Memory System [6].

With respect to cognitive functions, research in cognitive architectures focus more on higher-level cognitive abilities like learning, planning and general reasoning [7]. By applying cognitive architecture as a basic specification, it can provide a ready-made tool and theoretical basis for cognitive modeling, make the cognitive modeling normalizing, and achieve better results for cognitive modeling [8]. A hybrid framework developed based on different architectures takes advantage of all the relative strengths of the integrated components. Benjamin et al. designed cognitive architecture ADAPT (Adaptive Dynamics and Active Perception for Thought) for robotics [9] and Laird et al. develop autonomous robotic systems that have the cognitive abilities such as communication, coordination, adapting to novel situations, and learning through experience by integration of the Soar cognitive architecture with both virtual and physical robotic systems [10].

This study explores the cognitive architecture for cognitive robots, and conducts robot cognitive behavior simulations on a robotic simulation platform.

2 A Cognitive Robotics Framework

To be a cognitive robot, it must be a robot to meet with designed purpose firstly and then it possesses cognitive abilities. Therefor a cognitive robot must be composed of two main parts, actuators that like human body and cognitive abilities parts like human brain. Figure 1 shows the framework for a cognitive robot.

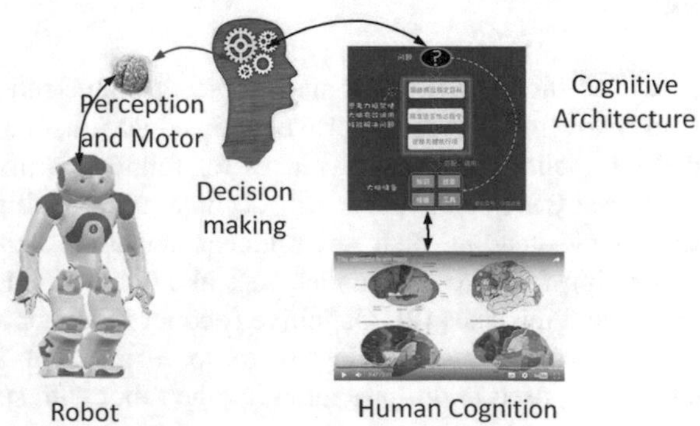

Fig. 1. Framework of cognitive robotic.

2.1 Cognitive Robotics

An industrial robot commonly refers to a robot arm used in a factory environment for manufacturing applications. Traditional industrial robots can be classified according to different criteria such as type of movement, different applications, architecture and brand, and there is new qualifier for industrial robots that can be collaborative or not. NAO is the world's leading and most widely used humanoid robot for education, healthcare, and research. NAO is also an autonomous and fully programmable robot that can walk, talk, listen to human, and even recognize human's face. Most important for research NAO is integrated in a sophisticated simulation platform v-rep, and therefore, it is chosen as the sample robot for this study.

Cognitive abilities are essential characteristic for cognitive robot to act appropriately in highly dynamic and complex environments involving attention, learning, and memory. Such robots need to exhibit a higher form of cognitive ability called cognitive control [11]. A cognitive robotic must have one or more cognitive abilities for its task purpose.

2.2 Cognitive Abilities of Robot

The definition of ability to describe attribute of an individuals is refers to the possible variations in the liminal levels of task difficulty that individuals perform successfully on a defined class of tasks. By adding the adjective cognitive, it means to limit the range of cognitive tasks to those that centrally involve mental functions not only in the understanding of the intended end results but also in the performance of the task, most particularly in the processing of mental information [12]. Cognitive abilities are supported by specific neuronal networks and brain-based. The human brain dominates most body functions of cognitive control include the ability of cognitive function such as perception, attention, memory, motor skills, language, visual and spatial processing, and motion functions. A cognitive robot should feature the abilities for most frequently mentioned cognitive functions include perception, attention, memory and learning.

Perception. A process that transforms raw input into the system's internal representation for carrying out cognitive tasks.

Attention. There are two major categories of the attention mechanisms - external and internal. External or perceptual attention selects and modulates information incoming from various senses, and internal attention modulates internally generated information, such as the contents of working memory or possible behaviors in a given context.

Memory. An essential part of any systems-level cognitive architecture for both studying human mind and solving engineering problems. Memory systems store intermediate results of computations, enable learning and adaptation to the changing environment. There are two approaches to modeling memory: (1) introducing distinct memory stores based on duration (short and long term) and type (procedural, declarative, semantic, etc.), and (2) a single memory structure representing several types of knowledge.

Learning. The capability of a system to improve its performance over time. Practically, any kind of learning is on the base of experience that is the knowledge.

Cognitive robot in most cases is designed to implement specific tasks. Once command for a task appointed, depending on the operations for performing the whole task, the robot firstly sets its goal being achieved in current state. The robot perceives the situation environment, and extracts important information relevant to its goal of the context. According to the goal having been settled and the knowledge being learned, the robot make decisions about what actions will be taken; and then the decision results are transferred to corresponding executors for conduction, after that the robot goes into a new working state waiting for new command. By doing like this, the robot completes a fragment cognitive task, and then the robot checks whether the new state is the indicating state of the whole task ends. If not, the robot will keep working until the task's final goal achieved. A robotic executor can be walking unit, moving unit, gripping arms, mechanical manipulation and so on depends on specific task.

3 A Framework of Cognitive Architecture for Robotics

Kotseruba et al. summarized applications in cognitive architectures as several major groups such as human performance modeling, games and puzzles, robotics etc. Furthermore, they indicated that there is an apparent gap between general research in robotics and computer vision and research in these areas within the cognitive architectures field. The statistical data confirms that the hybrid approach to cognitive modeling already dominates the field and will likely continue to do that in the future [8].

3.1 Module Components for Robot Cognitive Architecture

A cognitive architecture for robot (CAR) encompasses perception, information processing, knowledge and learning, decision-making and motor module in this study. Figure 2 shows the working principle of the cognitive architecture for cognitive robotics.

Benefiting a lot advantage of the knowledge representation from existing outstanding cognitive architecture, CAR split knowledge into two major kinds that is Declarative knowledge and Procedural knowledge.

Declarative Knowledge. It is definitions or conceptions about the world/tasks. They are facts that people make decisions depending on, Such as apple or there is an apple on desk.

Procedural Knowledge. It consists of conditional firing productions. Productions represent knowledge about why and how to do things: for instance, knowledge about why to type the letter "Q" on a keyboard, about how to drive etc.

Following human being's cognitive process, the study schematizes designs of cognitive architecture for robotic into five correlated functional modules to implement corresponding cognitive abilities, and they are perception, information processing, memory and learning, decision-making and motor modules respectively.

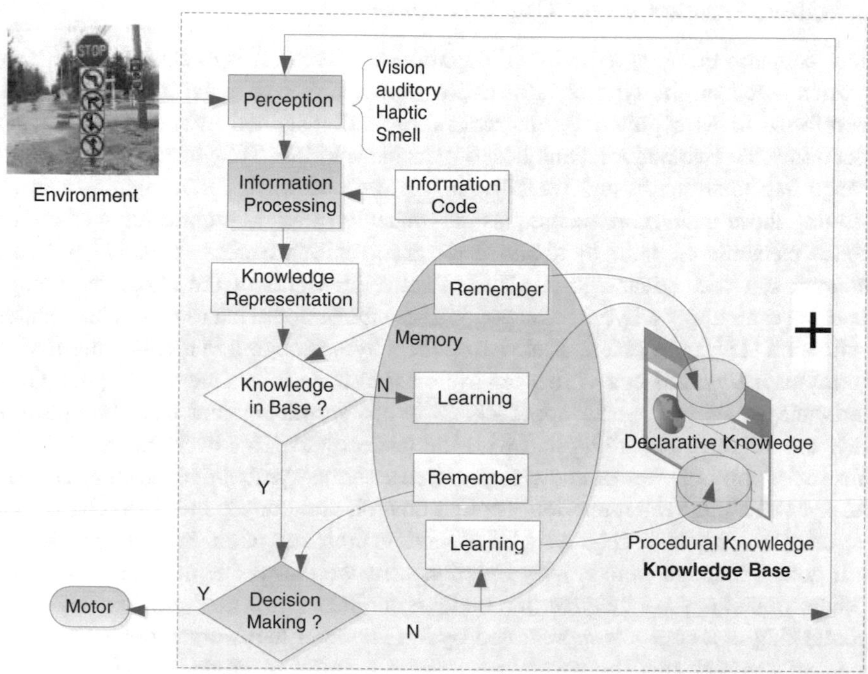

Fig. 2. Working principle of cognitive architecture for robotic

Perception Modules. Are the information entrance for the real world, the most well known perceptual modules are visual, auditory, haptic and smell modules. As for a robotic, the elements relating to perception are the sensors that robotic collect information from the environment.

Information Processing. Transform information collected from the outside into conception or definition.

Memory and Learning. The faculty by which the mind stores and remembers information. Inside the architecture, it learns and retrieves knowledge by using memory module. The memory activities may include learning from environment and remember knowledge from which it has learned before.

Decision-Making. Depending on knowledge that the robot collect from outside, the process hunts for procedural knowledge for matches and fires. The execution of the production will modify the buffers and thus change the state of the system.

Motor. The output of the system. It captures the motion result of a decision and transfers it to a control command to corresponding actuator.

Besides, buffer designed in CAR is to moderate the immediate impact of information on modules. The CAR accesses its modules through buffers and the buffer's content at a given moment represents the state of that module at that moment.

3.2 Hybrid Architecture for Cognitive Robotics

A more common grouping criterion of cognitive architectures is to classify three major categories based on the type of information processing they represent, it is symbolic (also referred to as cognitivist), emergent (connectionist) and hybrid. Symbolic and connectionist are two major techniques to machine learning. They began independently and each has its strengths and weaknesses. Researchers started investigating ways of integrating those even more techniques subsequently. Hybrid architectures attempt to combine elements of both symbolic and emergent approaches recently. A hybrid framework can take advantage of all the relative strengths of the integrated components. Lina et al. apply a hybrid connectionist-symbolic approach to real-valued pattern classification [13] and Lebiere et al. introduce a hybrid cognitive architecture SAL to autonomous navigation in a virtual environment to benefit the integration [14]. Given the advantages of the hybrid approach, it is not surprising that such architectures already are the most prevailing and with the tendency to grow even more.

From the point of view of knowledge-based systems, symbolic representations have advantages of human interpretation, explicit control, and knowledge abstraction, while any symbolic architecture requires a lot of work to create an initial knowledge base, but once it done the architecture is fully functional. As we know, one of the advantages of the connectionist approach is that the model is similar to how the brain actually works by postulating that concepts represented by, on the other hand, emergent architectures are easier to design, and however they require training to produce useful behavior. Furthermore, the existing knowledge may deteriorate with the subsequent learning of new behaviors. In the light of features of robotic different functional modules, the cognitive architecture applies symbolic technique to decision-making and knowledge representation module, and emergent technique to vision module, knowledge acquisition and object identification. As for the self-learning of the architecture, it employs one or more of single, multi-level or hybrid learning approaches. Figure 3 shows the framework of the hybrid architecture using in cognitive robotics.

By adopting hybrid approaches, the architecture can benefit various methods, and even burgeoning machine learning and deep learning technique.

3.3 Mapping Cognitive Functions to Brain's Anatomy Area

One of the main goals of a cognitive architecture is to summarize the various results of cognitive psychology in a comprehensive computer model. Not only cognitive architecture must account for apparent domain specificity, but also it must contrast the true cases with theory on conceptual grounds and support the more parsimonious hypothesis by showing that recent evidences and application studies.

The studies have proved that different parts of the brain control different human behaviors. Beyond, the more important for the behavioral outcome is interaction between brain areas than the efficient processing of a single brain region. Moreover, startling realization emerged that the cognitive theories were remarkably similar in the view of the overall architecture of human brain. Anderson postulated maps between cognitive architecture's modules and brain regions [15].

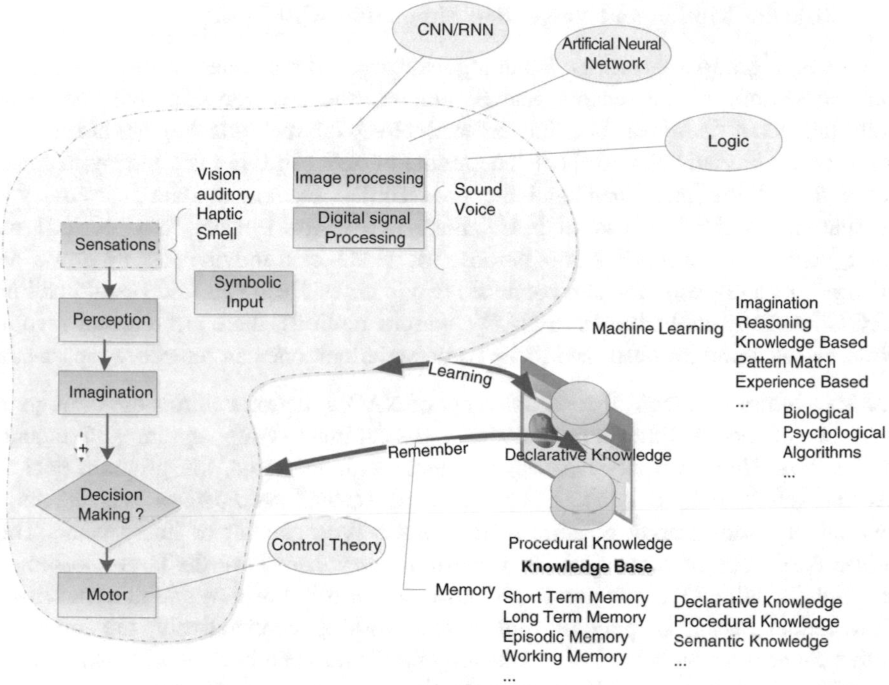

Fig. 3. Cognitive architecture for cognitive robotic

Based on knowledge on neuroscience and learned from merit of some exist sophisticated cognitive architecture, the different module in CAR is designed to map relevant brain region respectively either. Such as vision in perception is versus to fusiform, aural is versus to auditory, information processing is versus to cingulate cortex, memory is versus to hippocampus, decision-making is versus to frontal lobe, and vocal and motor is versus to cerebellum etc.

4 An Implementation of Cognitive Robotic

V-rep is a typical robot simulation platform, and there are varieties of popular robot model library and general robotic components. As a simulation of cognitive robotic, by applying robot NAO of v-rep it realizes motion controls for CAR's decision output. Python as a high-level programming language, it is the preferred language for data analysis and AI development. This study builds the CAR cognitive architecture on the python platform, and implements the robot's perception and motion behavior through v-rep python interface.

4.1 Robotics Motion and Perception Simulation with V-Rep

NAO has a group of sensors including cameras, microphones, sonar rangefinder, pressure sensors, tactile sensors, and IR emitters and receivers. Though one cannot really talk about "Artificial Intelligence" with NAO, the reality is they are already able to reproduce human behavior [16]. The model of NAO in v-rep is courtesy of Macro Cognetti, and the mesh data and the pre-recorded movement data is courtesy of Aldebaran. On the forehead of NAO, there two vision sensors NAO_vision1 and NAO_vision2 attached. For the purpose of NAO can move head up-down and left-right, joint collection which combines two joints of HeadYaw and HeadPitch links NAO's head and its body. As to NAO's various motions, there are relevant revolute joints installed and grouped, and theirs name and block one can refer to v-rep manual.

NAO's Motion Control. V-rep implement of NAO's motion controls are through the co-works of the revolute joint's motions. By calling a v-rep regular API function *simGetObjectHandle* it retrieves object handle with its name, the program gets the revolute joint's control handle. Function *simSetJointTargetPosition* sets the target position of a joint, target position of the joint maybe angular or linear value. Then calling API function *simGetJointTargetPosition*, it will retrieve the target position of the joint. By adjust the revolute joint's position the robot achieves joint's movement. With certain joints' keeping movement and working cooperatively, the robot will realize certain motion behavior, such as moving forward and back or making a turn etc. Therefore, the robot would implement the designed action as programs send specific motion command to the robotic to built-in corresponding joints according to the designer's control purpose.

NAO's Vision and Barrier Distance Perception. The NAO_vision1 and NAO_vision2 are two built-in head vision sensors for NAO robot that it is as the robotic vision system. They will render the objects that are in their field of view and a vision sensor's image content can be accessed via the API, and image-processing filters are available. By calling v-rep remote python API functions *simxGetVisionSensorImage* it retrieves the image of the vision sensor. Once the image data is available, in Python language there are many different ways to deal with it such as image processing or machine learning. By applying those methods, the robotic obtains vision information. Actually, Python platform are good at data analysis and rich in machine learning methods. For the reason that one cannot determine the distance from barrier in the range of the NAO view in current situation, hiring a proximity sensor here to obtain the distance from humanoid NAO to barrier. The function *simxReadProximitySensor* in python API can get coordinates of the detected point and change it to distance. Once the system gets to know that there is barrier come into its vision and knows the barrier's distance, it will make decision according to the barrier's situation. Finally, the system feedbacks the result to robotic and the robot take action depending on it.

4.2 Python Implementation for Perception and Control

V-rep allows a remote API to control a simulation (or the simulator itself) from an external application or a remote hardware (e.g. real robot, remote computer, etc.).

The v-rep remote API can be called from different ways including a python script, a Java application, or a Lua script etc. The remote API functions are interacting with v-rep via socket communication and all this happens in a hidden fashion to the user. The remote API can let one or several external applications interact with v-rep in a synchronous* or asynchronous* way, and even remote control of the simulator is supported (e.g. remotely loading a scene, starting, pausing or stopping a simulation for instance). As for this study, the v-rep will be the server side to conduct the robot's perception and motion, and the python part will be the client for completing the cognitive decision-making. Therefore, the v-rep remote API for python is the client side here to implement information processing, memory/learning and decision-making.

To carry out robot behavior control based on cognitive architecture, first step is to create a NAO humanoid model with v-rep and we call it server side. Moreover, it is essential to write python program to implement robot perception and decision-making that we name it client side. The program includes connection to the server side, image acquisition from vision sensor, image processing for vision perception, cognitive decision making based on cognitive architecture, changes from decision result to control command and feedback control command to robot etc. Finally, run the NAO humanoid model on the v-rep platform and run the python perception and decision-making program on the python platform. The NAO humanoid model, the python program that the client side connects and communicates to the server side, and the CAR cognitive architecture program in python are available at http://ccis.zstu.edu.cn/CogRobotic.

4.3 A Simulation Sample Case

As a study case, we conduct a simulation, mobile NAO humanoid robotic avoiding barriers, for the verification of behavior control based on cognitive architecture. For this purpose, The following three modules are indispensable, the simulation scenario – a NAO robot moves around an obstacle field and aims to avoid the obstacles which is built on v-rep, the cognitive architecture – for cognitive and motor function which is implemented in python, and communication and perception module. By adding a NAO robot model and adding a proximity sensor with v-rep, then adding whole bunch of viewable objects to the scene such as chairs, plants, tables and wooden floor etc. to create the scenario first. After the scene constructed, to make sure its joints are in 'Torque/force' mode, joint 'Motor enabled' property is checked, this will allow python control robotic joint using *simxSetJointTargetPosition* function, and the viewable objects' common properties collidable and detectible are enabled, this will allow the NAO robot to collide with them but also to detect them. The motor enabled property is emerge follows the 'Show dynamic properties dialogue' button, which it comes out when conduct a double click on the joint in the Scene Hierarchy. To construct knowledge for cognitive architecture to realize decision-making is prerequisite. The knowledge involves declarative and procedural knowledge. The declarative knowledge is definition or conception and the procedural knowledge are condition/fire rules such as if barrier and distance less than 0.2 m then makes turn.

The simulation is conducted on v-rep pro edu version 3.3.2 and python(x, y) 2.7.10.0 platform. As running the NAO robotic model on the server side and the python control module cognitive architecture based on the client side, the running result shows that the NAO humanoid can detect and avoid the barrier with the controls of cognitive model based on cognitive architecture as it moves in the designed scenario. The cognitive architecture can support the robotic perception and cognition.

5 Conclusion

This paper proposes a general idea for implementation of cognitive robot based on cognitive architecture and partially by simulation method. A cognitive architecture for robotic cognitive processing is implemented base on python platform. Robotic cognitive behavior controls for avoid obstacles are conducted and simulated on simulation platform v-rep. The simulation result shows the feasibility of the proposed idea in this study which to realize cognitive robot. The benefits of using python to implement cognitive architecture are not only it has advantage of rich resource, numerical algorithms and data processing solid infrastructure but also it is convenient that to implement various cognitive processes in cognitive architecture and to integrate cognitive architecture with true robotic development by using python.

Acknowledgments. This work is supported by the Feitian Foundation of China Astronaut Research and Training Center (No. FTKY201505), the China Advanced Space Medico-Engineering Research Project (No. 2012SY54A1705) and Zhejiang Provincial Natural Science Foundation (No. LY12C09005).

References

1. Edouard, R.: Reasons Why 2017 Will Be the Year of Robotics. https://www.roboticsbusinessreview.com/consumer/reasons-2017-will-year-of-robotics/. Accessed 23 Feb 2017
2. Liu, Y., Tian, Z., Liu, Y., Li, J., Fu, F., Bian, J.: Cognitive modeling for robotic assembly/maintenance task in space exploration. In: Proceedings of the AHFE 2017 International Conference on Neuroergonomics and Cognitive Engineering, 17–21 July 2017. The Westin Bonaventure Hotel, Los Angeles, (2015)
3. Samani, H. (ed.): Cognitive Robotics. CRC Press, Boca Raton (2015)
4. Wang, Y.: Cognitive learning methodologies for brain-inspired cognitive robotics. Int. J. Cogn. Inf. Nat. Intell. **9**(2), 37–54 (2015)
5. Merrick, K.: Value systems for developmental cognitive robotics: a survey. Cogn. Syst. Res. **41**, 38–55 (2017)
6. Ratanaswasd, P., Gordon, S., Dodd, W.: Cognitive control for robot task execution. In: ROMAN IEEE International Workshop on Robot and Human Interactive Communication, pp. 440–445. IEEE (2005)
7. Kotseruba, I., Gonzalez, O.J.A., Tsotsos, J.K.: A review of 40 years of cognitive architecture research: focus on perception, attention, learning and applications. arXiv preprint arXiv:1610.08602 (2016)
8. Liu, Y., Tian, Z., Liu, Y., Li, J., Fu, F.: Cognitive architecture based platform on human performance evaluation for space manual control task. In: Advances in Neuroergonomics and Cognitive Engineering. Springer International Publishing, pp. 303–314 (2017)

9. Benjamin, D.P., Lyons, D.M., Lonsdale, D.W.: ADAPT: a cognitive architecture for robotics. In: ICCM, pp. 337–338, July 2004
10. Laird, J.E., Kinkade, K.R., Mohan, S., Xu, J.Z.: Cognitive robotics using the soar cognitive architecture. Cognitive Robotics AAAI Technical report, WS-12 6 (2012)
11. Kawamura, K., Gordon, S.: From intelligent control to cognitive control. In: World IEEE Automation Congress, WAC 2006, pp. 1–8, July 2006
12. Carroll, J.B.: Human cognitive abilities: a survey of factor-analytic studies. Cambridge University Press, Cambridge (1993)
13. Ming, L.T., Kar, L.S.: A hybrid connectionist-symbolic approach for real-valued pattern classification. In: IFIP International Conference on Artificial Intelligence Applications and Innovations, pp. 49–59. Springer, Boston, September 2005
14. Lebiere, C., O'Reilly, R.C., Jilk, D.J., Taatgen, N., Anderson, J.R.: The SAL integrated cognitive architecture. In: AAAI Fall Symposium: Biologically Inspired Cognitive Architectures, pp. 98–104 (2008)
15. Möbus, C., Lenk, J.C., Özyurt, J., Thiel, C.M., Claassen, A.: Checking the ACT-R/brain mapping hypothesis with a complex task: using fMRI and bayesian identification in a multi-dimensional strategy space. Cogn. Syst. Res. **12**(3), 321–335 (2011)
16. Michieletto, S., Zanin, D., Menegatti, E.: Nao robot simulation for service robotics purposes. In: 2013 European Modelling Symposium (EMS), pp. 477–482. IEEE (2013)

Socio-Technical Health and Usage Monitoring Systems (HUMS)

Johan de Heer(✉) and Paul Porskamp

Thales Research and Technology - Hengelo, Thales Netherlands,
High Tech System Park: Gebouw N - Haaksbergerstraat 67,
7554 PA Hengelo, The Netherlands
{Johan.deHeer,Paul.Porskamp}@nl.thalesgroup.com

Abstract. This work in progress paper proposes a Socio-Technical Health Usage Monitoring Systems (HUMS). In addition, a lab setup is presented for empirical studies focusing on elucidating team performances.

Keywords: HUMS · Human behavior analytics

1 Introduction

Multiple technical solutions require human operators (individuals or teams) in- or on-the-loop. Therefore, the total capability of the envisioned solution is determined by technical system behaviors as well as interacting human behaviors [1]. Observed behaviors - either technical or human - that differ from expected behaviors may severely impact the capability of the total system solution. In order to make the capability of these socio-technical systems robust and resilient we envision a subsystem that detects shortcomings in technical and human behaviors and performances such that in-balances may be compensated or aligned. This requires a Socio-Technical Health Usage Monitoring Systems (HUMS) as depicted in Fig. 1.

However, the current focus in the domain of HUMS [2] is to focus only on the technical part of the capability solution. The reason seems obvious since its application is driven by condition based maintenance. Although, interacting humans are considered a failure type in this domain [3] to our knowledge HUMS functions focus only on the technical building blocks in need for maintenance, repair or replacement, and not on the degradation of human performances.

On the other hand, [4] traditional human factor evaluations are mostly constraint to a single criterion such as workload in relation to one subsystem such as the human interface. Although, in aviation virtually all aspects of human behaviors are analyzed based on measurements via video and audio recordings [5]. In addition, as these authors point out, it is necessary to determine human team performance as well, since it is becoming increasingly evident that degradation in human performance are due to breakdowns in the process of team coordination and communication. A potential interesting solution is provided by the concept of Human Performance Envelope (HPE) [6]. This paradigm is not focusing on one individual factor (e.g. fatigue, situation awareness, etc.), but considers their full range, mapping how they work alone or

H. Ayaz and L. Mazur (Eds.): AHFE 2018, AISC 775, pp. 394–399, 2019.
https://doi.org/10.1007/978-3-319-94866-9_40

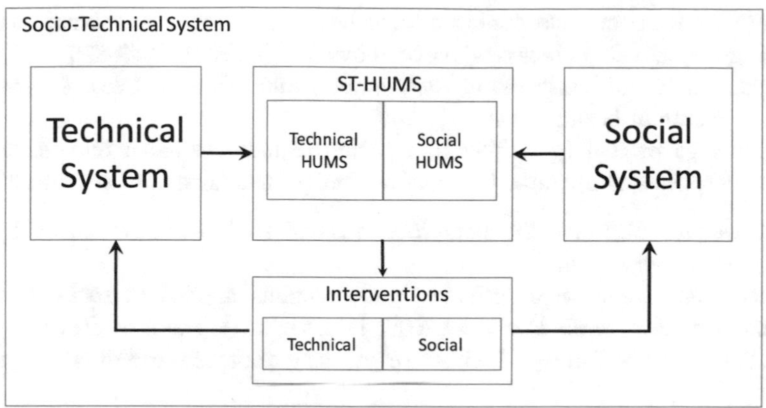

Fig. 1. Socio-Technical Health Usage Monitoring System (HUMS)

in combination leading to a decreased performance that could affect the capability of the total system [7]. As [5] indicates "*the fundamental problem however is still insufficient knowledge of the contribution of one factor to the outcome of the whole system or what factors contribute most*". In other words, how and in what way is the performance of the socio-technical system capability determined?

The outline of this work in progress paper is as follows. First, we focus on the Socio-Technical HUMS architectures. Second, we present a lab setup for empirical studies focusing on elucidating team performances.

2 Reference Architecture for Socio-Technical HUMS

Here we borrow from the Open System Architecture for Condition-Based Maintenance OSA-CBM [8]. The OSA-CBM specification is a reference architecture for moving information in a condition-based maintenance system. The OSA-CBM is designed as a "multi-technological implementation", meaning that it separates the information that can be exchanged in a condition-based maintenance system from the technical interfaces system integrators use to communicate the information. Vendors and integrators can implement the standard using the appropriate technology for their environment. The typical 'technical' HUMS functions by OSA-CBM, include:

- data acquisition (converting an output from a sensor to a digital parameter representing a physical quantity),
- data manipulation (performing signal analysis, computing meaningful descriptors, and deriving sensor readings from the raw functional measurements),
- state detection (facilitating the creation of normal baseline profiles, searching for abnormalities, whenever new data is required and determining in which abnormality zone, if any, the data belongs),

- 'health' assessment (determining current health state given functional parameters leading to diagnostic processes, preventative and corrective actions),
- prognostics 'health' assessment (determining future health states, failure modes based on current health states).
- Advisory generation (providing information required to make immediate operational changes and optimize lifecycle reliability, maintainability and availability).

Here we propose to use the same functions and apply them systematically to the social system as well (Fig. 2).

Further elaboration on describing these functions applied to social systems at individual and team levels is current work in progress. We aim to present an architecture such as depicted in Fig. 4 where we are using a simple sense-think-act paradigm (Fig. 3).

- Sensing level, including
  - Data acquisition
  - Data manipulation
- Thinking level, including
  - State detection
  - 'Health' assessment
  - Prognostics 'health' assessment
- Actuation level
  - Advisory generation.

A good starting point for further elaboration could be based on the concept for Human Performance Envelope (HPE) [7]. These authors are building on the current state of art on measuring and analyzing human factors aspects, including 'advisory generation functions' such as, mitigation and recovery rules for systems, organization, and individuals/teams. Possibly, HPE, which is currently defined for avionics can be applied to other domains as well.

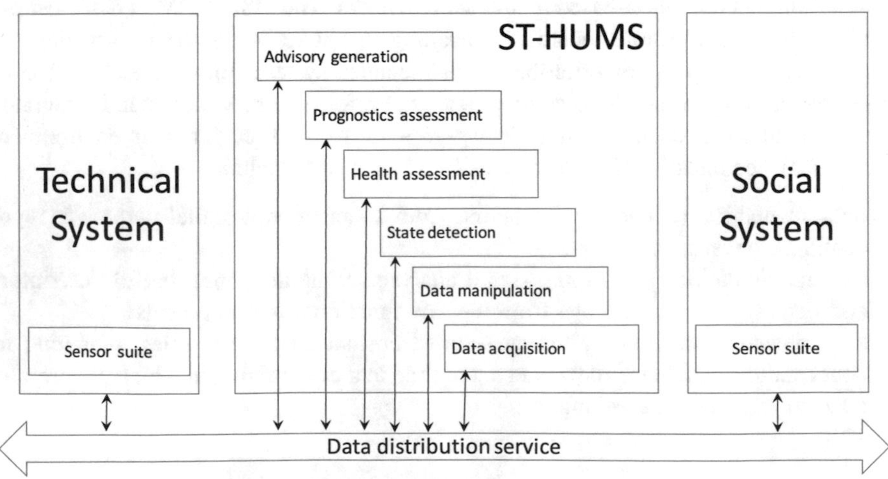

Fig. 2. Reference architecture Socio-Technical Health Usage Monitoring System (HUMS)

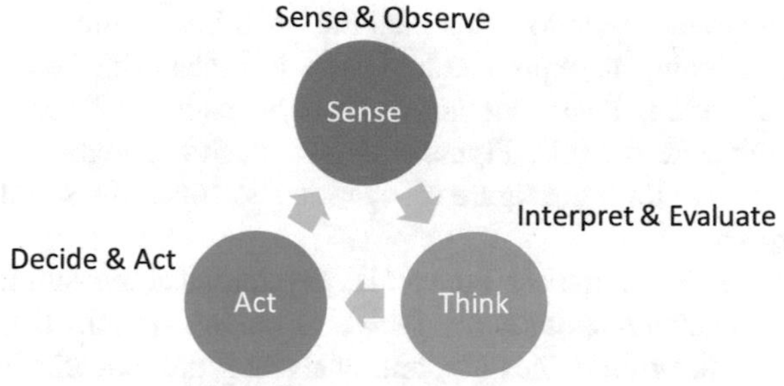

Fig. 3. Sense think act paradigm

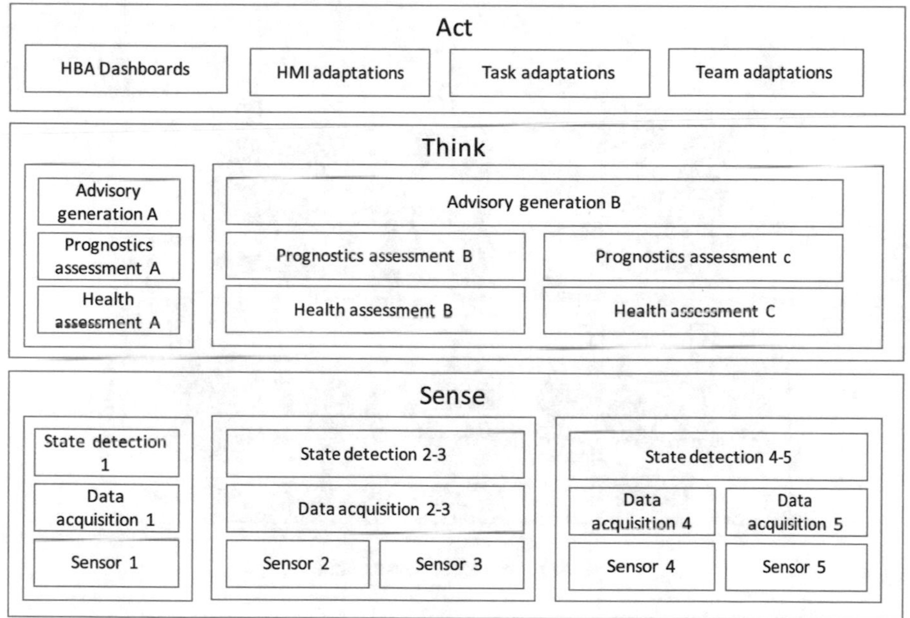

Fig. 4. Layered architecture Socio-Technical Health Usage Monitoring System (HUMS)

3 Experimental Lab for Empirical Studies

Our focus is on recording and analyzing (degradation in) performance at team level with command & control use cases in mind. Typically, 4 or 5 individuals working apart together make up the total team performance. We aim to measure several Human Performance Envelope factors such as mental workload, stress, mental fatigue-drowsiness, situational awareness, attention, vigilance, et cetera. These are measured at the individual level but aggregated to determine team performances. In other words, what is the overall situational awareness capability of this team at this point in time? Or, can we predict the stress level for this team at 10 min from now?

Together with Noldus Technologies[1] and the BMS Lab[2] (University of Twente) we are defining and designing an experimental lab environment for studying dynamic team performances in a working apart together type of applications. We are in the process of defining the sensor suite for this. Figure 5 depicts such a possible configuration. For illustration purposes we limit the figure to 2 people, however, the solution is scalable to more than 2 persons.

We include camera and audio systems (1), physiological measurements for cardiac activity and galvanic skin responses (2), facial expression systems (3), Brain computer interfaces (4), eye-tracking devices (5), and analyzing and visualization software (6). One of the challenges concerns defining the metrics and interpretations for teams. For example, which metrics to take for collaborative brain computer interfaces?

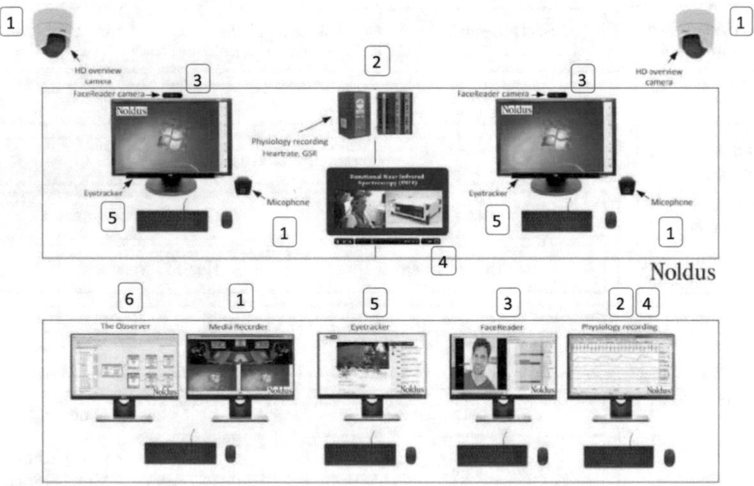

Fig. 5. Scalable sensor suite

4 Conclusion

The general goal of the present paper was to show some work in progress regarding our human behavior analytics activities. We foresee extending current HUMS architectures for condition based maintenance, including social systems to predict degradation in human performance, and sets of mitigation rules to optimize the balance for complex systems acting with complex humans.

Acknowledgments. Noldus technologies for defining and depicting Fig. 5.

---

[1] http://www.noldus.com.

[2] https://bmslab.utwente.nl.

References

1. Meister, D.: Basic principles of behavioral test and evaluation. In: Technical Proceedings of the NATO RSG.24 Workshop on Emerging Technologies in Human Engineering Testing and Evaluation. NATO R&T document AC/243 (Panel 8) TP/17 (1998)
2. https://en.wikipedia.org/wiki/Health_and_usage_monitoring_systems
3. Greaves, M.: Towards the next generation of HUMS sensor. ISASI 2014 Seminar, October 2014, Adelaide, Australia (2014)
4. Essens, P.: Interaction of individual and team performance in ship command centers. Paper presented at the RTO HFM Symposium on "Usability of Information in Battle Management Operations", held in Oslo, Norway, 10–13 April 2000, and published in RTO MP-57 (2000)
5. Pitts, P., Kayton, P., Zalenchak, J.: The national plan for aviation human factors. P 529–540 in Verification and validation of complex systems: human factors issues. In: Wise, J.A., Hopkin, V.D., Stager, P. (eds.) GmbH. Proceedings of the NATO Advanced Study Institute on Verification and Validation of Complex and Integrated Human-Machine Systems, Vimeiro, Portugal, 6–17 July 1992. Springer, Berlin, Heidelberg(1992)
6. http://www.stressproject.eu/wp-content/uploads/2017/02/stress-factsheet-extended.pdf
7. Silvagni, S., Napoletano, L., Graziani, I., Le Blaye, P., Rognin. L.: Human performance envelope, FSS_P6_DBL_D6.1, EU's Horizon 2020 Research and Innovation Programme, under Grant Agreement No. 640597 (2015)
8. http://www.mimosa.org/mimosa-osa-cbm

Author Index

H. Ayaz and L. Mazur (Eds.): AHFE 2018, AISC 775, pp. 401–402, 2019.
https://doi.org/10.1007/978-3-319-94866-9

Printed in the United States
By Bookmasters